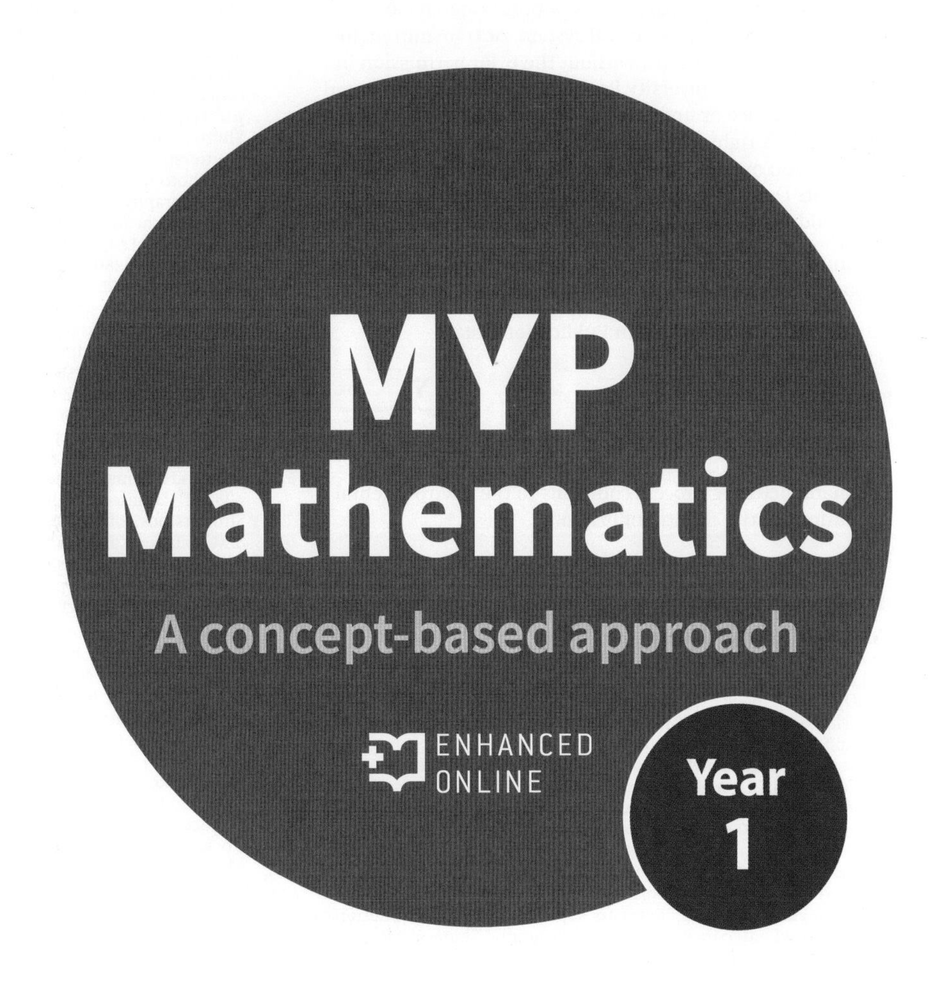


# MYP Mathematics

## A concept-based approach

David Weber
Talei Kunkel
Harriet Simand
Justin Medved

OXFORD

OXFORD
UNIVERSITY PRESS

Great Clarendon Street, Oxford, OX2 6DP, United Kingdom

Oxford University Press is a department of the University of Oxford.
It furthers the University's objective of excellence in research, scholarship, and education by publishing worldwide. Oxford is a registered trade mark of Oxford University Press in the UK and in certain other countries



First published in 2018



British Library Cataloguing in Publication Data
Data available

ISBN 978-0-19-835615-8

13

MIX
Paper | Supporting responsible forestry
FSC
www.fsc.org
FSC® C007785

The manufacturing process conforms to the environmental regulations of the country of origin.

Printed in Great Britain by Bell and Bain Ltd, Glasgow

## Acknowledgements

**Cover:** Gil.K/Shutterstock

p2:Alex Gontar/Shutterstock; p2:OUP_DAM; p3:Dominionart/Shutterstock; p3:James Brittain-VIEW/Alamy; p48:Trisha Masterson/Shutterstock; p49:OUP_DAM; p49:OUP_DAM; p49:Jochen Tack/Alamy; p84:Everett Historical/Shutterstock; p84:3Dsculptor/Shutterstock; p85:Alexander Nikitin/Shutterstock; p85:meunierd/Shutterstock; p120:icyimage/ Shutterstock; p120:nenetus/Shutterstock; p121:Jaromir Chalabala/ Shutterstock; p121:imagedb.com/Shutterstock; p160:Thomas La Mela/ Shutterstock; p161:Naypong Studio/Shutterstock; p161:marchello74/ Shutterstock; p200:sivanadar/Shutterstock; p200:Andriy Blokhin/ Shutterstock; p201:Arto Hakola/Shutterstock; p201:Ksy & Gu/ Shutterstock; p248:OlegD/Shutterstock; p248:Jose Ignacio Soto/ Shutterstock; p249:Sofiaworld/Shutterstock; p5:Oleg Znamenskiy/ Shutterstock; p6:Thipjang/Shuterstock; p44:MarcelClemens/Shutterstock; p33:Ekkapon/Shuterstock; p38:Thomas Siegmund/Shuterstock; p16:Rook76/Shutterstock; p30:Perlphoto/Shuterstock; p46a:Sergio Bertino/Shuterstock; p46b:Chungking/Shuterstock; p46c:Dmitry Rukhlenko/Shuterstock; p46d:Hhd/Shuterstock; p46e:Viacheslav Lopatin/Shuterstock; p46f:PerseoMedusa/Shuterstock; p46g:Sailorr/ Shuterstock; p7:Anton_Ivanov/Shuterstock; p14 (B):Andrea Izzotti/ Shuterstock; p15:Vectorfusionart/Shuterstock; p14 (T):Sidhe/Shutterstock; p24:Pavila/Shutterstock; p58:RikoBest/Shutterstock; p22:A_Lesik/ Shuterstock; p8:Shutterstock; p28:Julia Kuznetsova/Shutterstock; p45:Atdigit/Shutterstock; p70 (L):Mega Pixel/Shutterstock; p52:Rtimages/ Shutterstock; p52:Danny Smythe/Shutterstock; p51:Northfoto/ Shutterstock; p59:Glass and Nature/Shutterstock; p67 (MT):Brent Parker Jones/Oxford University Press ANZ; p67 (T):Diana Taliun/ Shutterstock; p66:Oriori/Shutterstock; p67 (MB):Ivonne Wierink/ Shutterstock; p67 (B):Bonchan/Shutterstock; p69:PaulPaladin/ Shutterstock; p71a:Surrphoto/Shutterstock; p71b:Kletr/Shutterstock; p71c:315 studio by khunaspix/Shutterstock; p71d:CYC/Shutterstock; p71e:LanKS/Shutterstock; p71f:Sarah2/Shutterstock; p71g:SunTime/ Shutterstock; p74 (TR):Bernd Juergens/Shutterstock; p74 (TL):Joe Gough/ Shutterstock; p75 (BR):Brulove/Shutterstock; p75 (CL):XiXinXing/ Shutterstock; p75 (T):Jaboo2foto/Shutterstock; p75 (CR):OUP_DAM; p74 (B):Sergey Peterman/Shutterstock; p70 (C):Africa Studio/Shutterstock; p75 (BL):Jacek Chabraszewski/Shutterstock; p70 (R):Cameramannz/ Shutterstock; p78 (T):Ruslana Iurchenko/Shutterstock; p78 (B):Wsphotos/ iStockphoto; p90:Eric Isselee/Shutterstock; p92 (B):Sergey Khachatryan/ Shutterstock; p88 (TC):StudioSmart/Shutterstock; p88 (TR):Crazy82/ Shutterstock; p91:Labrador Photo Video/Shutterstock; p92 (T):Cre8tive Images/Shutterstock; p99:PRILL/Sutterstock; p109 (B):Mariusz Switulski/ Sutterstock; p113 (T):irin-k/Shutterstock; p113 (B):Sophie Rohrbach/ The Organisation; p114:Luispa Salmon, James Walmseley; p112e:Marcus Cutler, Elisa Paganelli; p112f:Kotomiti Okuma/Shutterstock; p118 (TL):Jim Clark/Alamy Stock Photo; p118 (B):Bringolo/Shutterstock; p110:Tooykrub/ Shutterstock; p88 (TL):Patrick K. Campbell/Shutterstock; p87:Steve Scribner/Shutterstock; p96:Lukiyanova Natalia frenta/Shutterstock; p100:Plampy/Shutterstock; p103:Claudio Divizia/Shutterstock; p104a:Joao Virissimo/Shutterstock; p104b:Nrt/Shutterstock; p104c:Somchai Som/ Shutterstock; p104d:Alexdrim/Shutterstock; p104e:Mircea BEZERGHEANU/ Shutterstock; p104f:Dimedrol68/Shutterstock; p112a:Alberto Masnovo/ Shutterstock; p112b:Lightspring/Shutterstock; p112c:Tischenko Irina/ Shutterstock; p112d:Aboard/Shutterstock; p115 (T):Stepan Bormotov/ Shutterstock; p115 (C):Sergejus Byckovskis/Shutterstock; p115 (B):Andrey Armyagov/Shutterstock; p88 (BL):Hintau Aliaksei/Shutterstock; p88 (BC):Kondor83/Shutterstock; p88 (BR):Pal Teravagimov/Shutterstock; p109 (T):StudioSmart/Shutterstock; p117 (T):Stefan Petru Andronache/ Shutterstock; p117 (B):Lusoimages/Shutterstock; p119 (TL):Vitstudio/ Shutterstock; p119 (TR):Anna Omelchenko/Shutterstock; p119 (BL):Photo Image/Shutterstock; p119 (BR):Mycteria/Shutterstock; p252 (B):Artography/Shutterstock; p264:Aerial Archives/Alamy Stock Photo; p273:Jenoche/Shutterstock; p275 (T):Roman Samokhin/Shutterstock; p279:Olivier Le Queinec/Shutterstock; p280 (L):Astrid Hinderks/Alamy Stock Photo; p280 (R):Magnascan/iStockphoto; p252 (T):Emilio100/ Shutterstock; p255 (L):Katerina Sysoeva/Shutterstock; p255 (R):Andrea Willmore/Shutterstock; p256:Dim Dimich/Shutterstock; p253a:Rich Carey/Shutterstock; p253b:Ching/Shutterstock; p253c:Digitalbalance/ Shutterstock; p253d:Allween/Shutterstock; p253e:Cbpix/Shutterstock; p253f:Neirfy/Shutterstock; p261 (C):FotoRequest/Shutterstock; p261 (T):CreativeNature R.Zwerver/Shutterstock; p272:EcoPrint/Shutterstock; p275 (B):Cameramannz/Shutterstock; p254: National Council of Teachers of Mathematics; p257:MathPlayground.com; p261 (B): MrNussbaum. com; p265:Alberta Education. Math Interactives, Shape and Space, Area and Perimeter. www.learnalberta.ca . Accessed December 2017; p283 (T): LunaseeStudios/Shutterstock; p283 (B):Christopher Gardiner / Shutterstock.com; p251:Ariel Celeste Photography/Shutterstock; p253 (BKGD):Iakov Kalinin/Shutterstock; p275 (C):Urbanbuzz/Shutterstock; p163:Monkey Business Images/Shutterstock; p187 (B):Monkey Business Images/Shutterstock; p164 (B):Andresr/iStockphoto; p166:Andrey Lobachev/Shutterstock; p167:Tomo Jesenicnik/Shutterstock; p168:Sailorr/ Shutterstock; p169:Monkey Business Images/Shutterstock; p180:Lars Ove Jonsson/Shutterstock; p164 (T):Susan Law Cain/Shutterstock; p185:Olga Popova/Shutterstock; p170:Natalia Hubbert/Shutterstock; p175:Yongkiet Jitwattanatam/123RF; p182 (C):Christopher Elwell/Shutterstock; p182 (L):Sevenke/Shutterstock; p1782 (R):Africa Studio/Shutterstock; p195 (B):StanislavBeloglazov/Shutterstock; p195 (T):Dogi/Shutterstock; p181:Arina P Habich/Shutterstock; p188:SoumenNath/iStockphoto; p191:PzAxe/Shutterstock; p189 (B):Leenvdb/Shutterstock; p196:OUP_DAM; p171:TeddyandMia/Shutterstock; p186:Maxx-Studio/Shutterstock; p187 (T):Luchschen/Shutterstock; p189 (T):Michaeljung/Shutterstock; p194:G-stockstudio/Shutterstock; p197 (T):Monkey Business Images/Shutterstock; p190:Grekoff/Shutterstock; p197 (B):Bochkarev Photography/Shutterstock; p203:Syda Productions/Shutterstock; p206:NurPhoto/NurPhoto/Getty Images; p231:George Rudy/Shutterstock; p214 (L):Ttatty/Shutterstock; p214 (LC):Nd3000/Shutterstock; p214 (R):Gareth Boden; p214 (RC):Monkey Business Images/Shutterstock; p212:Alex Staroseltsev/Shutterstock; p216:Macumazahn/Shutterstock; p220:Norman Chan/Shutterstock; p238:Nigel Kitching/Sylvie Poggio; p240:Kokhanchikov/Shutterstock; p243:Maksim Kabakou/Shutterstock; p205 (BR):Podfoto/Shutterstock; p205 (TR):Serg64/Shutterstock; p205 (TL):Cameramannz/Shutterstock; p205 (BL): Oxford University Press ANZ; p215:Ifong/Shutterstock; p219:Kuzma/OUP Picture Bank; p221:Dragon Images/Shutterstock; p222:Maceofoto/Shutterstock; p223:Anna Kucherova/Shutterstock; p244 (T):Georgejmclittle/Shutterstock; p244 (B):Goodluz/Shutterstock; p246 (T):Flashon Studio/Shutterstock; p246 (B):Pixinoo/Shutterstock; p247 (T):Concept Photo/Shutterstock; p247 (B):MSPhotographic/Shutterstock; p124:Dpa picture alliance archive/Alamy Stock Photo; p128:Egle Lipeikaite/Shutterstock; p129 (T):Vadim Sadovski/Shutterstock; p129 (B):Exopixel/Shutterstock; p131:Erich Lessing/Art Resource; p134:By Gurgen Bakhshetyan/Shutterstock; p135:Prisma by Dukas Presseagentur GmbH/Alamy Stock Photo; p142: Alamy; p141:3Dsculptor/Shutterstock; p158 (CR):WitR/Shutterstock; p146:Farferros/Shutterstock; p148:Martinus Sumbaji/Shutterstock; p123:Kushch Dmitry/Shutterstock; p126:Petr Malyshev/Shutterstock; p158 (TR):imagefactory/Shutterstock; p158 (B):Hare Krishna/Shutterstock; p159 (TR):agsandrew/Shutterstock; p159 (B):Alesandro14/Shutterstock; p151:Valentyna Chukhlyebova/Shutterstock; p159 (TL):GOLFX/Shutterstock;

# Contents



Launch additional digital resources for this book

## MYP Mathematics 1: a new type of math textbook...

This is not your average math textbook. Whereas most textbooks present information (kind of like a lecture on paper), which you then practice and apply, this text will help you develop into a mathematician. Following the MYP philosophy, you will perform investigations where you will discover and formulate mathematical rules, algorithms and procedures. In fact, you will generate the mathematical concepts yourself, before practicing and applying them. Much like an MYP classroom, this text is supposed to be an active resource, where you learn mathematics by doing mathematics. You will then reflect on your learning and discuss your thoughts with your peers, thereby allowing you to deepen your understanding. You are part of a new generation of math student, who not only understands how to do math, but what it means to be a mathematician.

## How to use this book

MYP Mathematics 1 is designed around six global contexts. They are:

- identities and relationships
- orientation in space and time
- personal and cultural expression
- scientific and technical innovation
- globalization and sustainability
- fairness and development

Each unit in this book explores a single global context. However, a different global context could have been selected as a focus of study. What would the study of this content look like in a different context? The chapter opener gives you just a small taste of the endless possibilities that exist, encouraging you to explore and discover these options on your own.

7 Perimeter, area and volume

In this unit, you will see that being able to define and calculate the amount of space taken up by an object is an important step in figuring out how to minimize its impact on the environment. However, shapes and space also play an important role in understanding a wide range of other global contexts.

Fairness and development

Questions of power and privilege

How much space does a human being need in order to live? Why do some people have so much space while others have so much less? Perimeter, area and volume can be used to analyze housing settlements around the world, and to illustrate the stark differences between those who have power and privilege and those who do not.

At the same time, it has been said that less than 5% of the world's population uses at least 25% of its natural resources. What problems might this pose?

Calculating the volume of resources that are available and the extent of the damage caused to the planet in obtaining them can help to further demonstrate the inequality.

248

Orientation in space and time

Urban planning in ancient civilizations

We can study the great ancient civilizations through what remains of their architecture and the products that they created. The Ancient Egyptians and Chinese produced impressive structures that can be analyzed by looking at the space they occupy and shapes chosen. Whether it's an obelisk covered in ancient writing or the Great Wall of China, looking at perimeter, area and volume can help us understand the magnitude of a project and give important insight into the civilization that created it.

The Rama empire existed in what is now Pakistan and India. The empire was so advanced that its cities, including Mohenjo-daro, were carefully planned before their construction. The use of rectangular structures for the city grid and the housing predates evidence of the basic principles of geometry. More impressive still is the fact that the sewage system they developed was more sophisticated than those found in many cities today!

249

## Topic opening page

**Key concept** for the unit.

The identified **Global context** is explored throughout the unit through the presentation of material, examples and practice problems.

**Statement of Inquiry** for the unit. You may wish to write your own.

The Approaches to Learning (**ATL**) skills developed throughout the unit.

**You should already know how to:** – these are skills that will be applied in the unit which you should already have an understanding of. You may want to practice these skills before you begin the unit.

The **related concepts** explored in the unit. You may want to see if other related concepts could be added to this list.

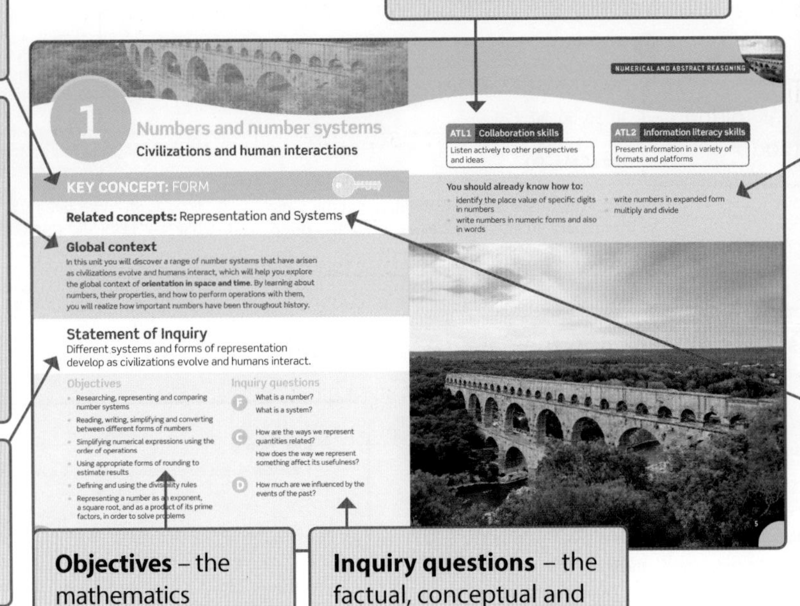


**Objectives** – the mathematics covered in the unit.

**Inquiry questions** – the factual, conceptual and debateable questions explored in the unit.

## Learning features

**Investigations** are inquiry-based activities for you to work on individually, in pairs or in small groups. It is here that you will discover the mathematical skills, procedures and concepts that are the focus of the unit of study.

### Investigation 2 – Divisibility rules

criteria B, C

You will be developing these rules for multiples of 2 to 10 (except 7).

1 **Determine the divisibility rule for multiples of 2.**
   - a Write down some small numbers (up to three digits) that are multiples of 2 and study the numbers to find a pattern.
   - b Describe these patterns and generalize the divisibility rule for 2.
   - c Test your rule with larger numbers (four digits and five digits) to see if it holds.

2 **Determine the divisibility rules for multiples of 5 and 10.**
   - a Write down some small numbers that are multiples of 5. Study the numbers, find a pattern and generalize your rule. Be sure to test it for larger numbers.
   - b Repeat for multiples of 10.

3 **Determine the divisibility rules for multiples of 4 and 8.**
   - a Write down all of the two-digit numbers that are divisible by 4. Now select some larger numbers that are divisible by 4. What do you notice about your lists of two-digit and larger numbers?
   - b Repeat these steps for multiples of 8.

4 **Determine the divisibility rules for multiples of 3 and 9.**
   - a Write down some small numbers (up to three digits) that are multiples of 3. Study the numbers, find a pattern and generalize your rule. Be sure to test it for larger numbers.
   - b Repeat this procedure for multiples of 9.

With 4b, you may have to look for other patterns, such as adding digits together.

5 **Determine the divisibility rules for multiples of 6.**

   Write down some small numbers (up to three digits) that are multiples of 6. Study the numbers, find a pattern and generalize your rule. Be sure to test it for larger numbers.

ATL2 6 Once you have determined all of the divisibility rules, make a reference chart with them all on one page so you can readily access them during this unit.

For question 5, you may want to look at combinations of other rules.

**Activities** allow you to engage with mathematical content and ideas without necessarily discovering concepts. They allow you to practice or extend what you have learned, often in a very active way.

### Activity 3 – Euclid's method

Find the GCF of each pair of numbers using both the ladder method and Euclid's method.

| | | |
|---|---|---|
| **a** 72 and 180 | **b** 24 and 132 | **c** 104 and 1200 |
| **d** 360 and 1024 | **e** 256 and 1600 | **f** 480 and 2040 |

**Hints** are given to clarify instructions and ideas or to identify helpful information.

In question **3**, be sure to justify your answers.

## Example 12

**Q** The eastbound bus leaves the depot once every 18 minutes. The westbound bus leaves once every 30 minutes. If the first bus on each line leaves at 3 pm, what time will it be when they again leave at the same time?

**A** What are you being asked to find?

You are trying to determine when they will next leave at the same time – so you are finding a larger number than both – so find the LCM.

18 : 18, 36, 54, 72, 90, 108

30: 30, 60, 90, 120

Find the LCM using your preferred method.

The LCM is 90.

Since the buses will leave the depot at the same time every 90 minutes, the next time they do so will be at 4:30 pm.

Look at what the question is asking to make sure you fully answer the question.

**Examples** show a clear solution and explain the method.

**Reflect and discuss** boxes contain opportunities for small group or whole class reflection on the content being learned.

### Reflect and discuss 13

Use the 3P strategy as you answer the following questions in pairs or small groups.

- Which comes first, multiplication or division? Explain.
- Try to create an acronym or a simple way to remember the order of operations.

**Practice** questions are written using IB command terms. You can practice the skills learned and apply them to unfamiliar problems. *Answers for these questions are included at the back of the book.*

### Practice 9

**1** Evaluate each of these calculations using the correct order of operations.

**a** $8+6\times 2$ **b** $\frac{18+6}{5+1}-4$ **c** $(7^2-4)(5^2-2)$

**d** $3(2^4-11)+3\times 2^2-4$ **e** $51\div(5^2-4\times 2)\times 7-\sqrt{100}$

**f** $\frac{5(3+4)-2\times 8-2^3}{3^2+(27\div 3)}$ **g** $48\div(4^2-4\times 2)\times 6-6$

**h** $\frac{2^4-3+12\div 4}{64\div 2^3}$ **i** $5\left(4^2-\sqrt{3^2+4^2}\right)-7\times 2^3$ **j** $\frac{3[(9-4)+6\times 3]-3^2}{48\div 4-[3(6-2)-10+2]}$

**2** Replace each □ with an operation to make the statement true.

**a** 1 + 15 □ 3 = 6 **b** (18 □ 2) ÷ 10 = 2 **c** 15 □ 3 + 2 □ 5 = 15

**3** Insert brackets where appropriate to make each statement true.

**a** 18 − 6 × 3 + 2 = 38 **b** 8 + 4 ÷ 2 + 2 = 3 **c** 5 + 3 × 6 − 10 = 38

**d** 16 □ 8 + 4 □ 4 = 3 **e** 12 □ 8 − 5 □ 4 = 1 **f** 5 × (3 + □) □ 6 = 9

**4** Change just one operation in each statement to make it true.

**a** 7 × 3 + 4 × 5 = 1 **b** (12 − 6) × 3 + 10 ÷ 5 = 4 **c** 24 − (10 − 6) ÷ 2 × 5 = 16

**d** 12 + 36 ÷ 9 ÷ 4 = 13 **e** 6 × 8 − 12 × 2 ÷ 7 = 15 **f** 14 ÷ 10 − 5 − 12 − 3 = 15

**5** The game '24' is a number game where players are given four numbers and must use them to create an expression that equals 24. Players can use any of the operations (+, −, ×, ÷) as well as exponents and parentheses. For each group of four numbers listed below, write down an expression that equals 24 using *each number* once.

**a** 1, 5, 7, 8 **b** 3, 7, 8, 8 **c** 2, 3, 6, 10 **d** 7, 6, 10, 12 **e** 2, 6, 6, 9

**Weblinks** present opportunities to practice or consolidate what you are learning online or to learn more about a concept. While these are not mandatory activities, they are often a fun way to master skills and concepts.

For practice with estimating fractions and decimals graphically, go to brainpop.com and search for 'Battleship Numberline'.

**Formative assessments** help you to figure out how well you are learning content. These assessments explore the global context and are a great way to prepare for the summative assessment.

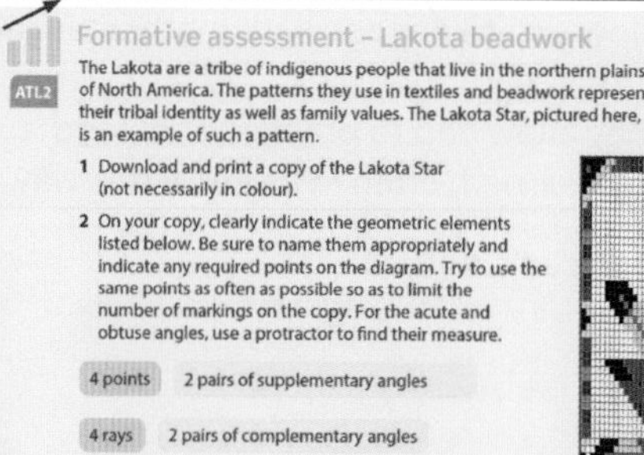

### Formative assessment – Lakota beadwork

ATL2

criterion C

The Lakota are a tribe of indigenous people that live in the northern plains of North America. The patterns they use in textiles and beadwork represent their tribal identity as well as family values. The Lakota Star, pictured here, is an example of such a pattern.

**1** Download and print a copy of the Lakota Star (not necessarily in colour).

**2** On your copy, clearly indicate the geometric elements listed below. Be sure to name them appropriately and indicate any required points on the diagram. Try to use the same points as often as possible so as to limit the number of markings on the copy. For the acute and obtuse angles, use a protractor to find their measure.

4 points | 2 pairs of supplementary angles

4 rays | 2 pairs of complementary angles

4 lines | 4 pairs of vertically opposite angles

4 line segments | 2 pairs of corresponding angles

4 acute angles | 2 pairs of alternate interior angles

2 pairs of parallel lines | 4 right angles | 4 obtuse angles

**3** Justify your decisions for each of the pairs of angles. Use a diagram to support your explanations.

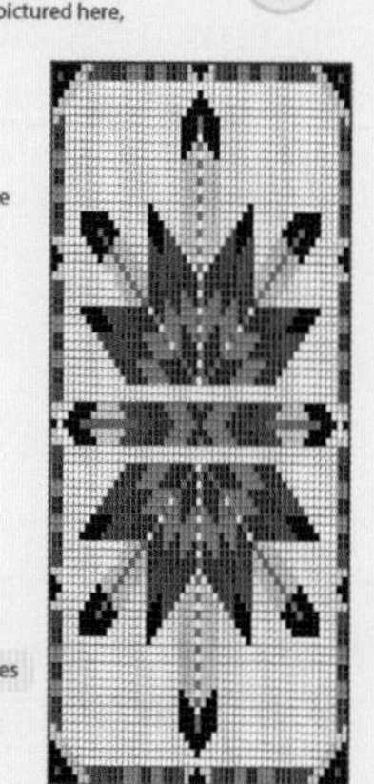

**Technology icon** indicates where you can discover new ideas through examining a wider range of examples, or access complex ideas without having to do lots of painstaking work by hand. This icon shows where you could use Graphical Display Calculators (GDC), Dynamic Geometry Software (DGS) or Computer Algebra Systems (CAS).

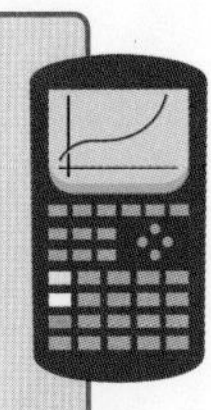

**ATL** icons highlight opportunities to develop the ATL skills identified on the topic opening page. ATL1

# Each unit ends with

**Unit summary** recaps the key points, ideas and rules/formulas from the unit.

## Unit summary

The number system that we currently use is the decimal number system. We use a place value system based on the number 10. The units place and the decimal point form the center of the place value system. The decimal system is made up of families (thousands, thousandths, millions, millionths, etc.) Each place value family has three members: units (or ones), tens, and hundreds.

ten billions, billions, hundred millions, ten millions, millions, hundred thousands, ten thousands, thousands, hundreds, tens, ones, • decimal point, tenths, hundredths, thousandths, ten thousandths, hundred thousandths

To write whole numbers from words:

1. Determine the place value name with the greatest value given in the number. This tells you how many places you need.
2. Draw a blank (a dash) for every place you will need, one per digit.
3. Fill in the blanks with the digits needed. Use zeros as place holders.

To read decimal numbers:

1. Ignoring the decimal point, read the decimal number part like a whole number.
2. At the end of the number, say the place name of the last digit.

To write decimal numbers:

1. Determine the place name of the last digit of the number. This tells you how many places there should be in the number.
2. Fill in the spaces so that the number reads correctly. If necessary, put zeros in as place holders.

**Expanded notation**

Examples:

$352.87 = 300 + 50 + 2 + 0.8 + 0.07$

or

$352.87 = (3 \times 100) + (5 \times 10) + (2 \times 1) + (8 \times 0.1) + (7 \times 0.01)$

40 1 Numbers and number systems

## Unit review

criterion A

Key to Unit review question levels:

Level 1–2 Level 3–4 Level 5–6 Level 7–8

1. **Write down** these numbers in words.
   a 32 560 042 b 17.08 c 1743.14 d 2.718
2. Represent these written descriptions as numbers using digits.
   a ten thousand four hundred fifty-five
   b twenty nine and thirty-two thousandths
   c one hundred ninety-thousand six hundred forty-five millionths
   d eighteen thousandths
3. Represent these numbers in words and in both expanded forms.
   a A blockbuster film made €15 342 807 over a single weekend.
   b At its furthest, the moon is 405 696.7 km from the Earth.
   c The length of an eyelash mite is 0.1375 mm.
   d In 2013, the population of India was 1 252 706 114 people.
   e The width of a Siberian Husky's hair is 0.000025 m.
4. Represent these numbers in both expanded forms.
   a 12.75 b 203.6 c 12 004.205 d 3.0019
5. **Write down** the given number from its expanded form.
   a $500 + 10 + 8 + 0.02 + 0.001$
   b $(3 \times 10\,000) + (6 \times 100) + (2 \times 10) + (7 \times 0.01)$
   c $9000 + 10 + 0.01 + 0.008$
   d $(4 \times 100\,000) + (6 \times 1000) + (2 \times 10) + (3 \times 0.01)$
6. Represent each given number in two other forms.
   a 8231.04 b $(4 \times 1000) + (3 \times 10) + (5 \times 0.1) + (4 \times 0.001)$
   c $8000 + 800 + 0.7 + 0.008 + 0.00002$ d 1400.12
7. Evaluate each number.
   a $8^3$ b $12^2$ c $6^3$ d $13^2$
   e $2^4$ f $3^3$ g $2^7$ h $5^4$

For a number such as 1047, some people say 'one thousand **and** forty seven' while others say 'one thousand forty-seven'. Can you think of a reason why the second way might be preferable?

42 1 Numbers and number systems

**Unit review** allows you to practice the skills and concepts in the unit, organized by level (criterion A) as you might find on a test or exam. You can get an idea of what achievement level you are working at based on how well you are able to answer the questions at each level. *Answers for these questions are included at the back of the book.*

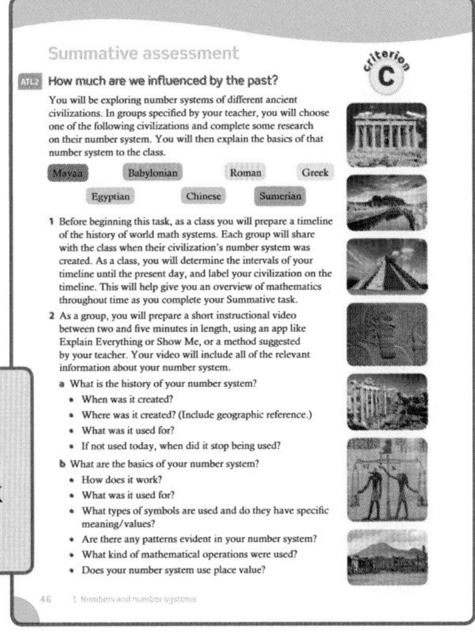

## Summative assessment

criterion C

ATL2 **How much are we influenced by the past?**

You will be exploring number systems of different ancient civilizations. In groups specified by your teacher, you will choose one of the following civilizations and complete some research on their number system. You will then explain the basics of that number system to the class.

Mayan Babylonian Roman Greek Egyptian Chinese Sumerian

1. Before beginning this task, as a class you will prepare a timeline of the history of world math systems. Each group will share with the class when their civilization's number system was created. As a class, you will determine the intervals of your timeline until the present day, and label your civilization on the timeline. This will help give you an overview of mathematics throughout time as you complete your Summative task.
2. As a group, you will prepare a short instructional video between two and five minutes in length, using an app like Explain Everything or Show Me, or a method suggested by your teacher. Your video will include all of the relevant information about your number system.
   a What is the history of your number system?
   - When was it created?
   - Where was it created? (Include geographic reference.)
   - What was it used for?
   - If not used today, when did it stop being used?

   b What are the basics of your number system?
   - How does it work?
   - What was it used for?
   - What types of symbols are used and do they have specific meaning/values?
   - Are there any patterns evident in your number system?
   - What kind of mathematical operations were used?
   - Does your number system use place value?

46 1 Numbers and number systems

The **summative assessment** task applies the mathematics learned in the unit to further explore the global context. This task is often assessed with criteria C and D.

# 1 Numbers & number systems

Numbers are an important part of everyday life. As you will discover in this unit, the systems that we use to represent numbers have developed over time as a response to human interactions. However, exploring numbers and number systems through a different context could open up worlds that are more personal or which push the boundaries of science and technology.

## Identities & relationships

### What it means to be a teenager

To understand what it means to be a teenager requires the ability to use numbers and number systems. From the number of 'posts' to 'likes' to 'followers', today's teenagers are used to seeing and understanding a much wider range of numbers than ever before.

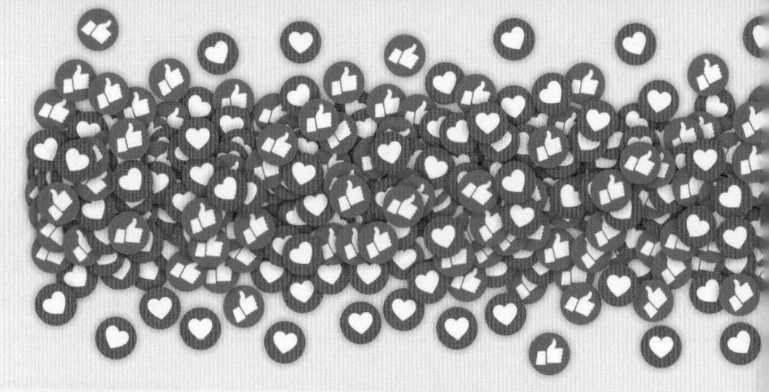

At the same time, the anatomy and chemistry of the adolescent brain is a study in small and large amounts. There are approximately 100 billion neurons in the brain, varying in size from 0.004 mm to 0.1 mm. The teenage brain undergoes both "pruning" and growth, where links between neurons are "pruned" depending on how often they are used, or where more neural pathways are formed.

## Ingenuity and progress

Throughout the centuries, humans have continuously innovated to produce more precise and more powerful tools. Amazingly, this ingenuity involves both very large and very small quantities and measures.

Very small...

Scientists have developed atomic clocks so precise that, after 15 billion years, they will not have lost or gained a single second. Before that, the most precise clocks lost up to 0.0000000033 seconds per year.

The Large Hadron Collider is the single largest machine in the world, housed in the CERN research headquarters in Geneva, Switzerland.

It is used to force particles to collide with one another as they travel at close to the speed of light (299 792 458 m/s).

The mass of each particle is so small that, to write it in grams would require 23 zeros after the decimal point, before writing the first non-zero number!

# 1 Numbers and number systems

## Civilizations and human interactions

## KEY CONCEPT: FORM

## Related concepts: Representation and Systems

## Global context

In this unit you will discover a range of number systems that have arisen as civilizations evolve and humans interact, which will help you explore the global context of **orientation in space and time**. By learning about numbers, their properties, and how to perform operations with them, you will realize how important numbers have been throughout history.

## Statement of Inquiry

Different systems and forms of representation develop as civilizations evolve and humans interact.

### Objectives

- Researching, representing and comparing number systems
- Reading, writing, simplifying and converting between different forms of numbers
- Simplifying numerical expressions using the order of operations
- Using appropriate forms of rounding to estimate results
- Defining and using the divisibility rules
- Representing a number as an exponent, a square root, and as a product of its prime factors, in order to solve problems

### Inquiry questions

**F** What is a number?
What is a system?

**C** How are the ways we represent quantities related?
How does the way we represent something affect its usefulness?

**D** How much are we influenced by the events of the past?

**ATL1 Collaboration skills**

Listen actively to other perspectives and ideas

**ATL2 Information literacy skills**

Present information in a variety of formats and platforms

## You should already know how to:

- identify the place value of specific digits in numbers
- write numbers in numeric forms and also in words
- write numbers in expanded form
- multiply and divide

# Introducing numbers and number systems

When did humans first use numbers? What did they use them for? Do animals use numbers? These are all questions that may never be answered definitively, but the discovery of the Ishango bone in 1960 in what is now the Democratic Republic of Congo is evidence that humans first developed a mathematical system more than 20 000 years ago. The bone is the fibula of a baboon with groupings of notches cut into it that could represent counting or perhaps even something more advanced. Some scientists believe it shows mathematical succession and can be thought of as a prehistoric calculator of sorts.

Cave paintings found in Lascaux, France, seem to demonstrate early humans' ability to distinguish between one animal and many animals. Although they did not have a system of numbers, they were able to communicate the idea of 'how many'.

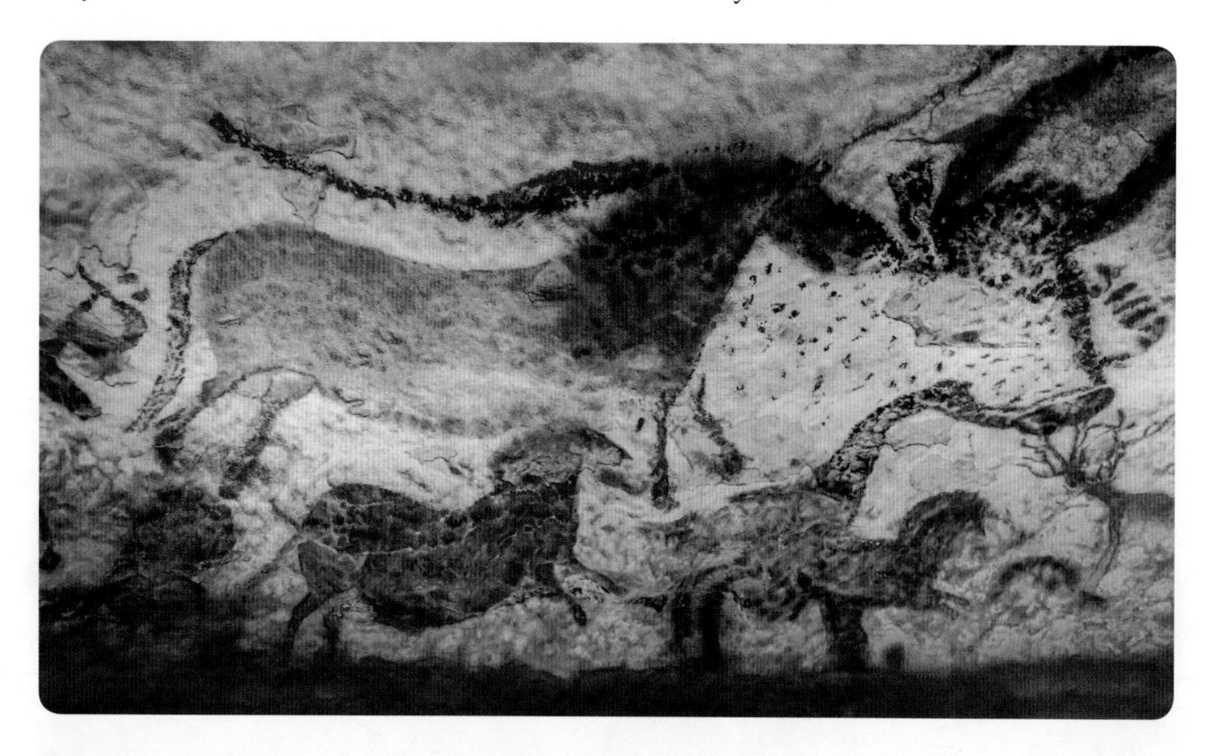

There is evidence that, in approximately 4000 BC, the Sumerians used numbers and counting. As one of the earliest civilizations, the Sumerians grew crops, raised livestock and even traded and acquired goods. As civilizations grew and became more complex, the need also grew for a number system that could handle tasks that required counting and performing operations with numbers.

## Reflect and discuss 1

In this unit, you will use the 'Pay attention, pause and paraphrase' (3P) strategy to build active listening skills. In this exercise, the emphasis is on paying attention. For each question below, have one person answer while everyone else simply listens. When others have their turn, they must add something new to the discussion rather than repeating what has already been said ('additive, not repetitive').

- Why do you think the Sumerians needed numbers? Explain.
- What evidence might have been found to support the idea that the Sumerians counted or used numbers?
- What evidence could there be to support the idea that some animals seem to have the ability to count?

# Representing quantities

## Activity 1 – The importance of a number system

Today you will go back in history, to a time when people did not speak English or any other modern language, and our current number system did not exist. You can do this activity in or outside of class. You will be in a group of four, and each group will be a family. One at a time, you will leave a message to the rest of your family telling them about the number of wild animals that you saw one day. (In North America, for example, that could include buffalo. In South America, it might be alpaca.)

One by one, with the rest of your family looking away, you will create a message for your family based on one of the scenarios below. You cannot use letters or words, nor can you use any of our current numbers. However, you can use any object(s) you see to help you. Once your message is ready, let your family look at it and try to guess what it represents. Once they have guessed correctly, the next person will leave their message.

**Person 1**

Pick a number between 1 and 5 and represent that as the number of people in your hunting party.

**Person 2**

Pick a number between 50 and 70 and represent that as the number of animals that were captured.

**Person 3**

Pick a number between 200 and 500 and represent that as the number of animals you saw.

**Person 4**

Pick a number between 1500 and 2000 and represent that as the number of animals you saw.

ATL1

## Reflect and discuss 2

You will again use the 'Pay attention, pause and paraphrase' (3P) strategy to build active listening skills. In this exercise, the emphasis is on pausing. For each question, have one person answer while everyone else simply listens. There must be a three-second pause before the next person speaks. Again, subsequent speakers' contributions must be additive, not repetitive.

- How did you figure out what was being communicated?
- Which person used a system that was the most effective? Explain.
- How did the system change so that large numbers could be represented?
- What is our number system based on? In other words, how do we group things? Why do you think we do this?

## Did you know...?

In many parts of the ancient world, clay tokens were used to keep track of items (jars of oil, amounts of grain, etc.) since there was no written number system. Some tokens represented groups of items while others represented single items. Because of this, humans now had a basic way of accounting and recording transactions. Of course, they would eventually require a more sophisticated way to store information.

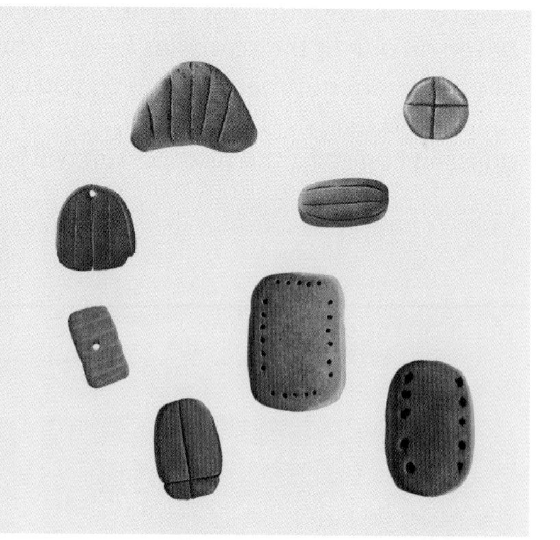

The number system that we currently use in mathematics is the *decimal number system*. Our number system is based on the number 10. In Greek, *dec* means 10, hence the word *decimal*. We use a place value system based on the number 10. Numbers are a way of representing an amount of something – a quantity – so it is important to be able to use our number system correctly and accurately.

The units place and the decimal point form the center of the place value system. The decimal system is made up of families (thousands, thousandths, millions, millionths, and so on), shown in bold in the following diagram.

Each place value family has three members:

- the units (or ones)
- the tens
- the hundreds.

Each family is separated by a space.

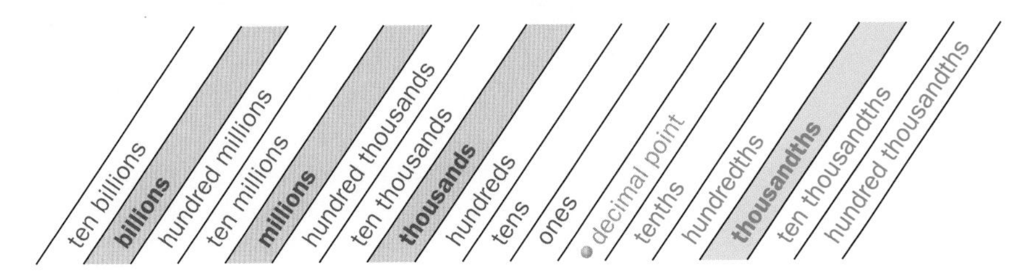


When reading a number you say the family name at the end of each family.

## Example 1

Read out the number 2 374 158 in words.

Two **million**, three hundred seventy-four **thousand**, one hundred fifty-eight

The family names are shown in bold.

To read decimal numbers:

- Ignoring the decimal point, read the decimal number part like a whole number.
- At the end of the number, say the place value family name of the last digit.

## Example 2

Q Read out the number 0.325 in words.

A Three hundred and twenty-five **thousandths**

Three decimal places puts this number into the thousandths place value family.

To write whole numbers from words:

- Determine the place value name with the greatest value given in the number. This tells you how many places you need.
- Put as many blanks as there are places in the number.
- Fill in the blanks with the digits needed; use zeros as place holders.

## Example 3

Q Represent the number *four hundred seventy-two million thirty thousand and sixteen* using digits.

A _ _ _   _ _ _   _ _ _

The greatest place value family is the millions, so you will need at most 9 digits.

4 7 2   _ 3 0   _ 1 6

Grouping digits by place value family name can help with large numbers. Fill blanks with zeros.

The number is 472 030 016

To write decimal numbers from words:

- Determine the place name of the last digit of the number. This tells you how many places there should be in the number.
- Fill in the spaces so that the number reads correctly. If necessary, put zeros in as place holders.

## Example 4

Q Represent *twenty-five thousandths* using digits.

A 0. _ _ _

The place value name family is *thousandths* so there are three places after the decimal.

0. 0 2 5

The first digit in the number is 0 as a place holder.

The decimal system makes it easy to represent numbers in different forms. Because each numeral has a specific value based on its place in the number, you can *expand* a number in several ways:

$164.34 = 100 + 60 + 4 + 0.3 + 0.04$

or

$164.34 = (1 \times 100) + (6 \times 10) + (4 \times 1) + (3 \times 0.1) + (4 \times 0.01)$

## Did you know...?

In the 6th and 7th centuries BC, Indian cosmologists needed to be able to represent large numbers, such as the age of the universe, so they used a set of ten symbols. This system was introduced to Arab mathematicians who would later introduce the system to Europe. This is the origin of the Hindu-Arabic number system and the decimal system we use today.

| Hindu | ० | १ | २ | ३ | ४ | ५ | ६ | ७ | ८ | ९ |
|---|---|---|---|---|---|---|---|---|---|---|
| Arabic | ٠ | ١ | ٢ | ٣ | ٤ | ٥ | ٦ | ٧ | ٨ | ٩ |
| Medieval | o | I | 2 | 3 | ߘ | ϥ | 6 | ʌ | 8 | 9 |
| Modern | 0 | 1 | 2 | 3 | 4 | 5 | 6 | 7 | 8 | 9 |

## Practice 1

**1** Read out these numbers in words.

**a** 401 302 059 677 **b** 496 812.03 **c** 0.0243

**d** 0.08570062 **e** 302.005 **f** 1.023

**g** 23 407.6 **h** 2 001.295 **i** 900 090 090

**2** Represent the following numbers using digits.

Eight hundred twenty-seven thousand ninety-six

Fifty thousand eight

Two hundred seventy-five thousand sixty-nine and ten millionths

Fourteen and six hundredths

Sixteen million eight hundred thousand six and seven ten thousandths

Five thousand and four tenths

Two and seven hundredths

Twelve thousandths

▶ Continued on next page

**3** Represent these amounts in words and in both expanded forms.

**a** A house costs \$640 700.

**b** At its closest, the distance from the Earth to the moon is 363 105.021 km.

**c** The diameter of a grain of sand is 0.0625 mm.

**d** At some time in 2017, the estimated global population was 7 492 115 325 people.

**e** The width of a human hair is about 0.0018 cm.

In part **3d**, you don't need to represent the year 2017. Also, putting it in expanded form would not serve much purpose. Or would it?

**4** Represent these numbers in both expanded forms.

**a** 385.901 **b** 6407.83 **c** 367 981.02 **d** 1 467 390.932

**e** 1003.2 **f** 12.5007 **g** 4.009 **h** 0.0104

**5** Write each number from its expanded form.

**a** $200 + 50 + 2 + 0.4 + 0.07$

**b** $(4 \times 1000) + (3 \times 100) + (8 \times 10) + (1 \times 0.1)$

**c** $3000 + 70 + 3 + 0.005$

**d** $(2 \times 10\,000) + (5 \times 100) + (7 \times 1) + (9 \times 0.0001)$

**e** $40 + 0.8 + 100 + 30\,000$

**f** $(6 \times 1\,000\,000) + (6 \times 1000) + (6 \times 1)$

**6** Represent each number in two other forms.

**a** 17 234.01

**b** $(3 \times 100) + (4 \times 1) + (7 \times 0.1) + (5 \times 0.01)$

**c** $2000 + 10 + 0.02 + 0.004 + 0.00005$

**d** 600.08

**e** $(2 \times 1000) + (5 \times 10) + (3 \times 0.01)$

**f** $(6 \times 10\,000) + (8 \times 100) + 8 + (7 \times 0.1) + (5 \times 0.001)$

Many decimal numbers end (or terminate), like the ones you have seen so far. There are others that never terminate and repeat themselves indefinitely. There are also decimal numbers that never terminate and never repeat themselves. In the next activity, you will research one of the more famous of these numbers: pi.

ATL2

## Activity 2 - Pi quest

In this activity, you will research the number pi and create a presentation (using software such as Prezi, Powerpoint, etc.) that includes the following information.

1 Pi is what kind of number? Write a brief definition for this kind of number.

2 What is the symbol that we use for pi? In which year was pi assigned this symbol, and who first used it?

3 What fraction is often used to approximate the value of pi? To how many decimal places is it a good approximation?

4 How many digits of pi have been discovered? Go to mypiday.com and enter your birthdate to find where your birthday occurs in the decimal representation of pi.

5 What day of the year is *Pi Day*? Why has this date been chosen?

6 Write one interesting fact that you have discovered about pi in your research, and discuss why you think it is interesting.

7 The Greek mathematician Archimedes (c 287 – c 212 BC) constructed a method for approximating the value of pi. Complete the 'Computing Pi' exploration on the NCTM Illuminations website to see how he did this. What fraction is known as the Archimedean value? Why? How did Archimedes use it?

8 Archimedes was not the first person to recognize pi. Research three ancient civilizations that used working approximations of pi and explain how they used it.

Could it really be possible that somewhere within the digits of pi *every* possible birthdate can be found? What do you think?

When the letter c is given before a date, it means the exact date is unknown. The c stands for *circa*, which is Latin for *around* or *about*.

# Other forms used to represent quantities

## Base-10 and base-60

Our system is base-10. There can be any one of 9 digits in any place value column. Once you go over 9, you need to start a new column. So our place value columns are 1, 10, 100, etc.

Ancient Babylonians used a base-60 system. The first digit is the 'ones' digit, as in our system, and it can be any number from 0 to 59. Once you go over 59, you need to start a new column. This means that the next column is the '60' digit and indicates how many 60s are required to write the number. Because base-60 numbers can be difficult to read, the digits are often separated from one another, for example: 2 15.

## Example 5

**Q** **a** Represent the base-10 number 68 in base-60.

**b** Represent the base-60 number 2  14 in base-10.

**A** **a** $68 \div 60 = 1$, remainder 8

$68 = 1 \quad 8$

Divide the number by 60. As long as the quotient is less than 60, place it in the second column and place the remainder in the ones or units column.

**b** 2  14

$2 \times 60 = 120$

$14 \times 1 = 14$

The number in the second column is the number of 60s. The number in the first column is the number of 1s. The total of this is the value of the number in base-10.

$2 \quad 14 = 134$

### Did you know...?

The Babylonian civilization replaced the Sumerian civilization in Mesopotamia. While their base-60 system would seem to be more complicated than our base-10 system, they wrote their numbers in *cuneiform*, which only had two symbols!

| Sumerian Cuneiform Numerals | | -1 |
|---|---|---|
| -2 | -3 | -4 |
| -5 | -6 | -7 |
| -8 | -9 | -10 |

ATL1

## Reflect and discuss 3

In this exercise, the emphasis is on paraphrasing. For each question, have one person answer while everyone else simply listens. When others have their turn, they must paraphrase what the previous speaker said before adding something new to the conversation. There must be a three-second pause between speakers.

- Why do you think the base-60 number system was appealing to ancient Babylonians?
- How is their system similar to ours?
- Can you think of where you use a system based on the number 60?
- Why should you have an understanding of how the base-60 number system works?

## Braille

The symbols in this section are numbers in Braille, which is a system of raised dots that can be 'read' with fingers by people who are blind or have reduced vision. Each symbol can be recognized by its position and the number of dots within a braille cell, which is two parallel vertical columns each having 0, 1, 2 or 3 dots in it. Braille is not a language in its own right, but rather a code which can be written or read in many languages around the world.

| This is the symbol for the word *number*. It appears before every number that is written so that when people read Braille they know that a number comes next. | 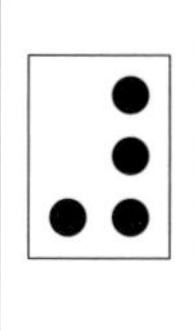 |
|---|---|

Numbers in Braille each have their own representation, which happen to match the first ten letters of the English alphabet. The *number* symbol placed in front of other symbols lets the reader know to interpret what follows as numbers instead of letters.

Here are the Braille symbols for the letters a to j:

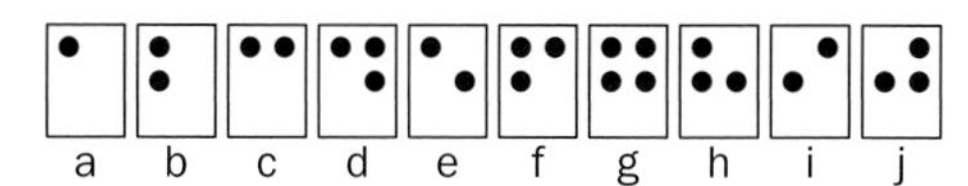


When preceded by the *number* symbol, these same ten symbols represent the ten digits:

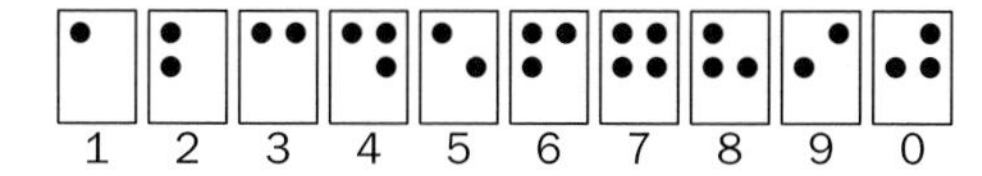


Therefore:

= hide

= 8945

| This is the Braille symbol for the decimal point. | |
|---|---|

Writing numbers in Braille requires putting the symbols for the digits in order.

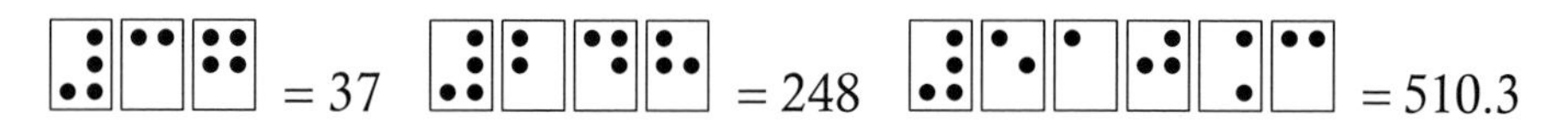


Note that the borders drawn around the symbols shown here do not actually feature in Braille. They are shown here only to separate the symbols from one another.

Writing expressions with numbers requires placing the following symbols between the values:

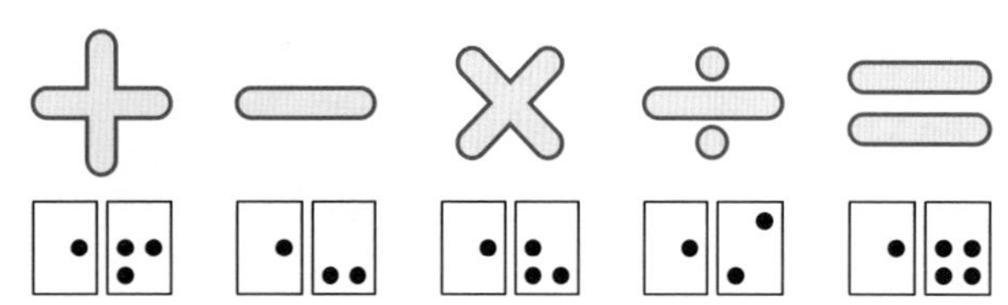

For example, $2 + 3 = 5$ is written as:

Note that the Braille symbol for 'number' precedes each of the three numbers in the statement.

## Reflect and discuss 4

- How difficult do you think it would be to learn Braille?
- How long do you think it would take you to read and write Braille without eyesight?
- Do you think it is important that you have a basic understanding of how Braille works?

## Did you know...?

Louis Braille lived in the 1800s in France. When he was young, he had an eye injury that got infected and left him blind in both eyes. Unable to read conventional books, he invented a system of raised letters and numbers that have allowed blind people all over the world to read. Braille symbols have been created for every known language.

## Practice 2

**1** Represent these base-10 numbers in base-60.

**a** 63 **b** 182 **c** 147 **d** 79

**e** 100 **f** 85 **g** 120 **h** 219

Continued on next page

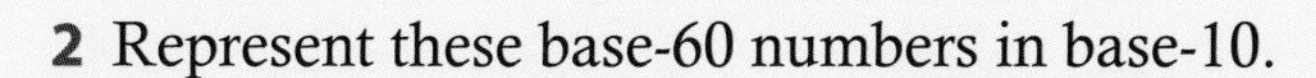

**2** Represent these base-60 numbers in base-10.

**a** 3 2 **b** 4 17 **c** 5 23 **d** 1 14

**e** 2 10 **f** 3 00 **g** 10 10 **h** 12 03

**3** Write down five base-10 numbers and five base-60 numbers. Then challenge a partner to convert them.

**4** Represent these numbers in Braille.

**a** 17 **b** 302 **c** 422 **d** 1705

**e** 981 **f** 2006 **g** 31.45 **h** 0.17

**5** Convert these Braille numbers to decimal form.

**a** **b** **c** 

**d** **e** 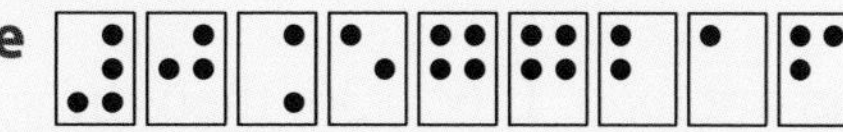

**6** Represent these Braille numbers in base-60.

**a** **b** **c** **d**

**7** The following two Braille lines are equations. Write them in decimal form.

**a**

**b**

**8** Find the value of the following expressions. Write your answer in Braille.

**a**

**b**

**c**

**d**

**e**

**f**

## Powers / indices

*Powers* or *indices* are a convenient form to use when writing a specific kind of quantity. Indices tell you how many times to multiply a number by itself. For example:

$$6^2 = 6 \times 6 = 36$$

The number 6 is called the *base* while the number 2 is called the *index* (or the *power*). The number $6^2$ can be read as '6 raised to the power 2'. This number is also referred to as 'a power of 6'.

When the power is 2, the base is often referred to as being *squared*, so $6^2$ is often read as 'six squared'. Likewise, when the power is 3, the term is *cubed*. Below is a representation of 'two cubed'.

$$2^3 = 2 \times 2 \times 2 = 8$$

Another word that is used to describe the index or power is *exponent*. You will often hear these terms used interchangeably.

## Roots

The opposite of powers are *roots*. If $6^2 = 36$, then the *square root* of 36 is 6. The square root symbol is $\sqrt{\ }$, so we write it as $\sqrt{36} = 6$.

Numbers whose square roots are whole numbers are called *perfect squares*.

Although square roots are the most common type of root, we can also work with *cube roots*, *fourth roots*, *fifth roots*, etc. When you have a square root, you ask yourself, 'What number multiplied by itself will equal the number under the root sign?' When you have a higher root, you are looking for the number that is multiplied by itself that many times. So, a fourth root (e.g. $\sqrt[4]{16}$) involves finding the number that is multiplied by itself *four* times to give the number under the root.

To denote a root that is not a square root we make a small modification to the symbol, for example: $\sqrt[3]{\ }$ and $\sqrt[4]{\ }$

$$\sqrt[3]{8} \rightarrow 8 = 2 \times 2 \times 2$$

$$\text{Therefore } \sqrt[3]{8} = 2$$

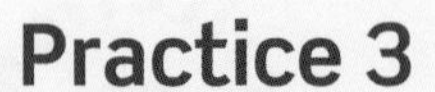

## Practice 3

**1** List the perfect squares up to 225. (They are good to remember so that you can calculate square roots quickly.)

**2** Express each number without indices.

| | | | |
|---|---|---|---|
| **a** $4^2$ | **b** $11^2$ | **c** $4^3$ | **d** $7^2$ |
| **e** $10^3$ | **f** $5^3$ | **g** $2^5$ | **h** $3^4$ |

**3** Evaluate these square roots.

| | | | |
|---|---|---|---|
| **a** $\sqrt{9}$ | **b** $\sqrt{81}$ | **c** $\sqrt{100}$ | **d** $\sqrt{324}$ |
| **e** $\sqrt{6400}$ | **f** $\sqrt{225}$ | **g** $\sqrt{2500}$ | **h** $\sqrt{4^2}$ |

**4** Evaluate the following roots.

| | | | |
|---|---|---|---|
| **a** $\sqrt[3]{27}$ | **b** $\sqrt[3]{125}$ | **c** $\sqrt[4]{16}$ | **d** $\sqrt[50]{1}$ |

**5** Find the square roots of these Braille numbers.

**a** ⠼⠊⠁ **b** ⠼⠃⠙⠁ **c** ⠼⠁⠊ **d** ⠼⠁⠙⠙ **e** ⠼⠊

## Classifying numbers

### Factors, multiples and prime numbers

A *multiple* is a number that can be divided *by* another number exactly (no remainder). A *factor* is a number that divides exactly *into* another number. For example, 4 is a factor of 12 and likewise 12 is a multiple of 4.

A *prime* number has only two factors: 1 and itself. Hence, 5 is a prime number since its only factors are 1 and 5. The number 6 is not prime since it has more than two factors: 1, 2, 3 and 6.

### Did you know...?

'The sieve of Eratosthenes' is a simple, ancient algorithm (mathematical procedure). Eratosthenes was a Greek scientist who devised an algorithm to filter out all of the prime numbers up to any given number. A sieve is a tool used to strain unwanted material from desired material, such as straining solids from liquids. In Eratosthenes' sieve, prime numbers are the desired material.

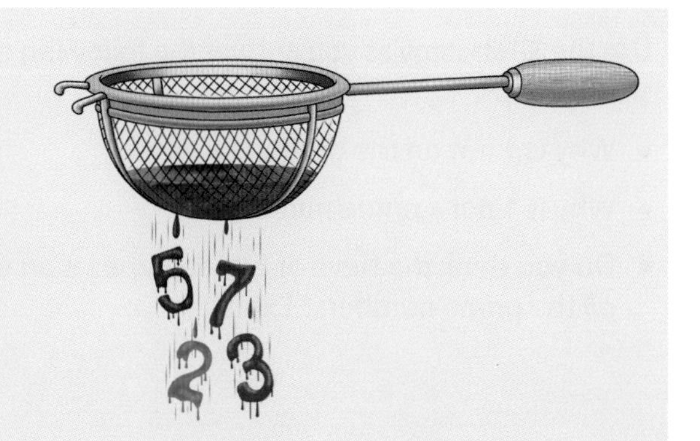

## Investigation 1 – The sieve of Eratosthenes

In this investigation you will determine what Eratosthenes' algorithm does.

|  | 2 | 3 | 4 | 5 | 6 | 7 | 8 | 9 | 10 |
|---|---|---|---|---|---|---|---|---|---|
| 11 | 12 | 13 | 14 | 15 | 16 | 17 | 18 | 19 | 20 |
| 21 | 22 | 23 | 24 | 25 | 26 | 27 | 28 | 29 | 30 |
| 31 | 32 | 33 | 34 | 35 | 36 | 37 | 38 | 39 | 40 |
| 41 | 42 | 43 | 44 | 45 | 46 | 47 | 48 | 49 | 50 |
| 51 | 52 | 53 | 54 | 55 | 56 | 57 | 58 | 59 | 60 |
| 61 | 62 | 63 | 64 | 65 | 66 | 67 | 68 | 69 | 70 |
| 71 | 72 | 73 | 74 | 75 | 76 | 77 | 78 | 79 | 80 |
| 81 | 82 | 83 | 84 | 85 | 86 | 87 | 88 | 89 | 90 |
| 91 | 92 | 93 | 94 | 95 | 96 | 97 | 98 | 99 | 100 |

You will need 4 colors (preferably highlighters) to do this investigation. Start with a grid of numbers from 2 to 100, like the one shown here.

1. The first number in the list is 2, but do not highlight the 2. Highlight every 2nd number in the grid *after* 2 (all the multiples of 2).
2. At this point, the lowest number not yet highlighted is 3. Do not highlight the 3, but highlight every 3rd number in the grid *after* 3 (all the multiples of 3).
3. The next lowest number not yet highlighted is 5, so highlight all the multiples of 5.
4. Now the next lowest number not yet highlighted is 7, so highlight all the multiples of 7.
5. Take a look at all of the numbers that are not highlighted. Describe the pattern of these numbers. What kind of numbers are they?

## Reflect and discuss 5

Use the 3P strategy as you answer the following questions in pairs or small groups.

- Why is 1 not on the grid?
- Why is 1 *not* a prime number?
- Do you think the sieve of Eratosthenes is an efficient way to find *all* the prime numbers? Explain.

## Factors and divisibility

A *factor* is a number that divides exactly into another number. For example, 3 is a factor of 6 since it divides exactly into 6, and so we say 6 is *divisible* by 3. Determining whether or not a number is divisible by another number can be time consuming when you have a large number to check. However, there are some rules that can help you determine this quickly. A *divisibility rule* is a faster way of determining if a given number is divisible by another number, using little or no division.

### Investigation 2 – Divisibility rules

criteria **B, C**

You will be developing these rules for multiples of 2 to 10 (except 7).

**1 Determine the divisibility rule for multiples of 2.**

- **a** Write down some small numbers (up to three digits) that are multiples of 2 and study the numbers to find a pattern.
- **b** Describe these patterns and generalize the divisibility rule for 2.
- **c** Test your rule with larger numbers (four digits and five digits) to see if it holds.

**2 Determine the divisibility rules for multiples of 5 and 10.**

- **a** Write down some small numbers that are multiples of 5. Study the numbers, find a pattern and generalize your rule. Be sure to test it for larger numbers.
- **b** Repeat for multiples of 10.

**3 Determine the divisibility rules for multiples of 4 and 8.**

- **a** Write down all of the two-digit numbers that are divisible by 4. Now select some larger numbers that are divisible by 4. What do you notice about your lists of two-digit and larger numbers?
- **b** Repeat these steps for multiples of 8.

**4 Determine the divisibility rules for multiples of 3 and 9.**

- **a** Write down some small numbers (up to three digits) that are multiples of 3. Study the numbers, find a pattern and generalize your rule. Be sure to test it for larger numbers.
- **b** Repeat this procedure for multiples of 9.

With **4b**, you may have to look for other patterns, such as adding digits together.

**5 Determine the divisibility rules for multiples of 6.**

Write down some small numbers (up to three digits) that are multiples of 6. Study the numbers, find a pattern and generalize your rule. Be sure to test it for larger numbers.

ATL2

**6** Once you have determined all of the divisibility rules, make a reference chart with them all on one page so you can readily access them during this unit.

For question **5**, you may want to look at combinations of other rules.

Go to aaamath.com. In the left-hand navigation bar, choose Division. Scroll down and click on Divisibility by 7.

## Reflect and discuss 6

Pairs

Use the 3P strategy as you answer these questions in pairs or small groups.

- Which divisibility rules are the easiest to remember?
- Did any divisibility rule surprise you in any way? Which rules are the most useful?

## Practice 4

1 Determine whether or not each number is divisible by 9 and explain your reasoning.
   **a** 135 **b** 1207 **c** 792 **d** 12 403 **e** 21 347 **f** 78 831

2 Determine whether or not each number is divisible by 4 and explain your reasoning.
   **a** 788 **b** 2140 **c** 3106 **d** 35 422 **e** 107 988 **f** 2 301 412

3 List all of the single-digit factors (1 to 9) of each of the following numbers.
   **a** 72 **b** 124 **c** 2505 **d** 1468 **e** 4096 **f** 12 240

4 Create two different four-digit numbers that are divisible by:
   **a** 2, 3 and 8 **b** 4, 6 and 9 **c** 2, 3, 5 and 9 **d** 4, 5, 6 and 9

5 Rugby is a game that is played internationally by both men and women. Points can be scored in a variety of ways. A 'kick' through a set of uprights is worth 3 points. A 'try' is scored when the ball is touched down inside the opposing team's End Zone and is worth 5 points. After scoring a try, a team can attempt a 'conversion', which is worth 2 more points.
   **a** Suppose the final score in a rugby match was 39–13. Could either team have scored on kicks alone? Explain.
   **b** Suppose the final score in a rugby match was 45–15. Could either team have scored on tries alone? Explain.
   **c** Determine one possibility for how the two teams scored their points if the final score was 41–22.
   **d** Determine two possibilities for how the two teams scored their points if the final score was 57–25.

## Greatest common factor (GCF)

Factors that are common to more than one number are called *common factors*. For example, 12 and 18 have 1, 2, 3 and 6 as common factors. Often we are interested in the *greatest common factor* (*GCF*), which is the largest factor that two or more numbers have in common. In the case of 12 and 18, the GCF is 6.

There are many ways to find the GCF of a group of numbers. Two such ways are described here.

### Example 6

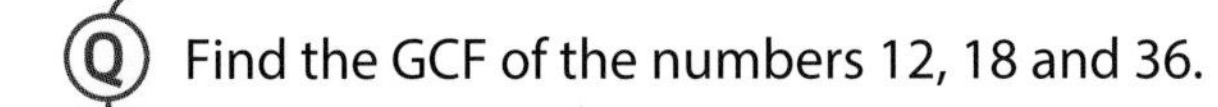

**Q** Find the GCF of the numbers 12, 18 and 36.

**A** 12: 1, 2, 3, 4, 6, 12
18: 1, 2, 3, 6, 9, 18
36: 1, 2, 3, 4, 6, 9, 12, 18, 36

The GCF is 6.

List all of the factors of each number and find the greatest one that occurs in all the lists.

Listing all of the factors can be time consuming and more difficult as the numbers get larger. In these cases, you can use the ladder method, demonstrated in Example 7.

### Example 7

**Q** Using the ladder method, find the GCF of the numbers 12, 18 and 36.

**A**

| | | | |
|---|---|---|---|
| **3** | 12 | 18 | 36 |

Find a factor of *all* the numbers. Write this factor in front of the three numbers, separated by a vertical line. Here, the factor identified is **3**.

| | | | |
|---|---|---|---|
| **3** | 12 | 18 | 36 |
| | 4 | 6 | 12 |

Divide each of the original numbers by this factor and write the quotients below the original numbers.

| | | | |
|---|---|---|---|
| **3** | 12 | 18 | 36 |
| 2 | 4 | 6 | 12 |

Now find a factor of this new set of numbers and write this in front (2). Divide.

| | | | |
|---|---|---|---|
| **3** | 12 | 18 | 36 |
| 2 | 4 | 6 | 12 |
| | 2 | 3 | 6 |

Continue until no more factors of *all* the numbers can be found.

$\mathbf{3} \times 2 = 6$

The GCF is 6.

Multiply all of the factors together (the numbers to the left of the vertical lines) to find the greatest common factor.

The Greek mathematician Euclid (c 300 BC) discovered a method for finding the GCF of two numbers, and first described this method in his famous mathematics book called *Euclid's Elements*.

In Euclid's method, you repeatedly subtract multiples of the smaller number from the larger number until you have a zero remainder. The last non-zero remainder is the GCF. The method is best understood by looking at a worked example.

## Example 8

**Q** Find the GCF of 490 and 1274.

**A** $1274 - (2 \times 490) = 294$

How many multiples of 490 will fit into 1274? Well, $3 \times 490$ would be greater than 1274. But $2 \times 490$ is less than 1274 by 294.

$490 - (1 \times 294) = 196$

How many multiples of 294 will fit into 490? Only 1, with 196 left over.

$294 - (1 \times 196) = \mathbf{98}$

196 is less than 294 by 98.

$196 - (2 \times \mathbf{98}) = 0$

98 is the last non-zero remainder.

The GCF of 490 and 1274 is 98.

## Activity 3 – Euclid's method

Find the GCF of each pair of numbers using both the ladder method and Euclid's method.

**a** 72 and 180 **b** 24 and 132 **c** 104 and 1200

**d** 360 and 1024 **e** 256 and 1600 **f** 480 and 2040

ATL1

## Reflect and discuss 7

Pairs

Use the 3P strategy as you answer the following questions in pairs or small groups.

- Which method do you prefer, the ladder method or Euclid's method? Explain.
- Describe the similarities and differences between the two methods.
- How does Euclid's method work?

## Practice 5

1 Find the GCF of each set of numbers. Make sure your chosen method is clear.

**a** 24, 36 **b** 48, 64 **c** 54, 81 **d** 15, 28

**e** 21, 35, 56 **f** 120, 56, 96 **g** 168, 252, 240 **h** 180, 300, 720

**i** 75, 200, 350 **j** 16, 48, 144 **k** 7, 14, 30 **l** 18, 28, 38

2 Another popular method for finding the GCF is the 'factor tree method', which you may have already seen. If not, research this method and try a few examples with it. Of the three methods (the ladder method, Euclid's method and the factor tree method), which do you prefer, and why?

### Lowest common multiple (LCM)

As with factors, different numbers will have common multiples. We are often interested in the *lowest common multiple* (*LCM*), which is the smallest multiple that the numbers have in common.

The next two examples show two methods for finding the LCM of a group of numbers.

## Example 9

Q Use the ladder method to find the LCM of the numbers 4, 8 and 10.

A
```
2 | 4   8   10
```
Find a factor of *at least* two of the numbers. The more of the numbers your factor divides into, the more efficient the process. Write this factor in front of the three numbers, separated by a vertical line.

```
2 | 4   8   10
    2   4    5
```
Divide each number by the factor (2, in this example) and write the quotients below the original numbers.

```
2 | 4   8   10
2 | 2   4    5
    1   2    5
```
Now find a factor (2 again, here) of at least two of the new numbers and write it in front. If a number does not have that factor simply carry it down to the next row. (Here, 5 does not have 2 as a factor.)

```
2 | 4   8   10
2 | 2   4    5
    1   2    5
```
Continue until no more factors can be found. Multiply all the factors and **remainders** together to find the lowest common multiple.

$2 \times 2 \times \mathbf{1} \times \mathbf{2} \times \mathbf{5} = 40$

The LCM of 4, 8 and 10 is 40.

## Example 10

Find the LCM of the numbers 4, 8 and 10.

4: 4, 8, 12, 16, 20, 24, 28, 32, 36, **40**, 44, 48, ...
8: 8, 16, 24, 32, **40**, 48, 56, 64, ...
10: 10, 20, 30, **40**, 50 , 60, ...

The LCM of the numbers 4, 8 and 10 is 40.

List multiples of each number (starting from the lowest multiple) until you find a common one.

## Activity 4 – Mystery numbers

**1** Read through the set of clues to determine the mystery number in each case.

**a** What number am I?

- I am a number between 40 and 70.
- I am a prime number.
- My digits add up to 8.

**b** What number am I?

- I am a four-digit number.
- My hundreds digit is 8.
- The difference between my hundreds digit and my ones digit is 1.
- My thousands digit is one quarter of my hundreds digit.
- My tens digit is triple that of my thousands digit.
- All my digits add up to 23.

**c** What number am I?

- I am a three-digit number.
- I am a multiple of 6.
- My tens digit is eight times my hundreds digit.

Continued on next page

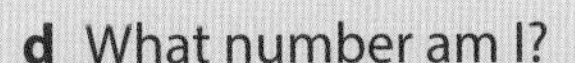

**d** What number am I?

- I am a five-digit odd number.
- The sum of my digits is 23.
- My ten thousands digit is twice my ones digit.
- My thousands digit is one third of my ten thousands digit.
- My tens digit is 4.
- If you multiply my hundreds digit by my ones digit, the product is 24.
- There is no 1 or 9 in my number.

Pairs

**2** Now your teacher will put you in groups of two or three, and each group will be given a whiteboard or paper. You will be given clues similar to the ones above and you will be asked to write down the number you think it is. You will hold up your answer when asked, and the groups with the correct number will proceed to the next round until there is one winning group.

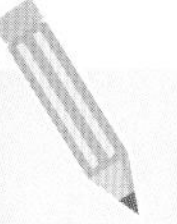

## Practice 6

**1** Find the LCM of each set of numbers.

**a** 4, 6 **b** 8, 12 **c** 12, 64

**d** 24, 32 **e** 35, 85 **f** 18, 54, 66

**g** 16, 20, 80 **h** 48, 68, 94 **i** 108, 180, 396

**2** Another popular method for finding the LCM is the 'factor tree method', which you may have already seen. If not, research this method and try a few examples with it. Of the three methods (the ladder method, writing a list and the factor tree method), which do you prefer, and why?

## Reflect and discuss 8

Pairs

- Finding the lowest common factor is not of interest to mathematicians. Explain why.
- Finding the greatest common multiple is not of interest to mathematicians either. Explain why.
- When using the ladder method, what are the differences between finding the GCF and LCM? Why can you not use the same method for both?

## GCF/LCM application

When faced with a word problem you need to determine what it is you have to find, the GCF or the LCM. Which strategy will you use to solve the problem? The next activity will help you form strategies for doing this.

### Activity 5 - GCF and LCM strategies

In groups of four, come up with specific strategies to follow when determining what you are finding, the GCF or the LCM. What will GCF word problems be asking you to do? What will LCM word problems be asking you to do?

Use the following questions to help you come up with your strategies, and then solve the problems.

1. Sliced ham comes in packages of 10. Cheese slices come in packages of 8. Bread rolls come in packages of 12. What is the least number of all three items you must buy in order to make the same number of ham and cheese sandwiches with nothing left over?
2. You have a piece of construction paper that measures 32 cm by 48 cm. You want to cut it into squares of equal size. What will be the dimensions of the largest possible square? How many squares will you have?
3. At a train station, the blue line has a train leaving every 15 minutes, the green line has a train leaving every 24 minutes, and the red line every 10 minutes. If the first train on each line leaves at the same time, how often will there be trains on all three lines departing the train station at the same time?

4. You have 54 hockey cards, 72 baseball cards, and 63 basketball cards and you want to put them in a binder. Each page of the binder should have cards from a single sport, and there should be the same number of cards on each page. What is the greatest number of cards you can put on a page? How many pages will you need for each sport?

As a class, come up with a list of key words/phrases to look for when solving these types of problems.

## Example 11

**Q** Your class is made up of 12 girls and 8 boys. Your teacher wants to split the class into groups, with each group containing the same combination of girls and boys, making sure that no one is left out. What is the greatest number of groups your teacher can make?

**A** What are you being asked to find?

You are breaking the class into specific groups – so you are finding a smaller number – so find the GCF.

8: 1, 2, 4, 8

12: 1, 2, 3, 4, 6, 12

The GCF is 4 .

Find the GCF using your preferred method. Since the numbers here are small, simply list the factors of each and identify the largest one in common.

There are 4 groups, with 3 girls and 2 boys in each.

Look at what the question is asking to make sure you fully answer the question.

## Example 12

**Q** The eastbound bus leaves the depot once every 18 minutes. The westbound bus leaves once every 30 minutes. If the first bus on each line leaves at 3 pm, what time will it be when they again leave at the same time?

**A** What are you being asked to find?

You are trying to determine when they will next leave at the same time – so you are finding a larger number than both – so find the LCM.

18 : 18, 36, 54, 72, 90, 108

30: 30, 60, 90, 120

The LCM is 90.

Find the LCM using your preferred method.

Since the buses will leave the depot at the same time every 90 minutes, the next time they do so will be at 4:30 pm.

Look at what the question is asking to make sure you fully answer the question.

## Practice 7

> ₽ is the symbol used for Russian rubles.

1. At the 2018 World Cup in Russia, one group of tickets cost ₽975 while another group of tickets in the same category cost ₽1170. What is the most a ticket could cost?

2. Saturn orbits the sun every 12 years, Jupiter every 30 years, and Uranus every 84 years. In how many years from now will all three planets have the same positions in the sky as they do today?

3. You are making emergency relief kits to give to bushfire victims. Bottles of water come in crates of 30; there are 24 cans of food in each carton; and beef jerky protein packs are in boxes of 80. You would like to create kits with one in each, with nothing left over. What is the smallest number of kits you can make?

4. Quilting (stitching together of layers of fabric) dates back to ancient Egypt, and the oldest quilt discovered is from about 5500 years ago. Today, many bedspread quilts consist of small squares sewn together. If you wanted to create a quilt for your bed measuring 42 inches by 78 inches, with squares of quilted fabric, what are the dimensions of the largest squares you could use? How many squares would there be?

5. The Mayans of south-eastern Mexico and northern Central America had two calendars of different lengths. The Haab' is 365 days and follows the solar year. There is also a sacred calendar called the Tzolk'in, which has a 260-day cycle. The *Calendar Round* is made from the connection between the two calendars. The duration of the Calendar Round is when both calendars start again on the same day.

   **a** Calculate the duration of the Calendar Round.

   **b** The Mayans believed that when a person lives for the duration of the Calendar Round, they will attain the special wisdom of an elder. At what age (in years) is this using the Haab' calendar?

## Calculating with numbers

Many math problems can be solved in your head using 'mental math' strategies. Some of these strategies take advantage of the place values in the decimal system.

### Addition

For addition, it is often easier to start by grouping numbers that sum to 10, or by finding multiples of 10.

## Example 13

**Q** Find the value of each summation.

**a** $6+2+4+7+8$

**b** $32+51$

**A** **a** $(6+4)+(8+2)+7$

$=10+10+7=27$

Rearrange to pair numbers that add to 10.

**b** $30+2+50+1$

$=(30+50)+(2+1)$

$=80+3=83$

Separate the numbers into tens and units.

Rearrange.

## Multiplication

Multiplying numbers in your head may seem difficult, but there are a few instances where it can actually be very easy, and even fun. You will discover two of these in the investigation below.

## Investigation 3 – Multiplying by 10 and 11

**1** Multiply the following without using a calculator.

**a** $10\times 24$ **b** $123\times 10$ **c** $574\times 10$ **d** $10\times 8105$ **e** $10\times 54203$

**f** $10\times 2.4$ **g** $3.19\times 10$ **h** $12.042\times 10$ **i** $100.5\times 10$ **j** $10\times 19.27$

**2** Generalize a rule for multiplying any number by 10. Be sure your rule includes whole numbers as well as decimal numbers.

**3** Multiply the following without using a calculator.

**a** $11\times 23$ **b** $11\times 35$ **c** $62\times 11$ **d** $81\times 11$ **e** $11\times 54$

**4** Do you notice any patterns in your answers to question **3**? How do your answers relate to the original numbers?

**5** Generalize a rule for multiplying a two-digit number by 11.

**6** Use your rule with each of these multiplications and then verify your answers without using a calculator.

**a** $11\times 16$ **b** $25\times 11$ **c** $11\times 44$ **d** $76\times 11$ **e** $11\times 49$

**7** Modify your rule based on your results in question **6**. What is a shortcut for multiplying by 11?

## Reflect and discuss 9

Pairs

- How does the multiplication rule for 10 work? Explain using the expanded form of a number.
- How does the multiplication rule for 11 work? Explain.
- Do you think your rule for 11 will work with three-digit numbers? Try an example to find out.

## Practice 8

**1** Find the value of each calculation.

**a** $24 + 62$ **b** $1 + 8 + 9 + 4 + 2$ **c** $5 + 3 + 7 + 6 + 5$ **d** $76 + 83$

**e** $340 + 220$ **f** $6 + 8 + 12 + 14 + 20$ **g** $400 + 1600$ **h** $98 + 45$

**i** $10 \times 34.2$ **j** $11 \times 61$ **k** $89 \times 11$ **l** $1230 \times 10$

**m** $8 \times 15$ **n** $10 \times 3.14159$ **o** $15 \times 16$ **p** $4001 \times 10$

**2** Find the sum of all the whole numbers from 1 to 20 inclusive. Explain your method. Do not add every single number! Think of a strategy to make it easier.

**3** Find the sum of all the whole numbers from 1 to 100 inclusive. Explain your method. Do not add every single number! Think of a strategy to make it easier.

### Activity 6 – Matho!

This game is like Bingo!, except that there are multiplications involved.

Your teacher will hold up a card with a multiplication on it, from $0 \times 1$ up to $9 \times 9$. You will do the multiplication in your head and, if you have that number on your grid, you cross it off. If you manage to cross off four numbers in one of the configurations listed below, shout out *Matho!* to win the game.

The teacher's set of cards contains no repeats. Hence, there will be one card for say $4 \times 5$ or $5 \times 4$, but not both.

Your grid will be four squares by four squares, like the example shown here.

You can choose any one- or two-digit numbers you want, but you may use each number only once.

| | | | |
|---|---|---|---|
| 18 | 32 | 25 | 44 |
| 12 | 72 | 14 | 9 |
| 65 | 16 | 8 | 64 |
| 81 | 42 | 54 | 19 |

Continued on next page

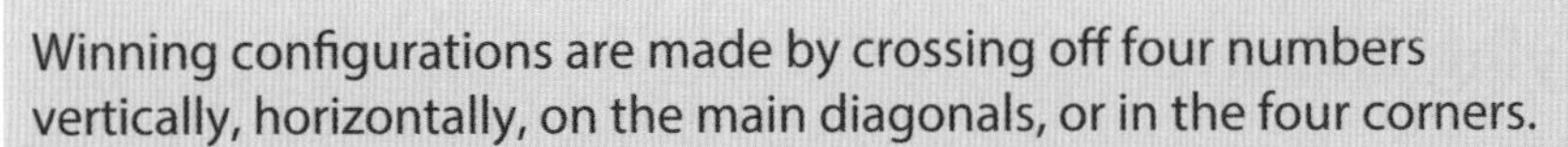

Winning configurations are made by crossing off four numbers vertically, horizontally, on the main diagonals, or in the four corners.

You will play this game for six rounds:

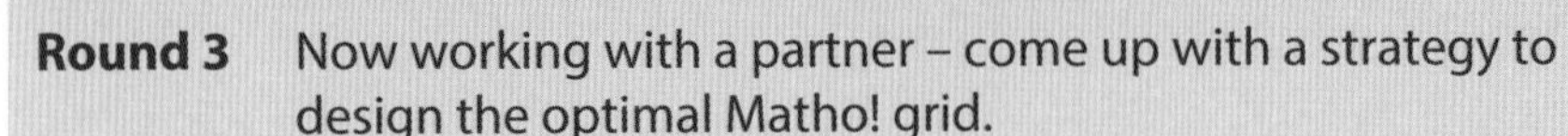

**Round 1** On your own – choose any numbers you wish.

**Round 2** On your own – do want to change any numbers?

Pairs

**Round 3** Now working with a partner – come up with a strategy to design the optimal Matho! grid.

**Round 4** Still with your partner – do either of you or both of you want to change any numbers?

**Round 5** Still with a partner (or perhaps for homework), – double-check that you have the optimal Matho! grid. Write down your strategy for choosing the numbers.

**Round 6** Final game – start with the winning group leading a discussion on what strategies they used to create the best grid.

## Reflect and discuss 10

Pairs

- What strategies helped create the optimal Matho! grid?
- How predictable were the numbers that were called?
- How effective was the strategy that you used to create your grid?
- What numbers in the example Matho! grid shown on page 30 were poor choices?

## Did you know...?

The abacus is an instrument first used by the Chinese in approximately 500 BC. It was used to do calculations with numbers involving addition, subtraction, multiplication, division and even square roots. The art of using an abacus is still taught in schools in some countries, and there are competitions for speed and accuracy. But that's not all. There are even competitions for students who can perform huge calculations at lightning speed, using only an imaginary abacus – one that they picture in their minds!

## Activity 7 – Multiplying visually

We can try to visually represent numbers in other ways in order to make them easier to work with. Take a look at this visual method for multiplying two numbers together which originated in Japan.

Example: Multiply 14 by 32.

**Step 1:** Represent the number 14 using 1 line and 4 lines:

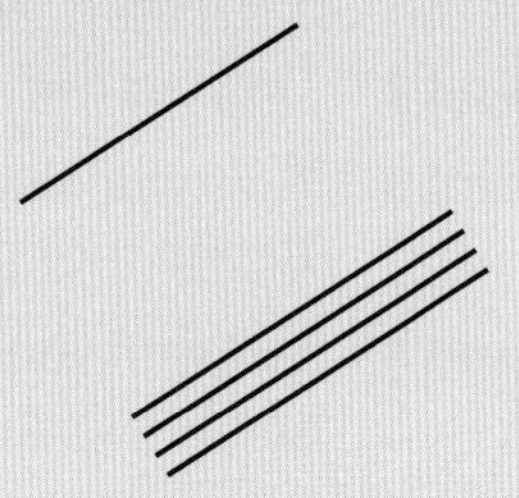

**Step 2:** Represent the number 32 using 3 lines and 2 lines, drawn perpendicular to the lines for 14:

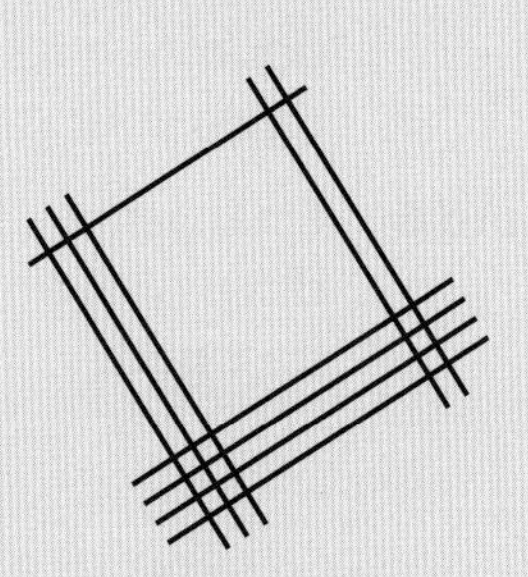

**Step 3:** Count the number of times the lines intersect at the three different vertical zones (in this case hundreds, tens and units). Note that there are two intersections in the top part of the vertical zone for tens, and twelve intersections in the bottom part.

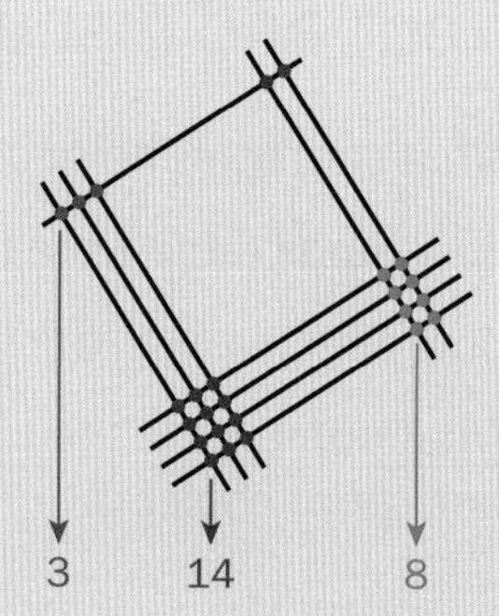


**Step 4:** If the number of intersection points in any zone is greater than 10, carry the number in the tens column into the next higher place value zone:

**Step 5:** Put the numbers together to read your answer!

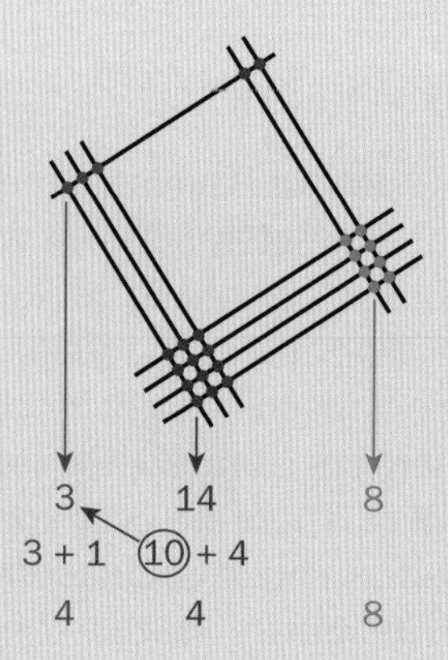


Therefore, $14 \times 32 = 448$.

You can watch an example on YouTube: search 'Japanese multiplication – how to multiply'.

Try using this method for the following calculations:

$32 \times 21$ $\quad$ $11 \times 18$ $\quad$ $46 \times 31$ $\quad$ $71 \times 25$

## Reflect and discuss 11

- Which step in this visual method is the most difficult?
- How would you represent a zero using this method? Explain your idea using an example.
- Will this method work for numbers greater than two digits? Try using this method to multiply a three-digit and a two-digit number together, and then two three-digit numbers. How many digits long would a number have to be before this method would not be efficient?
- Compare this method with traditional vertical multiplication. Which method is easier?
- Explain why this method works.

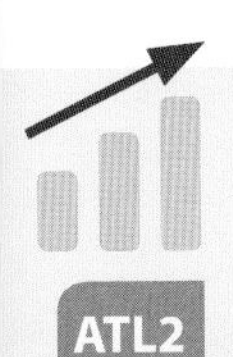

ATL2

## Formative assessment – The binary system

Civilization has come a long way since the abacus. Computers, laptops and tablets are now an important part of our lives. Instead of using the base-10 system, these devices use the binary or base-2 system. The simplest parts of most devices, including calculators, work like switches with just two positions or states: 'on' and 'off'. Because the binary system has only two numbers in it – zero and one – it was a perfect fit for an operating system for these devices.

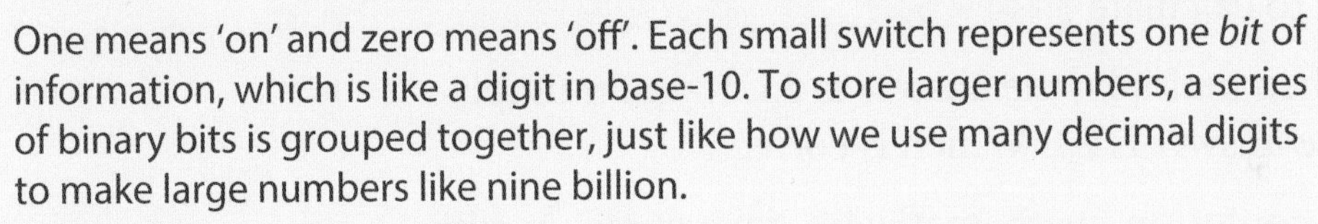

One means 'on' and zero means 'off'. Each small switch represents one *bit* of information, which is like a digit in base-10. To store larger numbers, a series of binary bits is grouped together, just like how we use many decimal digits to make large numbers like nine billion.

Your assignment:

**1** Research the binary system.

- When was it created?
- Where was it created? (Include a geographic reference)
- What was it used for?

**2** Find out about the basics of the binary system.

- How does it work?
- Does it use place value?
- How are numbers represented?
- How are 38, 95, 162 and 512 represented in binary?

**3** Find out about operations in binary.

- First, show an example of each mathematical operation (addition, subtraction, multiplication and division) using the decimal system.

Continued on next page

- Repeat the same calculations in binary.
- Generalize any rules for addition, subtraction, multiplication and division in binary.

You will create a short report (using word processing software) outlining these basic elements of the binary system. Be sure to use correct mathematical symbols, notation and vocabulary.

## Reflect and discuss 12

Why do you think computer and device storage capacities are given in sizes such as 32 Gb, 64 Gb, 512 Mb, and so on? What do you notice about these numbers?

## Performing many calculations

Mathematics is a language and, like any language, it follows certain rules. Mathematicians around the world follow an agreed order of operations so that when a calculation involves more than one operation, everyone gets the same answer.

Suppose you had to calculate $3 + 4 \times 5$.

If you add first, then multiply, the answer is 35. If you multiply first, then add, the answer is 23. Which is correct?

In order to avoid confusion, mathematicians have come up with rules about which operations come first. You will investigate this order of operations next.

## Investigation 4 – Order of operations

All the calculations below have been performed correctly. Look at the answers and determine the order in which the following operations should be done: addition, division, powers (indices), multiplication, parentheses (also called brackets), subtraction.

| | | | |
|---|---|---|---|
| $7+2\times4=15$ | $12-4\times2=4$ | $8-6\times0=8$ | $32\div2^2+3=11$ |
| $12\times2\div3=8$ | $3\times(8+2)=30$ | $(5-4)\times9=9$ | $15-4+7=18$ |
| $2+3^2=11$ | $(9-2^3)+15=16$ | $20\div10\times3=6$ | $2+24\div(2\times3)\times3=14$ |

ATL1

## Reflect and discuss 13

Pairs

Use the 3P strategy as you answer the following questions in pairs or small groups.

- Which comes first, multiplication or division? Explain.
- Try to create an acronym or a simple way to remember the order of operations.

Your teacher can share with you how he/she first learned how to remember the ordering of operations.

## Practice 9

**1** Evaluate each of these calculations using the correct order of operations.

**a** $8 + 6 \times 2$ **b** $\frac{18+6}{5+1} - 4$ **c** $(7^2 - 4)(5^2 - 2)$

**d** $3(2^4 - 11) + 3 \times 2^2 - 4$ **e** $51 \div (5^2 - 4 \times 2) \times 7 - \sqrt{100}$

**f** $\frac{5(3+4) - 2 \times 8 - 2^3}{3^2 \div (27 \div 3)}$ **g** $48 \div (4^2 - 4 \times 2) \times 6 - 6$

**h** $\frac{2^4 - 3 + 12 \div 4}{64 \div 2^3}$ **i** $5\left(4^2 - \sqrt{3^2 + 4^2}\right) - 7 \times 2^3$ **j** $\frac{3[(9-4) + 6 \times 3] - 3^2}{48 \div 4 - [3(6-2) - 10 \div 2]}$

**2** Replace each □ with an operation to make the statement true.

**a** 1 + 15 □ 3 = 6 **b** (18 □ 2) ÷ 10 = 2 **c** 15 □ 3 + 2 □ 5 = 15

**3** Insert brackets where appropriate to make each statement true.

**a** $18 - 6 \times 3 + 2 = 38$ **b** $8 + 4 \div 2 + 2 = 3$ **c** $5 + 3 \times 6 - 10 = 38$

**4** Change just one operation in each statement to make it true.

**a** $7 \times 3 + 4 \times 5 = 1$ **b** $(12 - 6) \times 3 + 10 \div 5 = 4$ **c** $24 - (10 - 6) + 2 \times 5 = 16$

**d** $12 + 36 + 9 \div 4 = 13$ **e** $6 \times 8 - 12 \times 2 + 7 = 15$ **f** $14 + 10 - 5 - 12 - 3 = 15$

**5** The game '24' is a number game where players are given four numbers and must use them to create an expression that equals 24. Players can use any of the operations (+, −, ×, ÷) as well as exponents and parentheses. For each group of four numbers listed below, write down an expression that equals 24 using *each number* once.

**a** 1, 5, 7, 8 **b** 3, 7, 8, 8 **c** 2, 3, 6, 10 **d** 7, 6, 10, 12 **e** 2, 6, 6, 9

▶ Continued on next page

**6** In Canada it is illegal to be awarded a prize for a 'game of chance'. So, when companies have contests, winners first have to answer a skill-testing question in order to claim their prize. One company's skill-testing question was to find the answer to this calculation:

$$100 + 3 \div 2 \times 5 + 6 \times 10 \div 2$$

**a** Find the correct answer.

**b** Find four other answers to this question that respondents are likely to give. Explain the errors that might be made.

## Activity 8 – flanking fifty!

For this activity, you will use part of one standard deck of playing cards. You will use nine cards, from 1 (Ace) to 9, from each of the four suits, making a total of 36 cards. Each team will be two or three students.

Teams are dealt their own set of five cards. Players use each card as a single digit to create any calculation they want, using any operations. The goal is for the answer to the calculation to be as close to 50 as possible. All five cards must be used and the correct order of operations must be followed. Your teacher will decide the length of time for each round.

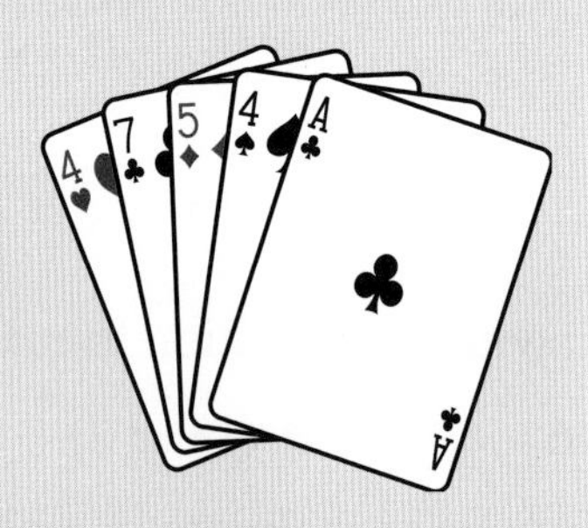

Points are scored by how many points you go over or under 50.

For each round, the deck is shuffled and five new cards are dealt to each team. The winners are the team with the least number of points at the end of five rounds.

## Did you know...?

Historically, as calculations became more difficult, tools were invented to do the computations more quickly than they could be done by hand. Until calculators became affordable and common items, the *slide rule* performed a wide range of calculations, but the user had to keep the order of operations in mind at all times. Today's *scientific calculators* follow the order of operations automatically.

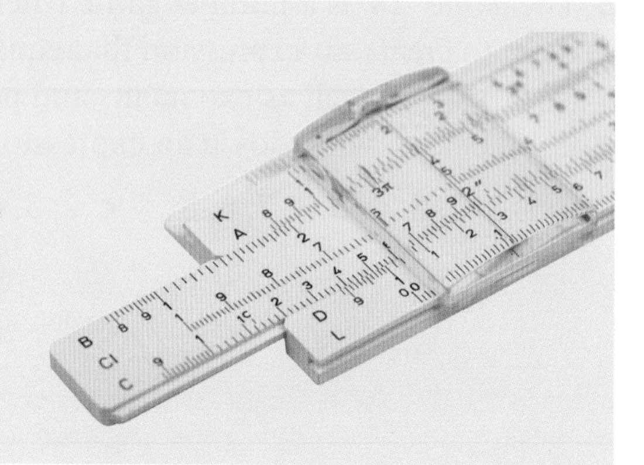

## Activity 9 - Bringing it all together

A mind map is a diagram used to visually organize information. It connects information around a central subject, which in this case is number sense. Create a mind map using all of the following words, showing all connections between concepts.

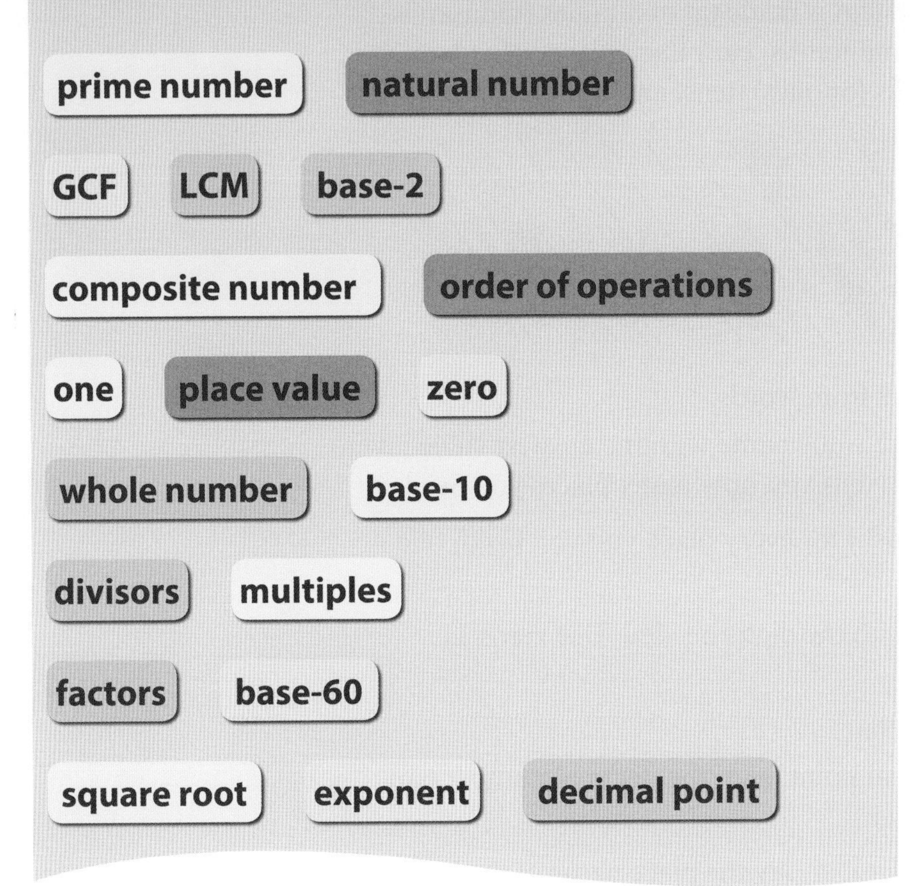

# Unit summary

The number system that we currently use is the decimal number system. We use a place value system based on the number 10. The units place and the decimal point form the center of the place value system. The decimal system is made up of families (thousands, thousandths, millions, millionths, etc.) Each place value family has three members: units (or ones), tens, and hundreds.

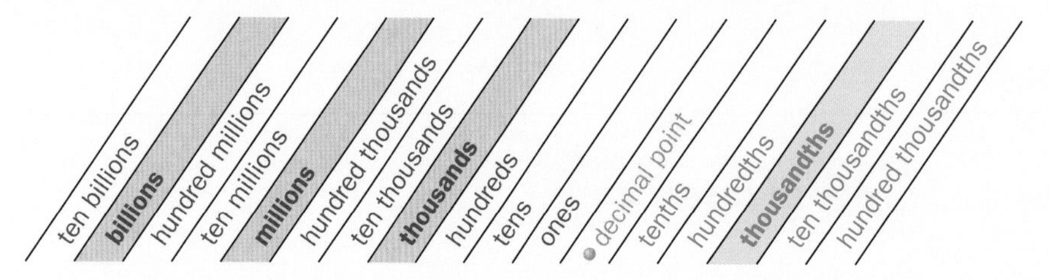


To write whole numbers from words:

1. Determine the place value name with the greatest value given in the number. This tells you how many places you need.
2. Draw a blank (a dash) for every place you will need, one per digit.
3. Fill in the blanks with the digits needed. Use zeros as place holders.

To read decimal numbers:

1. Ignoring the decimal point, read the decimal number part like a whole number.
2. At the end of the number, say the place name of the last digit.

To write decimal numbers:

1. Determine the place name of the last digit of the number. This tells you how many places there should be in the number.
2. Fill in the spaces so that the number reads correctly. If necessary, put zeros in as place holders.

**Expanded notation**

Examples:

$352.87 = 300 + [50 + 2 + 0.8 + 0.07$

or

$352.87 = (3 \times 100) + (5 \times 10) + (2 \times 1) + (8 \times 0.1) + (7 \times 0.01)$

## Divisibility rules

A number is divisible by a given number if the given condition is met:

2: the last digit is even
3: the sum of the digits is divisible by 3
4: the last two digits form a number that is divisible by 4
5: the last number is 5 or 0
6: the number is divisible by both 2 and 3
8: the last three digits form a number that is divisible by 8
9: the sum of the digits is divisible by 9
10: the last number is 0.

## Indices/Powers

Powers are a convenient way to write a number that is being multiplied by itself. For example: $7^2 = 7 \times 7 = 49$.

The opposite of powers are roots. If $7^2 = 49$, then the square root of 49 is 7. The square root symbol is $\sqrt{\ }$, so we write it as $\sqrt{49} = 7$.

Numbers whose square roots are whole numbers are called perfect squares.

To denote a root that is not a square root we make a small modification to the symbol. For example: $\sqrt[3]{\ }$ and $\sqrt[4]{\ }$ denote cube root and fourth root, respectively.

## Order of operations

Mathematicians have agreed upon the order of operations. The acronym **BEDMAS** is one of several acronyms that may help you remember the correct order:

**1** – **B**rackets
**2** – **E**xponents (or Indices)
**3** – **D**ivision and **M**ultiplication
**4** – **A**ddition and **S**ubtraction

Division and Multiplication are treated equally, so do whichever comes first when you read left to right. Likewise for Addition and Subtraction.

## Factors and multiples

A number is a *factor* of another number if it divides exactly into that number. For example, 5 is a factor of 60 and also a factor of 5, but not a factor of 24.

Factors that are common to more than one number are called *common factors*. The greatest common factor (GCF) is the largest of these factors.

A *multiple* is the result of multiplying a number by any other whole number.

*Common multiples* are multiples that are common to more than one number. The lowest common multiple (LCM) is the smallest multiple that two or more numbers have in common.

A prime number is a number whose only factors are 1 and itself.

## Unit review

Launch additional digital resources for this chapter

Key to Unit review question levels:

Level 1–2 | Level 3–4 | Level 5–6 | Level 7–8

1 **Write down** these numbers in words.

a 32 560 042 b 17.08 c 1743.14 d 2.718

2 Represent these written descriptions as numbers using digits.

a ten thousand four hundred fifty-five

b twenty nine and thirty-two thousandths

c one hundred ninety-thousand six hundred forty-five millionths

d eighteen thousandths

For a number such as 1047, some people say 'one thousand **and** forty seven' while others say 'one thousand forty-seven'. Can you think of a reason why the second way might be preferable?

3 Represent these numbers in words and in both expanded forms.

a A blockbuster film made €15 342 807 over a single weekend.

b At its furthest, the moon is 405 696.7 km from the Earth.

c The length of an eyelash mite is 0.1375 mm.

d In 2013, the population of India was 1 252 706 114 people.

e The width of a Siberian Husky's hair is 0.000025 m.

4 Represent these numbers in both expanded forms.

a 12.75 b 203.6 c 12 004.205 d 3.0019

5 **Write down** the given number from its expanded form.

a $500 + 10 + 8 + 0.02 + 0.001$

b $(3 \times 10\,000) + (6 \times 100) + (2 \times 10) + (7 \times 0.01)$

c $9000 + 10 + 0.01 + 0.008$

d $(4 \times 100\,000) + (6 \times 1000) + (2 \times 10) + (3 \times 0.01)$

6 Represent each given number in two other forms.

a 8231.04 b $(4 \times 1000) + (3 \times 10) + (5 \times 0.1) + (4 \times 0.001)$

c $8000 + 800 + 0.7 + 0.008 + 0.00002$ d 1400.12

7 Evaluate each number.

a $8^3$ b $12^2$ c $6^3$ d $13^2$

e $2^4$ f $3^3$ g $2^7$ h $5^4$

**8** **Use** a grouping strategy to find the value of each summation.

**a** $24 + 62$ **b** $1 + 8 + 9 + 4 + 2$

**c** $5 + 3 + 7 + 6 + 5$ **d** $76 + 83$

**e** $340 + 220$ **f** $6 + 8 + 12 + 14 + 20$

**g** $400 + 1600$ **h** $98 + 45$

**9** Represent these numbers in Braille.

**a** 28 **b** 362 **c** 12.3 **d** 90.12

**10** Convert these Braille numbers to decimal form.

**a** 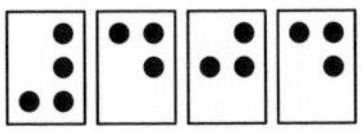 **b**

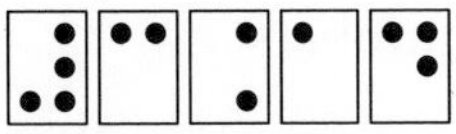

**c** 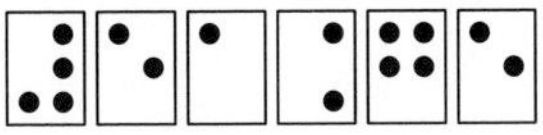 **d**

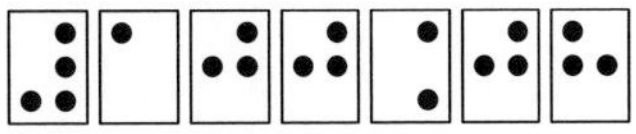

**11** Represent these Braille numbers in base-60.

**a** 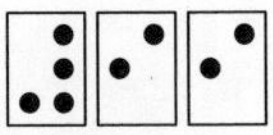 **b**

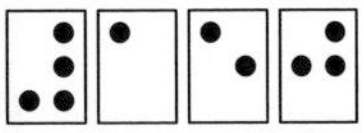

**c** 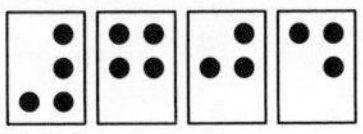 **d**

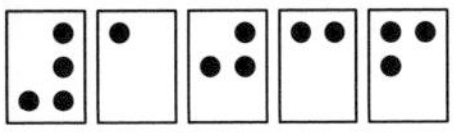

**12** Multiply these numbers using the visual multiplication method.

**a** $83 \times 21$ **b** $49 \times 16$

**13** Multiply these numbers using the visual multiplication method.

**a** $342 \times 17$ **b** $121 \times 213$

**14** Represent these base-10 numbers in base-60.

**a** 81 **b** 200 **c** 372 **d** 1248

**15** Represent these base-60 numbers in base-10.

**a** 1 14 **b** 3 05 **c** 2 43 **d** 5 00

**16** Evaluate these roots.

**a** $\sqrt{121}$ **b** $\sqrt{64}$ **c** $\sqrt[3]{1000}$ **d** $\sqrt[4]{16}$

**e** $\sqrt[3]{27}$ **f** $\sqrt{16}$ **g** $\sqrt{4^3}$ **h** $\sqrt{12^2}$

**17** Determine if each number is divisible by 3. **Explain** your reasoning.

**a** 204 **b** 1317 **c** 814 **d** 32 422 **e** 501 315 **f** 92 044

**18** Determine if each number is divisible by 8. **Explain** your reasoning.

a 1324 b 3080 c 71 256 d 4512 e 112 316 f 5 362 504

**19** List all of the single-digit factors (1 to 9) of each number.

a 64 b 256 c 1305 d 92 800 e 544 f 8280

**20** Find the GCF of each set of numbers. Make sure your chosen method is clear.

a 35, 42 b 144, 280 c 96, 240 d 18, 24, 42
e 12, 72, 156 f 126, 162, 180 g 120, 168, 216 h 90, 210, 3000

**21** Find the LCM of each set of numbers.

a 36, 54 b 45, 60 c 12, 30 d 9, 15, 30
e 8, 12, 30 f 18, 24, 36 g 11, 15, 21 h 14, 21, 28

**22** Celestial objects (like comets, planets and stars) have been observed by civilizations throughout history. Some people viewed these (especially comets and eclipses) as omens of future events while others interpreted them as signs from their gods.

Halley's comet orbits the Earth every 76 years while comet Swift-Tuttle has a 132-year orbit. If they were to pass by Earth in the same year, about how many years later would they again pass by Earth in the same year?

**23** Evaluate these expressions, using the correct order of operations.

a $10 + 12 \div 2$

b $12 - 3 \times 3 + 6 \times 2$

c $(3^3 - 12) \div (4^2 - 1)$

d $5(3^2 + 1) + 7 \times 3^2 + 8$

e $168 \div (2^4 + 2^3) + 11 \times \sqrt{169}$

f $2(5^2 + 2 \times 4) - 7 \times 8$

g $\dfrac{4(13-8)+3\times 4-4^2}{6^2 \div (2\times 9)}$

h $\dfrac{2^3+21\div 3-8\div 2}{9^2-7\times 10}$

i $2\left(8\times 3-\sqrt{8^2+6^2}\right)+3^2\times 2^3$

j $\dfrac{16[(12-4)-3\times 2]+4^2}{8\times 4-[2(10-7)+18\div 9]}$

**24** **Write down** an expression that equals 24 using the four numbers in each set below. You may use parentheses, addition, multiplication, division and/or subtraction, but each number *must* be used and *only once*.

**a** 2, 4, 5, 11 **b** 2, 3, 10, 12 **c** 3, 4, 9, 6 **d** 1, 3, 4, 8

In questions **24** and **25**, there are many possible answers.

**25** Arturo wrote down the equation below to make 24 using the numbers 3, 4, 6, 12.

$$3 + 6 \times 4 - 12 = 24$$

**a** Is Arturo correct? **Explain**.

**b** If Arturo is incorrect, make one change to his equation so that his answer is correct.

**c** Use the numbers 3, 4, 6 and 12 along with one multiplication and two divisions to **write down** another expression that equals 24.

**26** Using any numbers of your own choosing, **write down** an expression that uses four different operations (choosing from multiplication, division, subtraction, addition, exponent, and root) where the answer is 20. Remember to obey the correct order of operations.

**27** Using any numbers of your own choosing, **write down** an expression that includes at least three exponents, one set of parentheses, one multiplication and one addition, where the answer is 8.

**28** In a traditional Mayan medicine garden, herbs were planted for their healing powers, such as Bougainvillea for its antibiotic properties and Calendula for its anti-viral and anti-inflammatory properties.

**a** If a field was 54 land rods (ancient unit of measurement) by 30 land rods, what are the dimensions of the largest square plots of a single herb that could be planted?

**b** How many square plots would you have in this field?

**c** If the field was divided equally between both these types of herbs, how many square plots of Bougainvillea and Calendulas would there be?

# Summative assessment

## ATL2 How much are we influenced by the past?

You will be exploring number systems of different ancient civilizations. In groups specified by your teacher, you will choose one of the following civilizations and complete some research on their number system. You will then explain the basics of that number system to the class.

Mayan Babylonian Roman Greek

Egyptian Chinese Sumerian

1 Before beginning this task, as a class you will prepare a timeline of the history of world math systems. Each group will share with the class when their civilization's number system was created. As a class, you will determine the intervals of your timeline until the present day, and label your civilization on the timeline. This will help give you an overview of mathematics throughout time as you complete your Summative task.

2 As a group, you will prepare a short instructional video between two and five minutes in length, using an app like Explain Everything or Show Me, or a method suggested by your teacher. Your video will include all of the relevant information about your number system.

   **a** What is the history of your number system?

   - When was it created?
   - Where was it created? (Include geographic reference.)
   - What was it used for?
   - If not used today, when did it stop being used?

   **b** What are the basics of your number system?

   - How does it work?
   - What types of symbols are used and do they have specific meaning/values?
   - Are there any patterns evident in your number system?
   - What kind of mathematical operations were used?
   - Does your number system use place value?

c What were the positive aspects and limitations of the number system?

d Do we use any of the elements of this number system today?

e What are some interesting facts about your number system and the culture of the civilization that used it?

To accompany your video, you will create a one-page information sheet outlining the basics of your system and with explanations of the following:

- how to convert numbers from your number system to the base-10 number system (Your explanations must be clear enough so that the class can then determine how to write the following base-10 numbers using the symbols in your number system: 0, 8, 27, 74, 65, 1390.)
- how to add, subtract, multiply and divide in your number system, with an example of each
- how to find the value of the following: $6 + 8 \div 2 + 3 \times 5$.

## Reflect and discuss 14

After all of the presentations, answer the following questions in your group.

- Compare and contrast the number systems to the Hindu-Arabic number system we use today.
- Which were the easiest to learn? Which were the hardest? Why?
- How have the number systems you have studied here influenced the number system we use today?

### Enrichment activity

Computer programming is slowly becoming an important skill for most students to acquire. Computer programmers use the *hexadecimal* system to represent locations in memory. Like binary, it has practical advantages over the decimal system.

Create a short report on the hexadecimal system that includes:

- when and where it was created
- why it was created, and what purpose it serves
- how numbers are represented in hexadecimal
- two examples of how decimal numbers can be converted to hexadecimal
- one example of each mathematical operation in decimal, and the same calculation in hexadecimal.

# 2 Percentages

In this unit your use and understanding of percentages will give you a glimpse into the experience of refugees. However, the study of percentages could also be the gateway to discoveries in other contexts on a personal and global scale.

## Personal and cultural expression

### Entrepreneurship

Do you have an idea for your own business? What would it take to start your own company? What characteristics make for a successful entrepreneur? Answering these questions and interpreting the answers requires an understanding of percentages.

- Some research studies have shown that 90% of new companies fail within the first five years. Why does this happen?
- A start-up loan from the bank will charge interest, which is a percentage of the total amount. Could your new business afford to repay the interest?

Building a successful company also means understanding how to run a business. How do you know what price to charge for the product? How can you increase your profits without increasing the price? Percentages will help you analyze these questions and find optimal solutions.

## Globalization and sustainability

### Population and demography

Demography is the study of human populations. It can show us how the human population is broken down into different segments and how those segments change over time.

What is the current population of the planet? What was it fifty years ago? What percentage is male? Female? Teenage? Employed? Interested in American football? Religious? Likes pizza? Researchers look into different questions, for different reasons.

While some of these questions may not seem very important, looking at trends in the human population can help us to understand how the world might look in the future and explain differences between humans in different regions. The results of these questions may even point out where inequities exist.

2

# Percentages

## Inequality and difference

## KEY CONCEPT: FORM

### Related concepts: Equivalence and Quantity

### Global context

In this unit you will work with percentages, fractions and decimals to explore the global context of **fairness and development**. You will apply the relationships between percentages, fractions and decimals as you explore worldwide refugee displacement, working conditions, nutrition, and the availability of safe drinking water. You will also have opportunities to reflect on and discuss what it might be like to be a refugee in a foreign country.

## Statement of Inquiry

Inequality and difference become clearer through the use of equivalent forms of quantities.

### Objectives

- Representing a number in different forms – fractions, decimals and percentages
- Converting between equivalent forms of numbers – fractions, decimals and percentages
- Calculating percentage increase and decrease
- Applying mathematical strategies to solve problems involving percentages

### Inquiry questions

**F** What are different forms of numbers?
When are two things equal?

**C** How can different forms be equivalent?
When is it beneficial to use different forms?

**D** Can fairness be calculated?

**ATL1 Collaboration skills**

Practise empathy

**ATL2 Organization skills**

Plan short- and long-term assignments; meet deadlines

## You should already know how to:

- say and write decimal numbers
- round numbers correctly
- reduce fractions to their simplest form
- mentally multiply and divide by powers of ten
- find missing values in equivalent fractions

# Introducing percentages

Life seems to be full of percentages. They appear in advertising, in the news, even on our devices. But what do they mean? Why do we use them? Do they only describe a quantity or amount, or can they be used in other ways? Is it possible for percentages to help you learn important lessons, such as what it might be like to live in someone else's shoes?

DAILY NEWS

World - Business - Finance - Lifestyle - Travel - Sport - Weather

Est - 1965

THE WORLDS BEST SELLING NATIONAL NEWSPAPER

Issue: 240104

Monday 5th June

First Edition

Farmers say rats are eating up to 20% of their corn crops.

Prices rose by more than 10 percent in the summer.

70% off sale fuels mayhem at major department store.

The college acceptance rate is now just 14 percent.

Get 25% more storage when you sign a two-year contract.

Being comfortable with percentages will not only help you in your daily life, but may also help you understand someone else's life a little better.

## Reflect and discuss 1

Answer these questions individually before discussing them as a class.

- What do you think the word *percent* means?
- What is the definition of the word *cent*?
- In the following two examples, what does the word *per* represent?
  - o A bike courier works 7.5 hours per day.
  - o A cruise ship travels at 30 nautical miles per hour.
- Putting these together, what do you think the word *percent* means?

# Percentages

A percent, or percentage, is a ratio which can be written as a fraction out of one hundred. The symbol used to show a percentage is: %. For example, 20% is read as 'twenty percent', meaning twenty out of one hundred.

## Activity 1 – Uses of percentages

Percentages are a very useful way of expressing ratios and fractions. In pairs, find 10 real-life examples of where percentages are commonly used. You will score 1 point if several pairs have the same example, 3 points if only one other pair has it, and 5 points if yours is the only pair that has it.

The most common way to represent a percentage is by using notation, e.g. 75% or 13.5%. However, percentages can also be represented using models. These models can make percentages easier to understand by giving them a visual representation.

## Activity 2 – Visual representation of percentages

1 Write down the percentage represented by the shaded section of each grid (e.g. 60%).

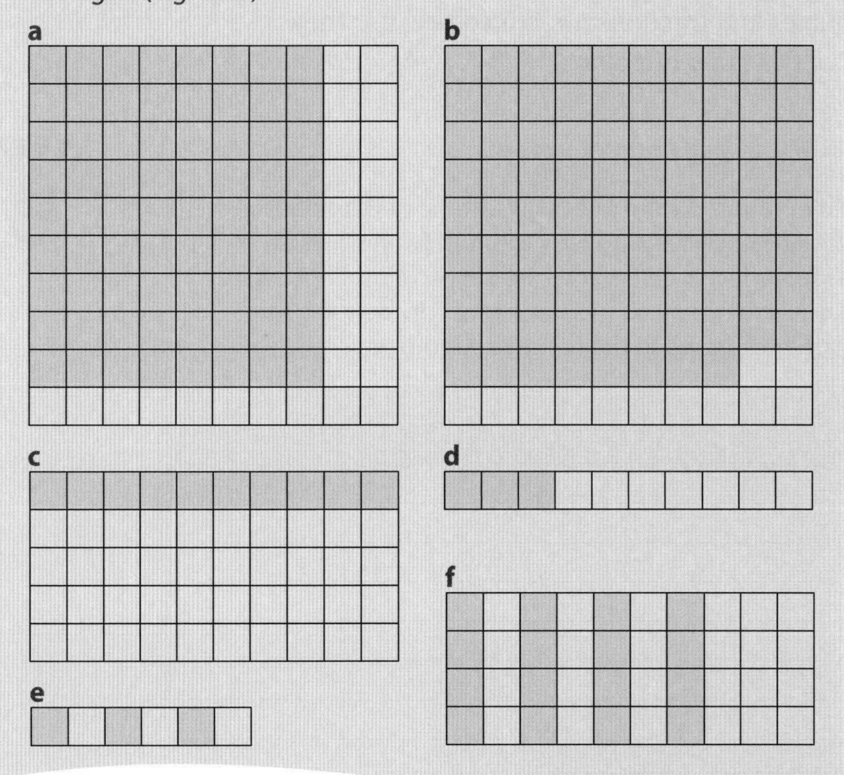


▶ Continued on next page

2 Describe how you were able to identify the percentage for the grids that do not have 100 squares.

3 Create two of your own grids to represent 80%.

4 Create two of your own grids to represent 30%.

5 Make a creative and unusual visual representation of 50%.

## Reflect and discuss 2

- Can you have a percentage that is larger than 100%? If so, how would you represent it visually?
- If you could have a percentage greater than 100%, what would it mean? Give an example to explain your reasoning.

In Activity 2, you saw that a percentage can be represented visually using shaded squares. Although using 100 squares makes representing and identifying percentages much simpler, percentages can also be shown using a different number of squares. Even though percentages are quantities or ratios written out of 100, you can sometimes simplify these ratios to produce equivalent fractions. In fact, percentages and fractions are linked because they can be equivalent forms of the same quantity.

## Investigation 1 – Key percentages

1 You have seen that a percentage is a ratio out of 100. Some percentages, like 50%, are easy to calculate since they are easy to simplify (in this case, one-half). Based on this idea, which percentages are easy to simplify? Copy the grid below and make a list of them. At the same time, divide the grid into the correct proportions, as shown in the first two examples.

<table>
<tr><td colspan="4">100%</td></tr>
<tr><td colspan="2">50%</td><td colspan="2">50%</td></tr>
<tr><td>25%</td><td>25%</td><td>25%</td><td>25%</td></tr>
<tr><td colspan="4"></td></tr>
<tr><td colspan="4"></td></tr>
<tr><td colspan="4"></td></tr>
<tr><td colspan="4"></td></tr>
<tr><td colspan="4"></td></tr>
<tr><td colspan="4"></td></tr>
</table>

▶ Continued on next page

**2** If you can calculate the percentages in step **1**, multiples of these should also be fairly easy to calculate. Copy the table on the right and make a list of these other percentages. If there is more than one way to calculate a percentage, list the easiest way. An example has been done for you.

| Percentage | How you can calculate it |
|---|---|
| 75% | multiply 25% by 3 |
| | |
| | |
| | |
| | |
| | |

**3** Based on the investigation, state a general rule for finding any percentage.

**4** Verify your rule for other percentages that you have not looked at yet.

## Reflect and discuss 3

- What makes some percentages easier to calculate than others?
- In some parts of the world, when people eat at restaurants, a common practice is to tip the server 10%, 15% or 20% of the cost of the meal. Can you think of an easy way to calculate these tip amounts?
- Why do you think servers receive tips? Suppose you ate at a restaurant, but you didn't know about the tipping custom and therefore you left no tip. How would it feel to be your server? What questions would your server ask himself/herself?

ATL1

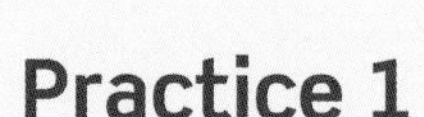

## Practice 1

**1** Write down the percentage represented by the shaded section of each grid.

**a**

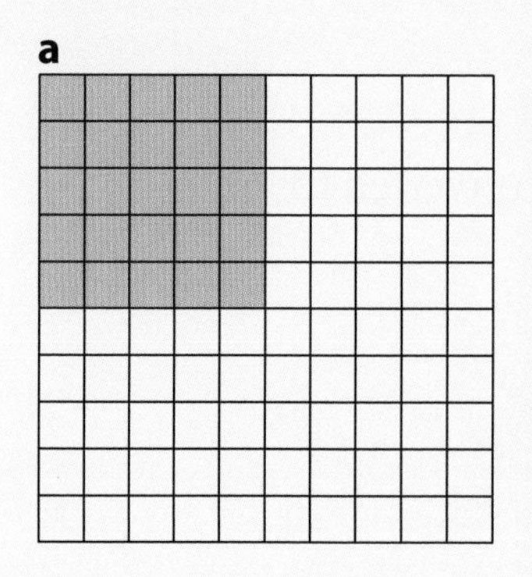

**b**

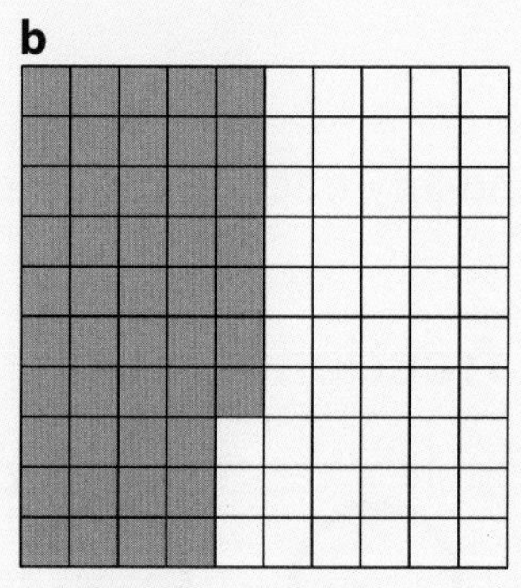

**c**

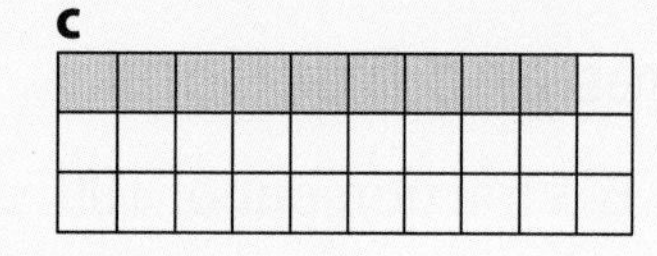

**d**

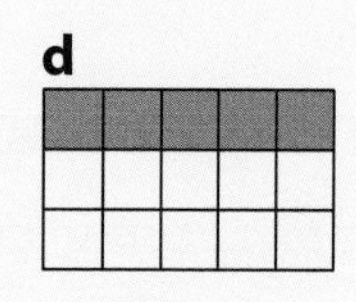

**2** For each percentage below, draw two grids of **different** sizes and then shade the percentage.

**a** 60% **b** 20% **c** 80% **d** 12% **e** 4%

**3** If 10% of a certain quantity is 12, find the following percentages of the same quantity.

**a** 20% **b** 30% **c** 70% **d** 100% **e** 5% **f** 25%

▶ Continued on next page

**4** If you knew what 4%, 5% and 6% of a certain quantity was, how could you calculate the following percentages of the same quantity?

**a** 12% **b** 9% **c** 10% **d** 30% **e** 11% **f** 14% **g** 17%

**5** In 2015, 6% of the people forced to flee their homes were hosted by countries in Europe. This represented approximately 4 million people.

**a** Using this data, approximate the number of people represented by the following percentages.

**i** 12% of the people fled to the Americas.

**ii** 30% fled to an African country (not including North Africa).

**iii** 15% fled to a country in the Asia-Pacific region.

**iv** 39% fled to a country in the Middle East or North Africa.

ATL1

**b** Imagine you had to leave your country because of a civil war. What would it feel like to move to a new place where you don't know anyone and don't speak the language? How would it feel to know that you may never go back home? What other feelings would you have?

**6** As of January, 2017, Canada had taken in approximately 40 000 Syrian refugees. Approximately 15 000 of these refugees were sponsored privately by individual Canadians who formed small groups of citizens to finance the refugees' first year in Canada. What percentage of the refugees were sponsored by private citizens?

# Equivalent forms

## Percentages and fractions

A percentage is a fraction out of 100, and any fraction can be converted to a percentage.

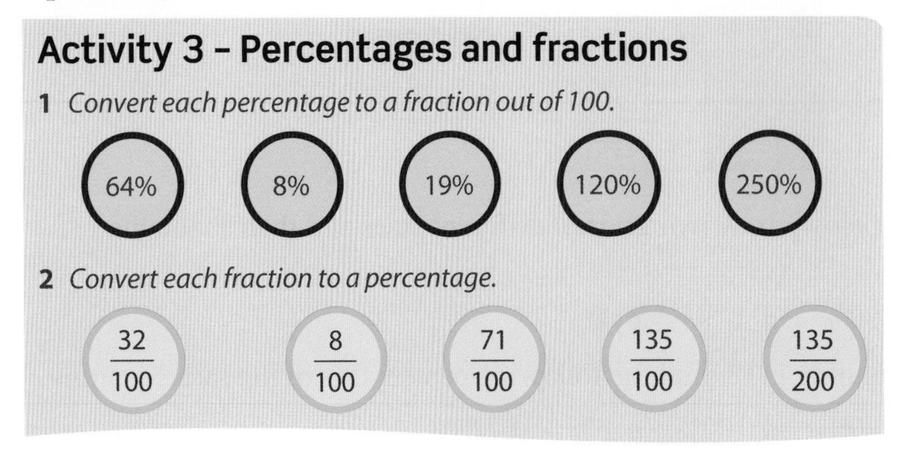

### Activity 3 – Percentages and fractions

**1** *Convert each percentage to a fraction out of 100.*

64% 8% 19% 120% 250%

**2** *Convert each fraction to a percentage.*

$\frac{32}{100}$ $\frac{8}{100}$ $\frac{71}{100}$ $\frac{135}{100}$ $\frac{135}{200}$

Being able to convert between equivalent forms is a very important skill in mathematics. When converting fractions to percentages, it is easiest to convert the quantity to an equivalent fraction out of 100 and then write the number as a percentage.

## Example 1

**Q** Convert $\frac{19}{25}$ to a percentage.

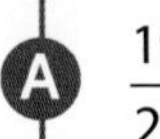

$$\frac{19}{25} = \frac{\square}{100}$$

$$\frac{19 \times 4}{25 \times 4} = \frac{76}{100}$$

Multiply the denominator by 4 to make an equivalent fraction out of 100. Multiply the numerator by the same amount in order to maintain equality.

$$\frac{76}{100} = 76\%$$

Write the fraction out of 100 as a percentage (since 'percent' means 'out of 100').

Therefore, $\frac{19}{25} = 76\%$.

## Reflect and discuss 4

What do you think you would do to convert $\frac{4}{80}$ to a percentage, since its denominator is not a factor of 100?

## Example 2

**Q** Write $\frac{7}{35}$ as a percentage.

$$\frac{7 \div 7}{35 \div 7} = \frac{1}{5}$$

Simplify the fraction.

$$\frac{1 \times 20}{5 \times 20} = \frac{20}{100}$$

Rewrite as an equivalent fraction out of 100.

$$\frac{20}{100} = 20\%$$

Rewrite the fraction as a percentage.

Therefore, $\frac{7}{35}$ as a percentage is 20%.

## Reflect and discuss 5

Sometimes the denominator of a fraction does not divide evenly into 100, yet you'd still like to convert it to a percentage.

- How could you use a calculator to help you find the percentage represented by $\frac{3}{17}$?
- Write down the steps you could use with your calculator to convert a fraction to a percentage.
- If a calculator is not available, what other option might you have? Use this method and round your answer to the nearest hundredth.

## Activity 4 – Refugees around the world

Research the statistics needed below and use a calculator to help you answer these questions. (Try unhcr.org as a statistical reource website.)

1 What is the approximate world population?

2 What is the approximate world refugee population?

3 What percentage of the world population consists of refugees? Round your answer to the nearest tenth.

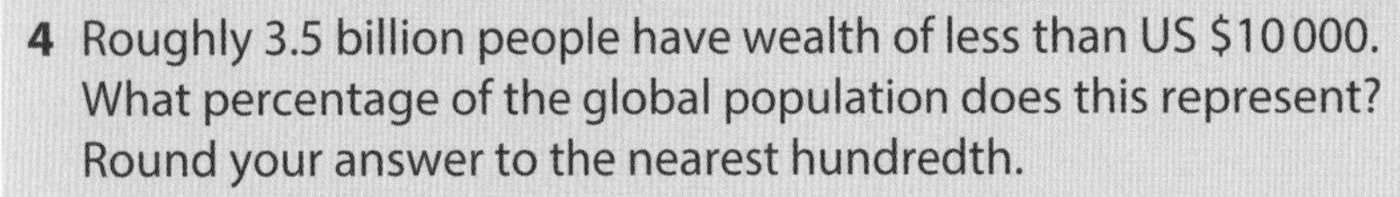

4 Roughly 3.5 billion people have wealth of less than US $10 000. What percentage of the global population does this represent? Round your answer to the nearest hundredth.

5 Approximately 50% of all refugees are children (under the age of 18). How many refugees are children? Round your answer to the nearest whole number.

## Reflect and discuss 6

- What would it mean to have a total wealth of less than US $10 000? What would life be like? How might your priorities be different than they are now?
- In Activity 4, why are some numbers rounded to one decimal place, but others to the nearest whole number?

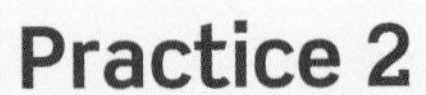

## Practice 2

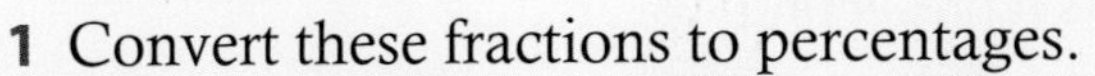

**1** Convert these fractions to percentages.

**a** $\frac{1}{2}$ **b** $\frac{17}{20}$ **c** $\frac{7}{10}$ **d** $\frac{3}{5}$ **e** $\frac{31}{50}$

**f** $\frac{3}{4}$ **g** $\frac{11}{25}$ **h** $\frac{3}{20}$ **i** $\frac{25}{25}$ **j** $\frac{47}{50}$

**2** Convert these fractions to percentages.

**a** $\frac{6}{12}$ **b** $\frac{12}{30}$ **c** $\frac{6}{40}$ **d** $\frac{7}{70}$ **e** $\frac{18}{45}$

**f** $\frac{21}{28}$ **g** $\frac{36}{60}$ **h** $\frac{81}{90}$ **i** $\frac{24}{32}$ **j** $\frac{9}{15}$

**3** Convert these fractions to percentages. Use two significant figures.

**a** $\frac{5}{18}$ **b** $\frac{2}{21}$ **c** $\frac{11}{43}$ **d** $\frac{71}{89}$ **e** $\frac{18}{19}$

**f** $\frac{24}{53}$ **g** $\frac{33}{92}$ **h** $\frac{50}{61}$ **i** $\frac{21}{17}$ **j** $\frac{1}{11}$

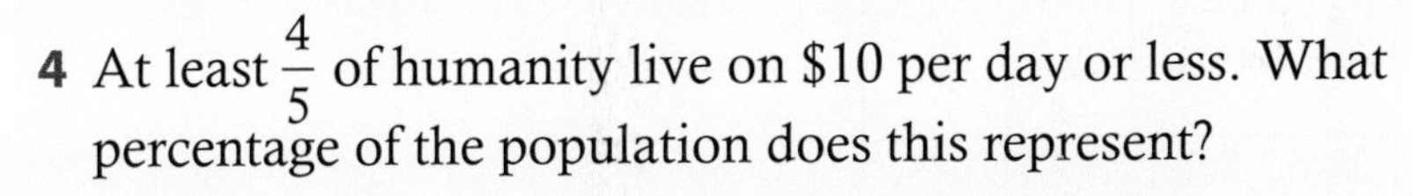

**4** At least $\frac{4}{5}$ of humanity live on \$10 per day or less. What percentage of the population does this represent?

**5** Of the estimated 2 billion children living in the developing world, approximately 400 million do not have access to safe drinking water.
Approximately what percentage of children in developing countries do not have safe water to drink?

### Activity 5 – The world's resources

Are the world's resources divided evenly? By looking at the population of each continent/region and its gross domestic product (GDP), you can get an idea of the distribution of the world's resources.

**1** Referring to the table on the next page, research the current population and GDP of each region and fill in the appropriate columns in a copy of the table.

**2** Calculate each region's share of the global GDP as a percentage and fill in the second column.

**3** Calculate the percentage of the world population that each region's population represents and fill in the fourth column.

▶ Continued on next page

Gross domestic product, or GDP, is a measure of the income of individuals and companies in a region.

4 Based on the percentage breakdown of the population, divide your class into groups that accurately represent the six regions. Fill in the last column with the number of students representing each region.

5 The world's GDP can be represented by 100 counters. Divide up the counters between the regions, based on the GDP percentage breakdown in the table. Give each group of students the counters for their region.

6 Take the counters for your region and divide them evenly among the group members. What do you notice? Make a general statement about the world's allocation of resources.

7 What if the counters were replaced with chocolate coins and each group member was allowed to eat them? Would that be fair?

| Continent or region | GDP | Percentage of global GDP | Population | Percentage of global population | Class size breakdown |
|---|---|---|---|---|---|
| Africa | | | | | |
| Asia | | | | | |
| Europe | | | | | |
| North America | | | | | |
| Oceania | | | | | |
| South America | | | | | |
| Total GDP: | | Total Population: | | | |

## Reflect and discuss 7

- Are the world's resources distributed evenly? Justify your answer.
- Do you think they should be distributed more equitably? If so, what can be done to make the distribution more equitable?
- As you have discovered, the world's resources are not distributed evenly. How would you feel if you were a citizen of a country that received a small percentage of the world's resources? How would you feel if you were a citizen of a country that received a large percentage of the world's resources?

ATL1

**equitable:** fair and impartial

## Percentages and decimals

A decimal can be converted to a percentage using a process that you are going to determine in the investigation below.

## Investigation 2 – Percentages and decimals

Copy the table below and complete it to discover the process of converting a decimal to a percentage.

| | 0.45 | 0.3 | 0.295 |
|---|---|---|---|
| Say the number (out loud) based on its place value and write it down. | | | |
| Write this number as a fraction. | | | |
| Convert to an equivalent fraction out of 100, if necessary. | | | |
| Write the fraction as a percentage. | | | |

1 How would you explain the process of converting a decimal to a percentage to a student who was absent for this lesson? Demonstrate and verify the process with two other decimal numbers.

2 How would you convert a percentage to a decimal?

## Reflect and discuss 8

- How does our decimal number system help with the process of converting decimals to percentages?
- How can you tell if a percentage and a decimal are equivalent?

## Example 3

Represent each percentage as a decimal.

**a** 17%  **b** 62.5%

**a** $\frac{17}{100} = 17 \div 100$

Think of the fraction as a division.

$17 \div 100 = 0.17$

Move the digits over two places when dividing by 100.

**b** $\frac{62.5}{100} = 62.5 \div 100$

Think of the line between the numerator and denominator as a division sign.

$62.5 \div 100 = 0.625$

Move the numbers over two places when dividing by 100.

You know how to convert fractions to terminating decimals (decimals that have a *finite* number of decimal places), but there are actually two other kinds of decimals:

- repeating decimals – decimals that have a sequence of digits that repeat *infinitely*
- non-terminating, non-repeating decimals – decimals that do not terminate and do not repeat. These are also called *irrational numbers* and you typically need a calculator to convert them between equivalent forms.

## Investigation 3 – Repeating decimals

criterion B

You will look at repeating decimals to ensure you can recognize these numbers in both decimal and fractional form.

A repeating decimal is often indicated with a line over the top of the digits to be repeated.

For example: $0.\overline{62} = 0.626262626262...$ $1.5\overline{34} = 1.5343434...$ $70.\overline{137} = 70.137137137...$

In the table below, the first column contains terminating decimals and their equivalent fractions, and the second column contains repeating decimals and their equivalent fractions.

| Terminating decimals | Repeating decimals |
|---|---|
| $0.7 = \frac{7}{10}$ | $0.\overline{7} = \frac{7}{9}$ |
| $0.1 = \frac{1}{10}$ | $0.\overline{1} = \frac{1}{9}$ |
| $0.23 = \frac{23}{100}$ | $0.\overline{23} = \frac{23}{99}$ |
| $0.017 = \frac{17}{1000}$ | $0.\overline{017} = \frac{17}{999}$ |

1 Looking at the denominators, what is the key difference between the fraction equivalents of terminating and repeating decimals?

2 Focusing on the denominators again, how do you know how many digits will be repeating infinitely?

3 What are the steps to convert a repeating decimal to a fraction?

4 Take a look at the following:

$0.\overline{3} = \frac{1}{3}$ $0.\overline{6} = \frac{2}{3}$ $0.\overline{72} = \frac{8}{11}$

Can you follow the same steps to convert the repeating decimals in these examples to fractions? Explain.

▶ Continued on next page

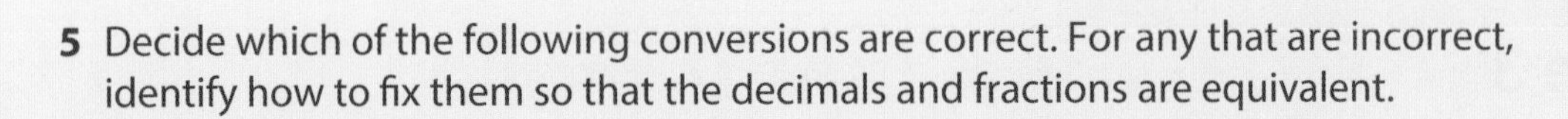

**5** Decide which of the following conversions are correct. For any that are incorrect, identify how to fix them so that the decimals and fractions are equivalent.

**a** $0.4 = \frac{4}{9}$ **b** $0.\overline{35} = \frac{35}{99}$ **c** $0.\overline{011} = \frac{11}{999}$

**d** $0.\overline{2} = \frac{2}{11}$ **e** $0.\overline{24} = \frac{8}{33}$ **f** $0.\overline{36} = \frac{4}{11}$

**g** $0.\overline{06} = \frac{3}{50}$ **h** $0.\overline{375} = \frac{375}{9}$ **i** $0.\overline{17} = \frac{7}{40}$

It is the opposite process to convert a fraction to a repeating decimal.

## Example 4

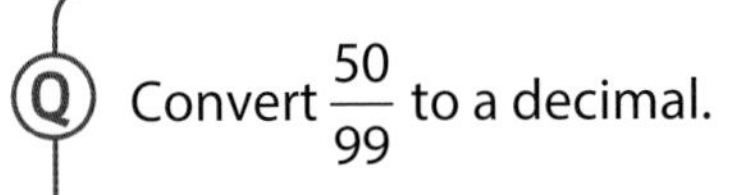

**Q** Convert $\frac{50}{99}$ to a decimal.

Determine the number of 9s in the denominator; this is the number of repeating digits in the decimal.

**A** 99 – so two digits repeat

NOTE: if the denominator does not contain only 9s you may have to convert the fraction first. If the denominator cannot be converted to contain only 9, then it is not a repeating decimal.

0.50

The numerator is placed in those decimal places.

$0.\overline{50}$

Draw a line over the top of the decimal to represent the infinite repeating of those digits.

$\frac{50}{99} = 0.\overline{50}$

To convert repeating decimals to percentages, it is best to simply look at the number in decimal form and convert using the process you already know (multiply by 100). You will typically round the number to the most appropriate number of decimals places, given the context of the question.

In Example 4, the repeating decimal was $0.\overline{50}$ so this would be 50.5050%. This number would round to 51%, accurate to the nearest percentage, and 50.5% if you wanted the number accurate to the nearest tenth.

## Practice 3

**1** Convert each decimal to a fraction and then a percentage.

**a** $0.87 = \frac{\square}{\square} = \square\%$ **b** $0.5 = \frac{\square}{\square} = \square\%$ **c** $0.732 = \frac{\square}{\square} = \square\%$

**d** $0.09 = \frac{\square}{\square} = \square\%$ **e** $0.006 = \frac{\square}{\square} = \square\%$ **f** $1.2 = \frac{\square}{\square} = \square\%$

**2** Convert each percentage to a decimal.

**a** 12% **b** 46% **c** 103% **d** 38.5% **e** 8.3% **f** 12.25%

**3** Write these numbers in all three forms: fraction, decimal and percentage. Write each fraction in its simplest form.

**a** $\frac{2}{25}$ **b** 0.8 **c** 35% **d** $\frac{4}{9}$

**4** Which is the largest number in each of the following groups? Justify your answer.

**a** $\frac{3}{10}$ or 28% or $0.\overline{28}$ **b** $\frac{17}{25}$ or 0.7% or $0.\overline{7}$ **c** 1.03 or 100%

**d** $\frac{3}{15}$ or $\frac{2}{9}$ **e** $\frac{24}{33}$ or 70% or 0.67 **f** $\frac{8}{40}$ or 20% or 0.2

**5** Which numbers are equivalent in each of these sets?

**a** $\frac{42}{150}$, 16%, 0.28, $\frac{21}{60}$, 28%, 0.286

**b** $\frac{17}{34}$, 4%, 0.4, $\frac{104}{260}$, 44%, 0.40

**6** Order each set of numbers from smallest to largest.

**a** $\frac{2}{5}$, 36%, 0.6, $\frac{2}{11}$, 14%, 0.186

**b** $\frac{1}{9}$, 9%, 0.16, $\frac{7}{20}$, 21%, 0.05

Continued on next page

**7** Order each set of numbers from largest to smallest.

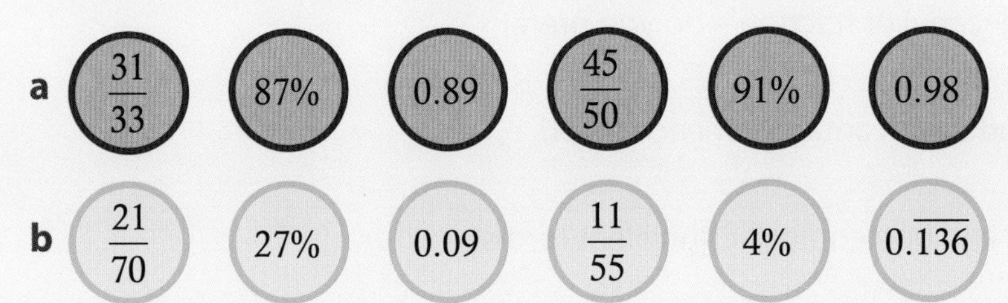


**8** In each of these scenarios, which option offers the greatest amount? Justify each answer using mathematics.

**a** 25% of one bag of popcorn, or 10% of three of the same sized bags of popcorn

**b** 50% of two mangos, or 40% of three of the same sized mangos

**c** 30% of one energy bar, or 5% of six of the same sized energy bars.

**9** The International Labour Organization (ILO) estimates that 60% of the 250 million children around the world who work in sweatshops live in Asia. Approximately how many Asian children work in sweatshops?

**sweatshop:** a workshop, often in the clothing industry, employing workers at low wages, for long hours, and under poor and often unsafe conditions

**10** The average hourly wage of a woman making clothes in Bangladesh is the equivalent of US \$0.13. The United States Federal minimum wage as of January 2017 was US \$7.25 per hour. Approximately what percentage of the US minimum wage is the Bangladeshi clothes worker's wage?

**11** In India, approximately 5% to 30% of the 340 million children under 16 are classified as child laborers. About how many children does this represent?

For **11**, give your answer as a range.

**12** The charity Oxfam is working to help refugees in Bangladesh. In September 2017, there were 470 000 Rohingya Muslim people from Myanmar living as refugees in Bangladesh. Only 50% of these refugees had access to clean drinking water.

**a** How many people does this percentage represent?

So far, Oxfam has been able to provide clean drinking water to 100 000 of the refugees who did not previously have access to clean drinking water.

**b** What percentage of the total number of people now have clean water?

## Reflect and discuss 9

- Which form – percentage, decimal or fraction – do you prefer when ordering numbers?
- What are the advantages and disadvantages of each form? Make a list for each one.
- How are fractions, decimals and percentages equivalent forms?

## Activity 6 – Conversion bingo

Draw a 4 by 4 grid and fill it in with a combination of fractions, decimals and percentages, at least four of each. Here is an example.

Your teacher will give you the specific guidelines for what numbers you can put in your grid. (For example, you may be told that you can use any percentages that are a multiple of 10. So you must come up with decimals, fractions and percentages that as a percentage are a multiple of 10.) You cannot include two equivalent numbers in different forms (e.g. 20% and 0.2).

Once you have written the numbers in your grid, the game will begin. Your teacher will call out a number in one of the three forms and you can cross it off your grid if you have that number in any of the three forms. The winner is the first student to get four in a row, column, or major diagonal.

| | | | |
|---|---|---|---|
| 20% | 0.4 | 30% | $\frac{17}{20}$ |
| $\frac{1}{4}$ | 0.75 | 0.05 | 45% |
| 10% | 16% | $\frac{1}{15}$ | 90% |
| $\frac{6}{35}$ | 70% | 0.68 | $\frac{3}{5}$ |

## Formative assessment – Food for refugees

The United Nations High Commissioner for Refugees (UNHCR) recommends that a refugee should consume at least 2100 calories per day. Yet in many refugee camps, refugees get less than this. For example, in one Tanzanian refugee camp, refugees were given only about 1400 calories per day.

1 What percentage of the minimum daily calories were the refugees in the Tanzanian camp given?

2 In many refugee camps, refugees end up trading their food for other non-food supplies that they need. If a refugee was eating only a third of the minimum daily calorie recommendation, how many calories per day would that be?

▶ Continued on next page

The table shows approximately what a Syrian refugee might eat *per day* in a Jordanian refugee camp. Remember that the UNHCR recommends that a person should have at least 2100 calories daily.

| Food | Grams (approx.) | Calories (approx.) | Type of nutrient |
|---|---|---|---|
| rice | 274 | 1100 | carbohydrate |
| flour | 57 | 230 | carbohydrate |
| lentils | 24 | 97 | protein |
| dried chickpeas | 12 | 49 | protein |
| sardines | 18 | 71 | protein |
| canned kidney beans | 57 | 229 | protein |
| vegetable oil | 43 | 386 | fat |

**3** What is the number of total daily calories in this refugee diet?

**4** What percentage of a refugee's calories come from protein?

**5** What percentage of a refugee's calories come from carbohydrates?

**6** What percentage of a refugee's calories come from fat?

**7** Make a similar list of the foods that you eat in a typical day.

**8** Using an online source, calculate the approximate number of calories for every item on your list and state the type of nutrient.

**9** Calculate the total number of calories you consume in one day. Calculate how many calories you consume daily of fat, protein, and carbohydrates.

**10** What percentage of your calories come from each of these three nutrient groups?

**11** Explain the degree of accuracy of your results in step **10**. Describe whether or not your results in step **10** make sense.

**12** How does your diet differ from the refugee diet? How is it similar? (Compare the number of calories, breakdown of food groups, etc.)

ATL1

**13** How do you think refugees feel, having to follow this diet? Explain.

## Applications of percentages

There are many practical applications involving percentages which you will see on a regular basis, such as finding the discount on an item for sale in a store or calculating the tax when purchasing an item. Remembering that percentages and fractions are really just equivalent forms can help you to solve these types of problems.

## Example 5

**a** What percentage of 80 is 8?

**b** What is 30% of 25?

**c** 50% of what number is 13?

**a** 80 is the whole and 8 is the part, so the fraction is $\frac{8}{80}$.

$$\frac{8}{80} = \frac{8 \div 8}{80 \div 8} = \frac{1}{10}$$

$$= \frac{1 \times 10}{10 \times 10} = \frac{10}{100} = 10\%$$

Convert the fraction to a percentage.

Therefore, 8 is 10% of 80.

**b** $\frac{30}{100} = \frac{\square}{25}$

In this case, 25 is the whole and the part is unknown.
We need to find the equivalent to the fraction represented by 30%.

$$\frac{30 \div 4}{100 \div 4} = \frac{7.5}{25}$$

Since $100 \div 4 = 25$, we need to divide the numerator by 4 as well.

Therefore, 30% of 25 is 7.5.

**c** $\frac{50}{100} = \frac{13}{\square}$

In this case, 13 is the part but the whole is unknown.
We need to find the equivalent to the fraction represented by 50%.

$$\frac{1}{2} = \frac{13}{\square}$$

Simplify the first fraction.

$$\frac{1 \times 13}{2 \times 13} = \frac{13}{26}$$

Therefore, 50% of 26 is 13.

## Reflect and discuss 10

- How would you write 30% as a decimal?
- What does the word 'of' mean in mathematics?
- What is another way of solving problems like 'Find 70% of 25'?

## Practice 4

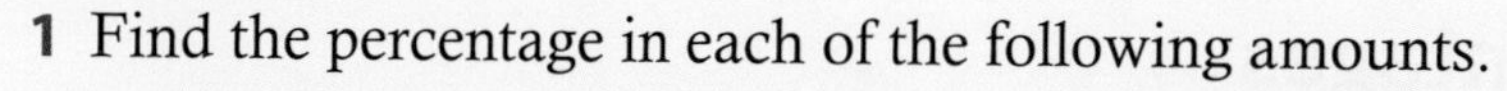

**1** Find the percentage in each of the following amounts.

**a** What percentage of 100 is 75?

**b** What percentage of 48 is 36?

**c** What percentage of 300 is 75?

**d** What percentage of 620 is 124?

**e** What percentage of 50 is 18?

**f** What percentage of 220 is 132?

**g** What percentage of 96 is 24?

**h** What percentage of 850 is 340?

Continued on next page

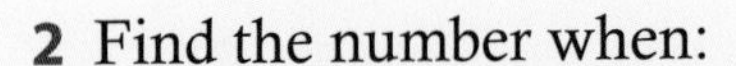

**2** Find the number when:

**a** 50% of the number is 20
**b** 20% of the number is 8
**c** 10% of the number is 12
**d** 25% of the number is 38
**e** 80% of the number is 208
**f** 14% of the number is 49
**g** 62% of the number is 124
**h** 130% of the number is 78.

**3** Find the number when given the percentage.

**a** 25% of 100
**b** 40% of 100
**c** 20% of 50
**d** 80% of 100
**e** 10% of 1000
**f** 70% of 200
**g** 15% of 200
**h** 5% of 60
**i** 100% of 80
**j** 35% of 200
**k** 8% of 50
**l** 12% of 32
**m** 60% of 55
**n** 45% of 80
**o** 90% of 20

**4** In a store, an item has a new price of $63 after a reduction of 30%. What was the original price?

**5** In Spain, the average annual income of a male is approximately €26 000 and the average annual female income is approximately €20 000. Approximately what percentage of the average male income is the average female income?

**6** In Canada, the average annual income of a female aged 25 to 34 is 77% of a male in the same age group. If the average annual income of a female is approximately $50 000, how much, rounded to the nearest thousand dollars, does the average male in the group earn?

**7** In Australia, the average full-time weekly salary for a woman is approximately 84% of the full-time weekly salary for a man.

If an Australian man earns approximately $1600 per week, how much does an Australian woman earn?

ATL1

**8** What feelings would you experience if you earned less than someone else for doing the same work? Explain.

**9** During a six-week period in 2012, the organization Doctors Without Borders managed to vaccinate approximately 75% of a population at imminent risk of a disease called cholera in Guinea and Sierra Leone, averting a potential disaster. What was the total population if 170 000 people were vaccinated?

Many countries around the world add a sales tax (sometimes called a value added tax) onto most of the goods that are sold. In some countries, the tax is included in the listed price, while in others the tax is added on to the listed price. To make sure you have enough money, you need to be able to quickly determine what that final price will be. You can do this easily in your head without a calculator.

## Activity 7 - More than one way

You want to buy an item that costs £60. However, a tax of 15% will be added. What is the amount of the tax?

One way to solve this is to divide the problem in two:

$15\% = 10\% + 5\%$

To find 10% of 60, you can multiply $0.1 \times 60$.

$0.1 \times 60 = 6$

Since 5% is half of 10%, find half of £6 ,which is £3.

Therefore, the total tax is £6 + £3 = £9.

**a** Solve the problem by using one calculation involving a decimal.

**b** Solve the problem by using two equivalent fractions.

**c** Solve the problem in as many other ways as you can think of.

## Reflect and discuss 11

- Which method did you prefer to use when solving the problem in Activity 7?
- Will your choice of method depend on the question, or will you always use the same one?
- How can different methods produce equivalent results?

## Practice 5

**1 a** Calculate the tax on each item, then calculate the total cost of each item. Use the average sales tax worldwide, which is 15%.

**b** You have $100 to spend. Can you afford the painting (including sales tax)? Explain.

**c** List three items you could buy (including sales tax) with your $100. Write down their total cost.

**d** What combination of items can you buy that is the closest to your $100, without going over it? Remember to include sales tax on each item.

**2 a** Find the price of each of the following items after a discount is applied (often called the 'sale price').

**i** A shirt that was originally $62 is discounted by 50%.

**ii** A pair of $120 shoes are now marked 30% off.

**iii** A $40 concert ticket now has 40% off.

**b** What is the total price of each item in part **a** if the worldwide sales tax of 15% is applied after the discount?

> You can solve these problems in a similar way to the tax problems, but this time subtracting the discount from the original price.

**3** Carlos buys a backpack for $30. There is a 10% discount followed by a 15% tax. How much will the backpack cost in total?

**4** The war in Syria, which led to millions of refugees, also increased prices of goods dramatically.

**a** A shawarma sandwich used to cost 100 Syrian pounds (SYP), but has now increased by 100%. What is the new price of the sandwich?

**b** A tax of 10% was then added by the government on the shawarma sandwich. What is the new price of the sandwich?

▶ Continued on next page

**c** The price of six eggs was 120 SYP and increased by 550%. What is the new price of six eggs?

**d** The price of a kilogram of butter used to be 520 SYP. However, that has increased by 250%. What is the new price of a kilogram of butter?

**e** Butter could be replaced by margarine, the cost of which is only 350 SYP per kilogram. Using the new price of butter, what percentage of the price of a kilogram of butter is this?

**f** Imagine you are living in Syria and the price of food is increasing as you have seen. How would it feel to be the head of a family trying to provide for them, despite your salary not increasing? What would you consider as your options?

## Percentage change

Often, we are interested in finding by what percentage a quantity changes, which is a comparison of a change in value to the original value.
The percentage change is found by:

$$\text{Percentage increase or decrease} = \frac{\textit{new amount} - \textit{original amount}}{\textit{original amount}} \times 100\%$$

### Example 6

**Q**

**a** In 2008, the number of undernourished people was 1 billion worldwide. In 2010, approximately 900 million people were undernourished globally. Find the percentage decrease.

**b** Despite the scenario highlighted in part **a**, in 2008, the United States wasted 126 billion pounds of food. In 2010, that amount climbed to 133 billion pounds. Find the percentage increase, rounded to the nearest hundredth.

**A**

**a** To find the percentage change (decrease):

$$\frac{900\,000\,000 - 1\,000\,000\,000}{1\,000\,000\,000} \times 100\% = -10\%$$

It represents a 10% decrease.

Use the percentage change formula.
The negative value here indicates a decrease.

**b** To find the percentage change (increase):

$$\frac{133\,000\,000\,000 - 126\,000\,000\,000}{126\,000\,000\,000} \times 100\% \approx 5.56\%$$

It represents a 5.56% increase.

Give your answer to the degree of accuracy stated in the question.

## Practice 6

1 Calculate the percentage change of the following examples, clearly stating if it is an increase or a decrease. Round to 1 decimal place if necessary.

  a The percentage change from 270 to 60
  b The percentage change from 34 to 91
  c The percentage change from 580 to 2479
  d The percentage change from 8452 to 7203
  e The percentage change from 54 to 540
  f The percentage change from 3160 to 718

2 There are currently approximately 3 400 000 births per year by girls who are under 17 years of age in sub-Saharan Africa and south and west Asia. It is predicted that the number would drop to approximately 3 100 000 births if all girls in those areas were to receive a primary school education. Calculate the predicted percentage decrease in births if all the girls receive a primary school education.

3 It is predicted that the figure given in question **2** would drop even further, to 1 400 000 births, if all girls in those areas also had a secondary school education. Calculate the predicted percentage decrease in births if all the girls receive a secondary school education.

Search online for Math Playground Percent Shopping and play the game that involves calculating discounts and sale prices.

## Reflect and discuss 12

Discuss these issues regarding teenage pregnancies in small groups.

- What would be the implications of such a drop in births? (Practice 6, question 3)
- Why do you think education would have such an impact on the number of births by girls under 17 years of age? How would it feel to be an educated female in sub-Saharan Africa or south and west Asia?
- Do you think that in developed countries 'education for all' is taken for granted?
- If you were to write a headline right now that captured the most important aspect of your discussion, what would that headline be?
- Share your headlines with the class.

ATL1

## Activity 8 – What the world eats

The quantity, quality and variety of foods that people eat around the world differ significantly between regions. You are going to look at the percentage breakdown of the daily diet of different countries and compare them to the world average.

Search online for 'what the world eats' to answer these questions: https://www.nationalgeographic.com/what-the-world-eats/

### World consumption changes

1. What was the average worldwide daily calorie intake in 1961?
2. What was it in 2011?
3. Calculate the percentage increase. Do you think this is a *significant* increase? Why do you think there has been an increase?
4. What category, as a percentage of the total, has increased the most over the years? Why do you think this is?

### Your country compared to the world

1. Find the data for your country. If your country is not listed, your teacher will tell you a country that has a similar demographic.
2. Is your country's calorie intake higher or lower than the world average? Calculate your country's daily calorie intake as a percentage of the world average.
3. Discuss with a partner why your country's calorie intake is higher or lower than the world average. Summarize three reasons to share with the class.
4. Look at the changes of your country's calorie intake over the last 50 years. Pick two categories and calculate the percentage increase or decrease over the 50 years. What is significant about your calculations?

## Your country compared to other countries

1 Look at the 22 countries listed on the right side of the webpage. Which country has the smallest calorie intake per day? Which has the highest? Calculate the percentage difference between the two. Give two explanations for why there is such a difference.

2 Find the country that has the highest percentage of each category in their daily diet. Create a table for your results. Did you find any of these surprising? Why?

3 Select two other countries that are different to your own country. Create a table to compare these countries in the way that you think would best allow the data to be compared. Record all necessary information in your table.

4 Based on their daily diet, how do you think the daily life of people in these countries differs from yours?

## What will the future look like?

1 Calculate the average annual percentage change of daily calorie intake in a country of your choice.

2 In 20 years, if your country follows the same trend it has followed over the past 50 years, what will the daily calorie intake be?

3 In 20 years, what could be the health implications of this?

# Unit summary

A *percent* or *percentage* is a ratio which can be represented as a fraction out of 100.

Fractions, decimals and percentages are all equivalent forms that express the same quantity.

Some percentages, like 50% or 25%, are easier to calculate because you know the fraction that they represent.

## Conversions

**A fraction to a percentage**

Write an equivalent fraction out of 100. Then, write the number using the percent (%) symbol.

or

Divide the numerator by the denominator and then multiply by 100%.

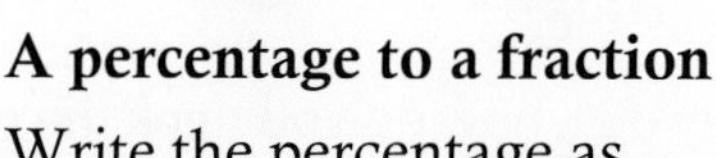

**A percentage to a fraction**

Write the percentage as a fraction out of 100. Be sure to simplify the fraction whenever possible.

**A decimal to a fraction**

Say the decimal number using its place value (e.g. '35 hundredths') and then write the corresponding fraction. Be sure to simplify the fraction whenever possible.

**A percentage to a decimal**

Divide the percentage by 100 and write without the percent symbol.

Calculate percentage increase by dividing the amount of increase by the original quantity. This decimal can then be converted to a percentage by multiplying by 100%.

Calculate percentage decrease by dividing the amount of decrease by the original quantity. This decimal can then be converted to a percentage by multiplying by 100%.

# Unit review

Launch additional digital resources for this chapter

Key to Unit review question levels:

Level 1–2 | Level 3–4 | Level 5–6 | Level 7–8

**1** **Write down** the percentage represented by the shaded section of each grid.

a 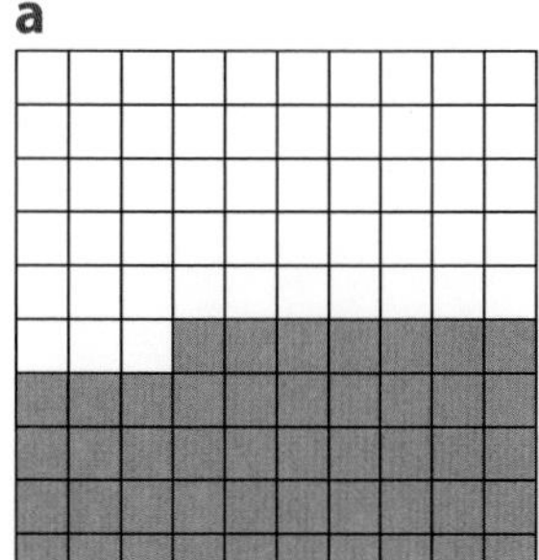

b 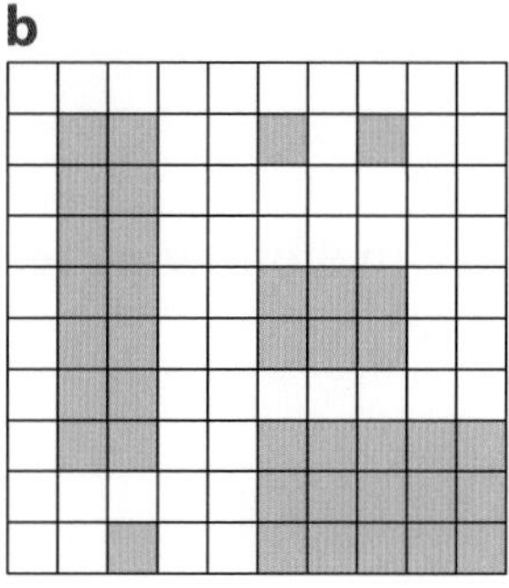

c

d 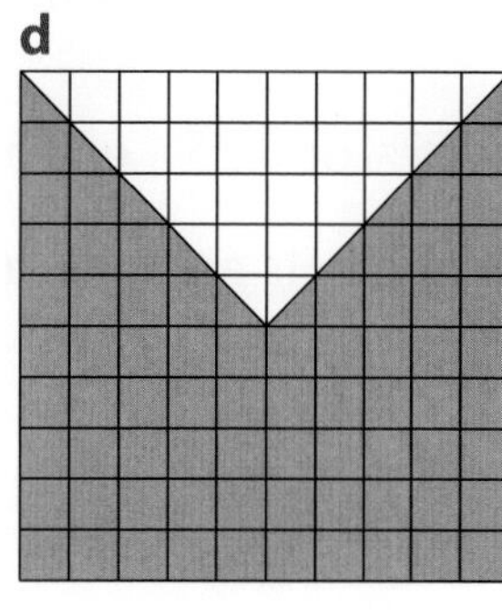

e

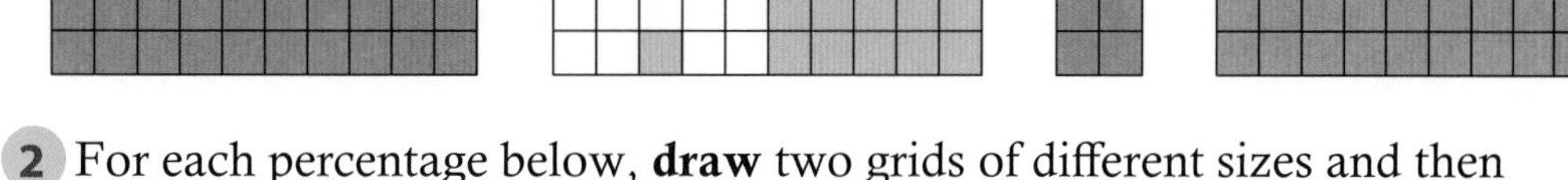

**2** For each percentage below, **draw** two grids of different sizes and then shade the given percentage.

**a** 30% **b** 50% **c** 75% **d** 18% **e** 12%

**3** If you knew what 3%, 8% and 10% of a quantity was, how could you **calculate** the following percentages of the same quantity?

**a** 9% **b** 11% **c** 21% **d** 18% **e** 16% **f** 26% **g** 77%

**4** If 20% of a quantity is 30, find the following percentages of the same quantity.

**a** 10% **b** 40% **c** 100% **d** 30% **e** 4% **f** 24%

**5** Convert these fractions to percentages.

**a** $\frac{9}{10}$ **b** $\frac{17}{20}$ **c** $\frac{14}{7}$ **d** $\frac{3}{12}$ **e** $\frac{22}{55}$

**f** $\frac{20}{50}$ **g** $\frac{12}{60}$ **h** $\frac{8}{99}$ **i** $\frac{37}{37}$ **j** $\frac{40}{11}$

**6** Convert these fractions to percentages, and round to the nearest whole percentage.

**a** $\frac{3}{13}$ **b** $\frac{8}{35}$ **c** $\frac{14}{47}$ **d** $\frac{61}{66}$ **e** $\frac{7}{19}$

**f** $\frac{31}{51}$ **g** $\frac{12}{73}$ **h** $\frac{153}{88}$ **i** $\frac{36}{35}$ **j** $\frac{212}{100}$

7. **Write down** the following in all three forms: fraction, decimal and percentage. Give each fraction in its simplest form.

   a 28%  b 0.8  c $\frac{13}{20}$  d 0.32  e 55%

   f $\frac{7}{15}$  g 92%  h $\frac{4}{20}$  i $0.\overline{60}$  j 110%

8. Which is the largest number in each of the following lists? **Justify** your answer.

   a $\frac{4}{5}$ or 75% or 0.81  b 0.92 or 90% or $\frac{17}{19}$  c $\frac{2}{3}$ or 65% or 0.63

   d $\frac{1}{4}$ or 26% or $0.\overline{25}$  e 0.55 or 53% or $\frac{6}{13}$

9. Which numbers are equivalent in each of the following sets?

   a $\frac{12}{50}$, 25%, $\frac{1}{4}$, 0.24, $\frac{3}{8}$, 2.4%, 24%

   b 5%, $\frac{2}{20}$, 0.5, $\frac{30}{600}$, 50%, $\frac{1}{2}$, 0.05

10. In the 2016 Olympics, the International Olympic Committee created a 'Refugee Olympic Team' to draw attention to the worldwide refugee crisis. Ten athletes were selected from four countries to take part in this team. **Calculate** the percentage of athletes from the following countries of origin.

    a 5 athletes were originally from South Sudan.

    b 2 athletes were originally from Syria.

    c 1 athlete was originally from Ethiopia.

11. There are an estimated 700 million children living in the world who are of primary school age. Approximately 70 million of them still do not have access to education. Approximately what percentage of primary-age children do not have access to education?

12. In the United States Senate, 20 senators out of 100 are women. In Canada, 88 women have been elected to the House of Commons out of 338 members. Which country has the higher percentage of women in its government house?

**13** **a** What percentage of 50 is 10?

**b** What is 30% of 20?

**c** 15 is 45% of what number?

**d** What is 60% of 90?

**e** 40 is 25% of what number?

**f** What percentage of 80 is 60?

**14** Due to the efforts of many charitable organizations, enrollment in primary education has reached approximately 90% worldwide. If there are approximately 60 million children still not attending schools, approximately how many are enrolled in primary schools?

**15** Many charitable organizations have worked hard to reduce child mortality. Since 1990, the number of deaths worldwide in children under five years old declined from approximately 13 million to approximately 6 million.

**a** What is the approximate percentage decrease?

**b** Assuming a constant rate of change, by approximately what percentage did the number of deaths drop per year?

**c** If this trend continues at the same rate, what would you predict to be the number of deaths in children under five years old in ten years' time?

**16** Approximately 780 million people in the world are illiterate. What percentage is that of the total population? (You found this statistic earlier in this unit). Of the 780 million, almost 70% are female. Approximately how many females are illiterate in the world?

**17** **a** Before 2011, a typical Syrian could expect to live to be about 80 years old.
In 2014, during a civil war where millions of people died, life expectancy in Syria decreased to 56 years old. **Calculate** the percentage decrease in life expectancy.

**b** Due to the civil war, 4 million refugees fled Syria. Some of these refugees fled to neighboring Jordan, where the population swelled from approximately 5 850 000 to approximately 6 500 000. **Calculate** the percentage increase in the population of Jordan.

**c** Lebanon, another country that took in many refugees, saw its population increase from roughly 3 million to 4 million. **Calculate** the percentage increase in the population of Lebanon.

# Summative assessment

criteria C, D

## 'You are a refugee' assignment

Pairs

You and your partner are refugees and have just arrived in your current city.
You have no money, but the government gives you 'social assistance' funding to survive.
Your teacher will tell you the monthly amount that you will receive.
This amount needs to cover all living expenses. Your teacher will also tell you how many members are in your family and the ages of any children.

You must find a place to live and prepare a monthly budget for rent, food, transportation, living expenses, clothes and other expenses. As a class, you will create a list of all of these expenses and every pair will use this list.

ATL2

This assignment has many different tasks that will require a lot of organization. You and your partner will need to collaborate well. It will be very important to set up a good organizational framework to ensure that you can both access all your information easily, and that it will not get lost.

Once you have read through the assignment with your partner, your first task will be to create a graphic organizer/diagram or other method to organize all of the different components in this assignment. The second task will be to create a to-do list along with a calendar of when each item will be completed.

Think about the following questions:

- How will you organize your information?
- How will you record and keep track of the websites you have used?
- How will you break up the tasks into manageable sections?
- How will you create a to-do list?
- What headings will your to-do list have?
- How will you communicate with each other?
- How will you manage your time so that you do not leave too much work to the last minute?

Once you have submitted this graphic organizer/diagram and to-do list, your teacher will provide you with the appropriate platform on a web-based application (connection through link or email) for you both to work on. Then you will set up the organization framework that you outlined. If technology cannot be used as the self-management tool for this assignment, you must still explain how you will organize and store your documents and information, and how you will manage your time.

### Part A – Preparation

1 **Find a place to live.** You will look for an apartment in a local newspaper or online. Remember your budget and keep in mind the other expenses that you have!
Once you have found your accommodation, locate it on a map of your city/town.

2 **Find food to eat.** Write down a nutritious meal plan for breakfast, lunch and dinner for seven days. Be specific – don't just write 'sandwich' – consider all the parts of the sandwich you will need: bread, mustard, tomato, cheese, etc. Use a chart like the one below to record your meal plan.

| | Monday | Tuesday | Wednesday | Thursday | Friday | Saturday | Sunday |
|---|---|---|---|---|---|---|---|
| Breakfast | | | | | | | |
| Lunch | | | | | | | |
| Dinner | | | | | | | |
| Snack | | | | | | | |

a Once you have an idea of what you will eat in a week, you will have to calculate the cost of the items. First, create a grocery list for the food you will need to make the meals listed in your chart for one week.

Then put the list in a chart with three columns: the food item, the estimated cost and the actual cost. You will then estimate how much you think each item costs and calculate the total. You could create your chart using a spreadsheet so that you can do the calculations efficiently. To find the real costs, you can either search a grocery store's prices online, or go to the grocery store.

Consider a field trip to a supermarket as a class, or individually with your parents. Alternatively, use an online shop to find the costs of grocery items.

b Calculate the actual total cost of the groceries.

c Calculate the percentage difference between what you estimated the total cost of the groceries to be and their real cost.

d Explain the degree of accuracy of your results in part **c**. Describe whether or not your results make sense.

3 **Money for other expenses:** You will set up a monthly budget spreadsheet. A budget tracks your income and your expenses to help you see how much money you will have left to save or how much you will be in debt. The budget is best created using a spreadsheet, so you can set up a formula for 'Income less total expenses'. This number should be positive so that you do not go into debt. Here is an example budget spreadsheet.

| MONTHLY BUDGET | |
|---|---|
| **CATEGORY:** | **Amount:** |
| *Income: (how much you earn)* | |
| *Expenses: (what you have to pay)* | |
| Rent | |
| Groceries (multiply weekly amount by 4) | |
| [Create a row for each expense] | |
| **Total expenses: (add up all expenses)** | |
| **Net Income: (Income less total expenses)** | |

4 As you complete this assignment, you will be recording your thoughts and reactions. Discuss what struggles you are facing, and your thoughts about how refugees might survive on this money. You will submit this reflection with your assignment.

## Part B – Analysis

1 Using the grocery list based on your meal plan for the week, approximate what percentage of the grocery list is:

**a** carbohydrates  **b** protein  **c** fat

2 Compare your grocery list with your country's nutrition guidelines. How close does your grocery list match the recommended daily allowances of your country's nutrition guidelines?

3 Select one of the days of the week from Part A question **2** and calculate the number of calories one person will eat in a day as a refugee moving to a new country. To do this, divide the total by the number of people in your family. Compare this amount to what a person in a refugee camp eats in a day and what you actually eat in a day. (You calculated these in the formative assessment task earlier in the unit.)

**a** Who consumes the greatest number of calories in a day?

**b** Who consumes the greatest percentage of carbohydrates, protein and fat in a day: you, someone in a refugee camp, or the refugee moving to a new country?

**c** Did any data in the comparison surprise you? Why or why not?

4 Looking at your monthly budget, calculate each expense as a percentage of the total. Did any of the expenses surprise you? Why or why not?

## Reflect and discuss 13 – Empathy for a refugee

- Were you able to stay on budget? Describe any sacrifices you had to make.
- Were you able to eat nutritiously on the budget? Explain.
- How did the life you created as a refugee differ from the life you live now? What would it feel like to be a refugee? Explain your answers.
- What recommendations would you make to your city/country to make life better for refugees who move to your country?

ATL1

# Self-assessment

## ATL2 Self-management

Throughout this assignment, you have worked on developing your self-management skills to stay on task and meet the deadlines outlined by your teacher. Using a copy of the checklist below, reflect on how you were able to manage your organizational skills and affective skills throughout the assignment. In the first column, rank yourself as 1 (emergent), 2 (capable) or 3 (exemplary) to indicate how each skill has developed. In the next column, discuss what you did well and what you would do next time to improve each skill.

| Skill | Emergent (1)<br>Capable (2)<br>Exemplary (3) | What would you do next time to improve this skill? If you think you developed this skill well, list the things you did that helped your development and would do again next time. |
|---|---|---|
| Managing time and tasks effectively | | |
| Meet deadlines of the assignment<br>• Used a calendar effectively<br>• Tasks were well spaced out – no cramming/rushing to complete tasks on time | | |
| Create plans to prepare for summative assessments<br>• Created to-do lists, followed and updated them<br>• Prioritized tasks on the to-do list<br>• Broke down the assignment into manageable tasks | | |
| Keep an organized and logical system of information files/notebooks<br>• Created a logical system to organize all documents/tasks in an effective way (e.g. using Google Docs effectively)<br>• Used this system throughout the assignment so all information was easily accessible to both partners | | |
| Select and use technology effectively and productively<br>• For organization and collaboration (Google Docs)<br>• For researching information | | |
| Managing state of mind | | |
| Practise strategies to overcome distractions<br>• Turned off social media and email notifications while working on computer/tablet<br>• Kept workspace tidy so resources were readily accessible | | |
| Demonstrate persistence and perseverance<br>• Kept trying and revised plans when things did not go right<br>• Learned from making mistakes instead of seeing them as setbacks | | |

# 3 Algebraic expressions and equations

Nature is full of patterns. In this unit, you will begin to understand how these patterns can be described using mathematics, particularly algebra. From art, to history, to urban planning, many other aspects of life also involve patterns. Where might an exploration of algebraic patterns in these other areas take you?

## Orientation in space and time

### Eras, epochs and turning points

The history of our planet is sometimes observed as a series of turning points that can be described and analyzed with patterns. The Industrial Revolution, for example, brought with it the invention and application of the combustion engine where pistons follow a pattern of intake, compression, power and exhaust. Each phase within the pattern has its own sub-pattern, which can be described using algebra.

Two centuries later came what is known as The Space Age. The Space Age began in the 1950s and focused on space exploration and space technology. Our ability to send a human to the moon relied on understanding the relationships between the forces on a rocket and it's size and mass. Patterns can also be observed when studying the distances travelled by a rocket at different speeds. Being able to describe these relationships algebraically has helped humans not only look at, but also visit and study, objects in the night sky.

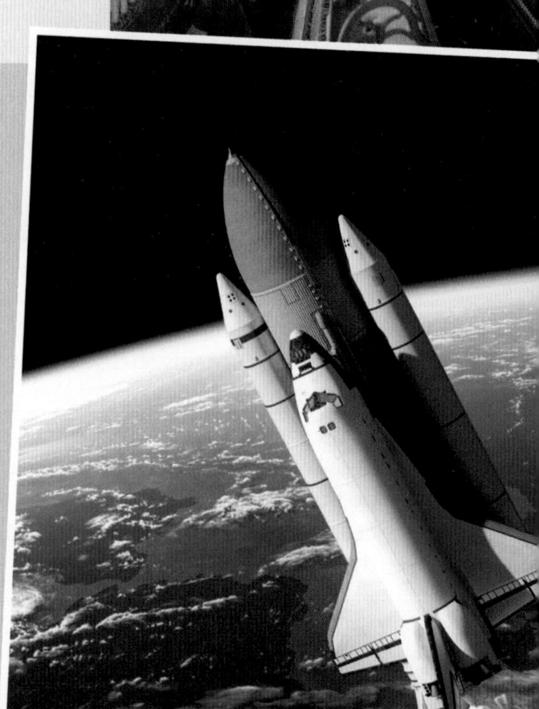

## Scientific & technical innovation

### Personal and cultural expression

Music is full of patterns. The tempo of a song is the number of beats per minute, and this usually follows a consistent pattern. Drummers and other percussionists are experts at following complex patterns. Even converting music to a digital form requires representing the music mathematically. Algebra is used to capture, analyze and even create music.

Artistic beadwork is found in many different cultures around the world. In parts of Kenya and Tanzania, the Maasai people use beads to represent aspects of their culture and then create ornate pieces of jewelry with them. These necklaces, headbands and bracelets are composed of patterns that can be described algebraically.

# Algebraic expressions and equations

## Patterns in nature

## KEY CONCEPT: LOGIC

## Related concepts: Generalization, Models and Patterns

## Global context

In this unit you will use algebra to explore patterns in the natural world, which is one aspect of the global context **scientific and technical innovation**. In order to better understand the patterns around us, you will need the tools of algebra, such as writing expressions and solving equations. By the end of the unit, you may begin to wonder if the natural world is actually more mathematical that it initially appears.

## Statement of Inquiry

A logical process helps to model and generalize patterns in the natural world.

### Objectives

- Using correct terminology when analyzing algebraic patterns and sequences
- Representing patterns in different forms – diagrams, sequences, tables, words
- Creating and simplifying basic algebraic expressions
- Generalizing a mathematical pattern using algebra, and solving applications involving algebraic patterns
- Solving single-step and basic two-step algebraic equations

### Inquiry questions

**F** What is a pattern?

What are the different types of patterns?

**C** How do we model patterns?

How is it possible to model a pattern and make predictions?

**D** Is there a mathematical order to our natural world?

**ATL1 Reflection skills**

Consider content:

- What did I learn about today?
- What don't I yet understand?
- What questions do I have now?

**ATL2 Creative-thinking skills**

Practise visible thinking strategies and techniques

## You should already know how to:

- convert between different units
- perform calculations following the order or operations
- evaluate basic exponents and perfect squares

# Introducing patterns in nature

Can you see what the following pictures all have in common?

While they all represent something in nature, they also have a pattern that repeats itself. Some patterns are decorative while others serve an important purpose, like strength or camouflage or protection. Patterns that you can identify and describe can then be studied and understood better. Much of the mathematics that you learn involves establishing patterns and then describing them using appropriate mathematical notation and vocabulary.

## Reflect and discuss 1

- When does something become a pattern, as opposed to just a series of events?
- What patterns do you notice in your daily life?
- What patterns do you see in nature?
- Give two examples of mathematical patterns in your life.

## Sequences

An ordered list of items (numbers, shapes, etc.) is called a *sequence*. Items in a sequence are called *terms*.

ATL2

Pairs

### Activity 1 – Looking for patterns

Take a minute to look at each part of question **1**. Then, in pairs, discuss possible answers to each one. Be sure to explain your reasoning.

**1** What are the next two terms in each of these sequences?

**a** 1 3 5 7 ___ ___

**b** 2 4 8 16 ___ ___

**c** M T W T F ___ ___

**d** J F M A M J ___ ___

**e** 1 4 16 64 ___ ___

**f** O T T F F S S ___ ___

**g** M V E M J S ___ ___

**h** Α Β Γ Δ Ε Ζ ___ ___

**i** 100 80 60 40 ___ ___

**2** Make up your own sequence and trade with a classmate. See if you can correctly identify the next two terms in your classmate's sequence and explain your reasoning.

Some of the sequences in question **1** are words you will see in everyday life.

In question **1** of Activity 1, you may have found that there was sometimes more than one answer. Whether an answer is correct or not depends on if it follows the *pattern* of the letters or numbers.

## Patterns

A *pattern* is a set of items or numbers that follow a predictable rule. The *core* of a pattern is the part of the pattern that keeps repeating.

In *algebraic patterns,* the same mathematical procedure is applied each time. For example, you might add or multiply by the same number to determine the next term, or even perform a combination of different operations.

## Activity 2 - The core of a pattern

1 Which of the sequences in question **1** of Activity 1 follows an algebraic pattern? Explain the pattern using words.

2 Identify the core of the following patterns.

3 Birds make sounds to define territory, call a mate, scold other members of the flock, and even beg for food. Their songs and calls follow patterns that have allowed people to identify birds without even seeing them. Identify the core of the following patterns in bird calls.

**a** *meeoopeeoomeeoopeeoo…* (buzzard)

**b** *jeditjeditjeditjedit…* (ruby crowned kinglet)

**c** *zzzzzdzzzzzdzzzzzd…* (lazuli bunting)

**d** *pekpekcackcackcackpekpekcack…* (Cooper's hawk)

**e** *tseeewtseeewtseeew…* (American robin)

**f** *urrREEErrrurrREEErrrurrREEErrru…* (common pauraque)

Is it possible for one pattern to lead to a different pattern? In the following investigation, you will analyze a pattern of shapes and look for an algebraic pattern within it.

## Investigation 1–Sequences

criterion B

**a** Draw the next two shapes in the sequence on a copy of the grid below.

**b** Explain the pattern in words.

**c** Write this pattern as a list of 5 numbers.

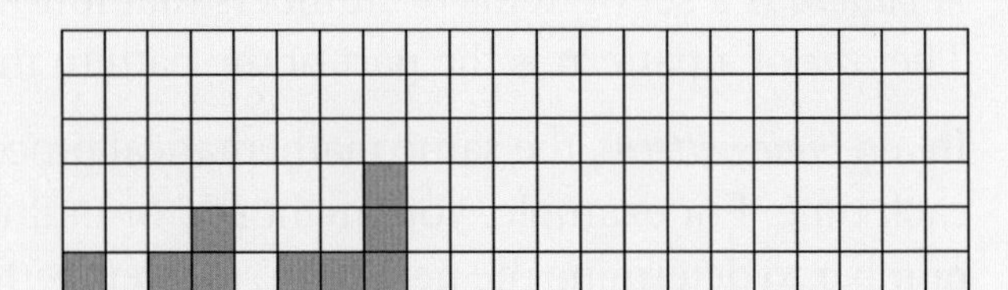

**d** Describe the number pattern in words. How do you find the next number in the sequence?

**e** Determine what the 10th term in the sequence would be, without drawing it. How do you know?

**f** What would the 100th term in the sequence be? How do you know?

**g** Will the number 20 be in this sequence? Explain why or why not.

**h** Draw a different sequence of three shapes that follow the same pattern as the one above.

## Reflect and discuss 2

- In Investigation 1, which was easier: finding the pattern in the shapes (**b**) or finding the pattern in the numbers (**d**)? Explain.
- What is the minimum number of shapes needed to establish a definite pattern? Explain.

Recognizing a pattern and being able to describe it are two very important skills in mathematics. You can describe a pattern using words or even mathematical symbols; mathematicians call this 'generalizing a pattern'. It is better to be specific in your description, so instead of saying 'the numbers get bigger', try to say something like 'the numbers increase by 2 every time'.

When you generalize a pattern, try to be very precise about how the numbers relate to one another. For example, when the difference between terms is always the same, it is called a *linear pattern*.

## Practice 1

**1 a** Draw the next two shapes in the sequence below.

**b** Explain the pattern in words.

**c** Write this pattern as a sequence of five numbers.
How do you find the next number in the sequence?

**d** Is this a linear pattern? Explain.

**e** If one of the shapes has $L$ leaves in it, how many leaves will there be in the next shape?

**f** Determine what the 10th term in the sequence would be, without drawing it.
How do you know?

**g** Will the number 30 be in this sequence? Explain why or why not.

▶ Continued on next page

**2 a** Draw the next two shapes in the sequence below.

**b** Explain the pattern in words. How do you find the next number in the sequence?

**c** Is this a linear pattern? Explain.

**d** Determine what the 30th term in the sequence would be, without drawing it. How do you know?

**e** If one of the shapes has $w$ water droplets in it, how many droplets will there be in the next shape?

**f** Will the number 40 be in this sequence? Explain why or why not.

**3 a** Draw the next two shapes in the sequence below and explain the pattern in words.

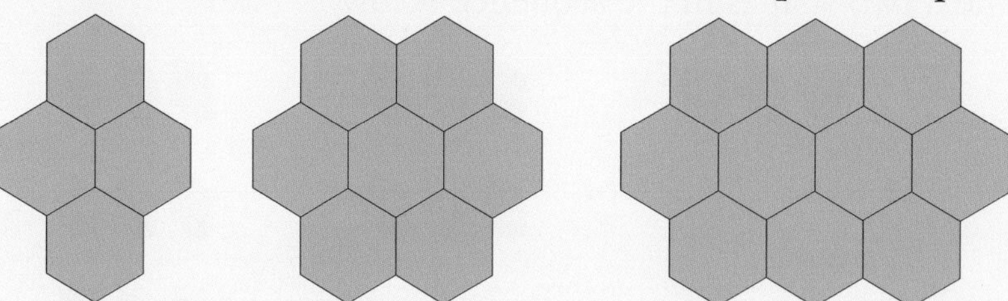

**b** Determine what the 26th term in the sequence would be, without drawing it. How do you know?

**c** Will the number 409 be in this sequence? Explain why or why not.

**d** Write a rule to find the next term of the sequence if you know the one before it.

**4** Draw a pattern for the first three terms in each of the following sequences. Then draw the next two shapes in each pattern.

**a** 3, 7, 11 **b** 1, 9, 25

**5** Maria starts the pattern below. She stays up all night and falls asleep after finishing the 914th term. What shape was it?

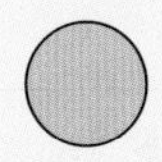 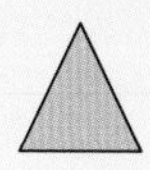 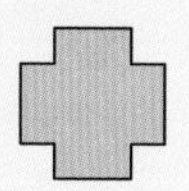 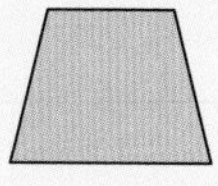 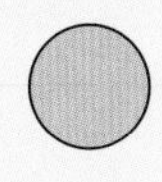 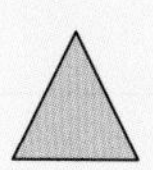

## Reflect and discuss 3

- What did I learn about today?
- What don't I yet understand?
- What questions do I have now?

Go to learnalberta.ca. Choose English as your language. Under the Find Resources heading, select Mathematics, Grade 6 and enter the keyword 'Exploring Patterns'. Click on the Exploring Patterns link on the right of the screen to open the resource 'It's a Bit Nutty!'. Follow the instructions and complete the activities.

## Representing patterns with expressions

When a pattern is generalized in mathematics, an *expression* is written that uses operations and numbers instead of words. This requires being able to use the correct mathematical notation, vocabulary and/or symbols. Translating between words and mathematical symbols is another important part of algebra.

Pairs

### Activity 3 – The language of operations

Create a table with five columns to represent each of the five mathematical operations below (in blue), and then put the words in the correct column. Add one extra word for each column if you can.

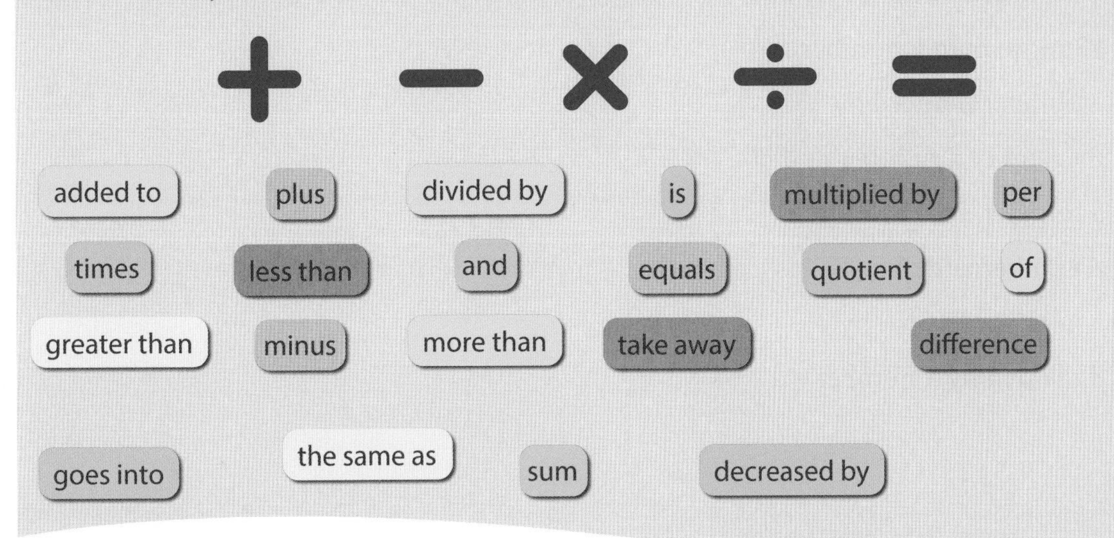

An *algebraic expression* consists of variables (typically letters), numbers and operations. Examples of algebraic expressions are:

$$4y \qquad t - 10 \qquad n \div 2 \qquad 3w + 5$$

When using algebraic letters, the multiplication symbol × is not generally used, to avoid confusion with the letter $x$. To describe multiplication, you simply write a number and a letter together and put the number first with no operation symbol (e.g. $3w$).

Instead of the division symbol (÷), a fraction bar is used $\left(\text{e.g. } \frac{m}{8}\right)$.

Some algebraic expressions and their meanings are:

$y + 7$ = 'a number plus seven'

$3m$ = '3 times a number'

$4a - 8$ = '8 less than four times a number'

$\frac{d}{5}$ = 'a number divided by 5'

$\frac{k}{6} + 2$ = 'two more than a number divided by 6'

## Practice 2

**1** Translate each of the following phrases into a mathematical expression. Use whatever letter or symbol you like for the original number.

**a** 3 times a number

**b** a number increased by 7

**c** a number multiplied by 5

**d** 12 more than a number

**e** a number reduced by 13

**f** the sum of a number and 60

**g** a number divided in half

**h** a number added to 17

**i** a number divided by 43

**j** 8 divided by a number

▶ Continued on next page

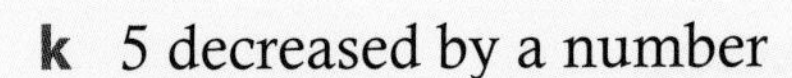

**k** 5 decreased by a number

**l** the product of 15 and a number

**m** a number multiplied by 4, then decreased by 26

**n** 8 times a number, then added to 3

**o** 7 added to a number, then the total multiplied by 4

**p** 20 divided by a number, then reduced by 11

**2** Translate the following symbols back into words (For example: 'seven times a number'). Try to use different mathematical words as much as possible.

**a** $h - 1$ **b** $p + 11$ **c** $14y$ **d** $\frac{z}{3}$ **e** $q - 12$

**f** $20 - c$ **g** $\frac{12}{m}$ **h** $\frac{j}{20} + 7$ **i** $8w - 9$ **j** $2 - 3t$

ATL1

## Reflect and discuss 4

- What did I learn about today?
- What don't I yet understand?
- What questions do I have now?

In order to generalize a pattern, you need to figure out the relationship between successive terms. It is sometimes helpful to draw circles between the numbers to study the pattern. You can then fill in the circles with the pattern that you notice.

## Example 1

**Q** Find the next term in the sequence that begins 1, 5, 9, 13, and generalize the pattern using a mathematical expression.

**A**

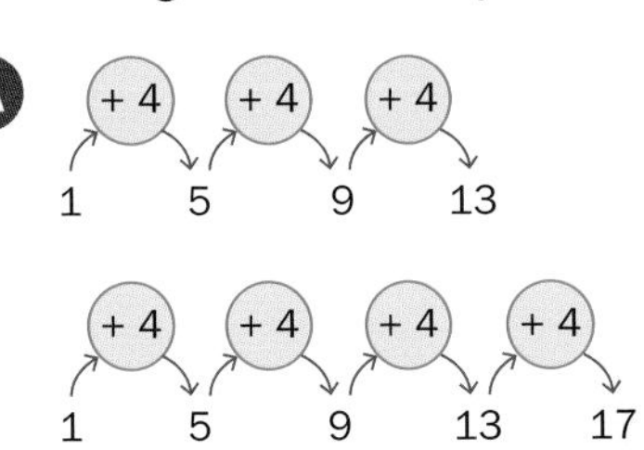


The numbers go up by 4 each time.

Since this linear pattern adds 4 to each term, the fifth term would be 17.

The general pattern is $n + 4$, where $n$ is the previous term.

We can generalize how to find the next term in a sequence by writing an expression called a 'pattern rule'. One type of pattern rule describes how to find a term based on the term before it. In Example 1, the pattern rule is $n + 4$, since every term comes from adding four to the term before it ($n$).

## Practice 3

ATL2

Pairs

1 For each of the following sequences, find the next term and the pattern rule based on knowing the term before it. (For example: $n - 2$ or $6n$.)

> You can use any of the mathematical operations +, −, and ×, but make sure you represent division as a fraction.

**a** 32, 38, 44, . . .

**b** 103, 98, 93, . . .

**c** 3, 9, 27, . . .

**d** 250, 50, 10, . . .

**e** 2, 8, 32, . . .

**f** 17, 26, 35, 44, . . .

**g** 200, 100, 50, 25, . . .

**h** 40, 400, 4000, . . .

2 Your body is made up of many cells. As you grow, your body makes new cells by the process of cell division. So, you start with one cell and it divides to become two. Then each of the two cells divide to become four, and then they divide to become eight, and so on.

**a** After 1, 2, 4, 8, how many cells will there be in the next three stages of cell division?

**b** If there are 512 cells, how many stages have occurred?

**c** Is this a linear pattern? Explain.

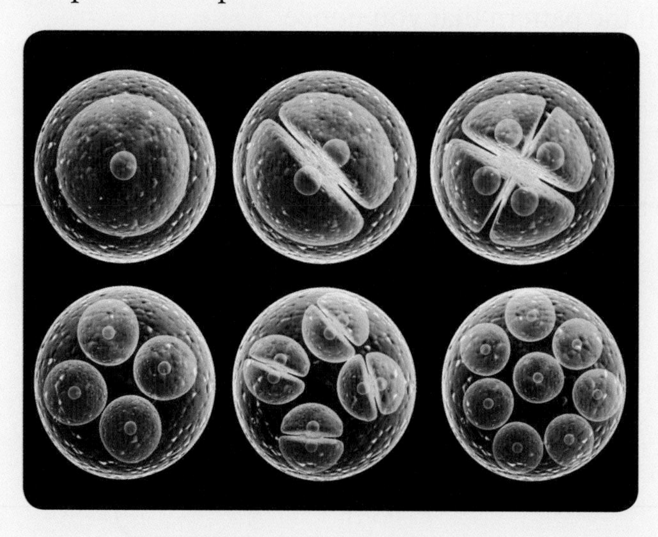

▶ Continued on next page

**3** There is a common belief that animals get an average of 1 billion heartbeats in their lifetime. You will use the table below to verify whether or not this belief is true.

| Animal | Heart rate (beats per minute) | Average approx. lifespan (years) | Lifetime heartbeats |
|---|---|---|---|
| Pig | 70 | 25 | |
| Rabbit | 205 | 9 | |
| Horse | 44 | 40 | |
| Hamster | 450 | 3 | |
| Monkey | 190 | 15 | |
| Whale | 20 | 80 | |
| Elephant | 30 | 70 | |

**a** Find the lifetime heartbeats for each animal. Is the common belief true, that animals get approximately 1 billion heartbeats in a lifetime?

**b** Find the *total* of the lifetime heartbeats and divide this by 7, the number of animals in the list. This will give you the *average* lifetime heartbeats for this group of animals.

**c** Based on the number you found in part **b**, if a cat lives for approximately 15 years, what is its approximate heart rate?

**d** Humans have a heart rate of 60 beats per minute and live for approximately 70 years. Do they follow a similar pattern in terms of lifetime heartbeats?

**4** The San Andreas fault is the boundary between two tectonic plates that are part of the Earth's crust. When the plates move, it causes earthquakes, such as the devastating San Francisco earthquake of 1906.

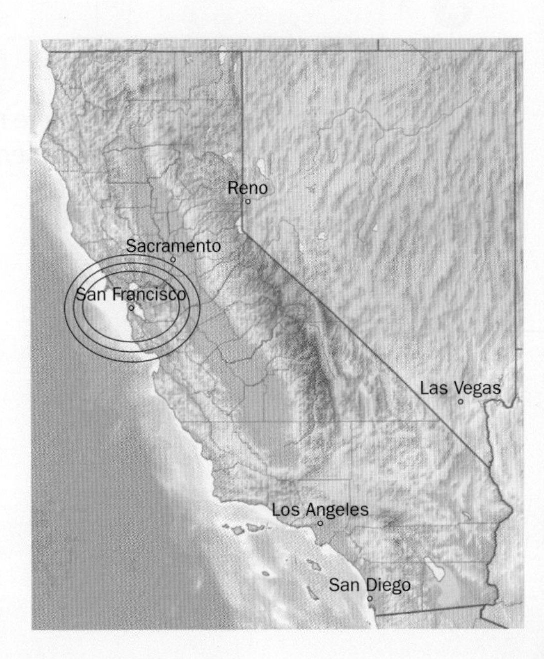


**a** If the fault moves roughly 3.5 cm every year, write a sequence that represents the total movement of the fault over the first 5 years.

**b** Write the pattern rule to represent how to find a term, given that you know the one before it.

**c** Is this a linear pattern? Justify your answer.

**d** Scientists believe that the fault was formed approximately 28 million years ago. How far has the fault moved in total since it was formed?

ATL1

## Reflect and discuss 5

- What did I learn about today?
- What don't I yet understand?
- What questions do I have now?

There is a type of pattern rule which allows you to find any term in a sequence without knowing the previous term. Instead you substitute the term's position number into the rule. For example, if you want to know the 12th term in a sequence, you substitute the number 12 into the rule. Substituting a number into an expression is sometimes referred to as 'finding the value of an expression'.

## Example 2

Here is the beginning of a sequence.

2, 4, 6, 8, 10, . . .

**a** Find the pattern rule based on the position of the term in the sequence.

**b** Determine the 12th term of the sequence, without writing more terms.

**c** Determine the 150th term of the sequence.

**a**

| Term position ($p$) | 1 | 2 | 3 | 4 | 5 |
|---|---|---|---|---|---|
| Term ($t$) | 2 | 4 | 6 | 8 | 10 |

Draw a table that includes the term positions and the terms.

The pattern rule is $t = 2p$, where $t$ is the term and $p$ is its position in the sequence.

You could use any letters that you like.

**b** $t = 2p$
$= 2 \times 12$
$= 24$

To find the 12th term, substitute 12 into your rule.

The 12th term is 24.

**c** $t = 2 \times 150$
$= 300$

The 150th term is 300.

To find the 150th term, substitute 150 into the rule.

## Practice 4

**1** Substitute the given number into each expression to find the value of the expression.

**a** $x + 7$, when $x = 3$

**b** $5a$, when $a = 12$

**c** $c - 10$, when $c = 23$

**d** $9e$, when $e = 4$

**e** $\frac{y}{6}$, when $y = 42$

**f** $\frac{g}{2} + 13$, when $g = 8$

**g** $6w - 1$, when $w = 11$

**h** $4x - 2$, when $x = 15$

**i** $2e + 9$, when $e = 4$

**j** $15 + \frac{z}{8}$, when $z = 64$

**k** $12 + 7b$, when $b = 9$

**l** $4x - 2$, when $x = \frac{1}{2}$

**m** $\frac{f}{4} + \frac{1}{2}$, when $f = 2$

**n** $\frac{m+7}{2}$, when $m = 19$

**2** Orb spiders create webs that are nearly circular, with every new row almost exactly the same distance away from the previous one. No one is sure why they use such precise measurements, since they don't seem to catch more prey than spiders that spin webs that don't follow this pattern.

As an orb spider spins its web, the number of segments increases. After one complete trip around (cycle), there are 13 segments. After the second cycle there is a total of 26 segments. By the end of the third cycle there are 39 segments in total.

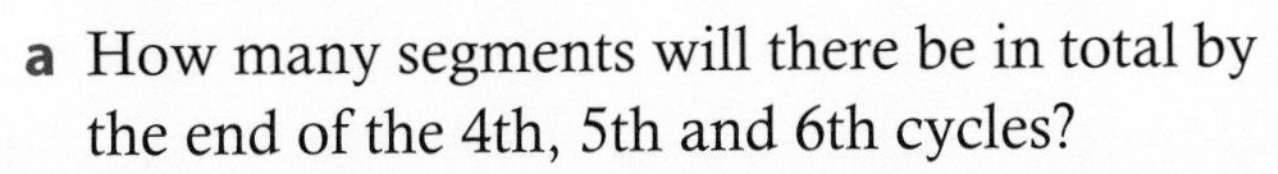

**a** How many segments will there be in total by the end of the 4th, 5th and 6th cycles?

**b** Is this a linear pattern? Explain.

**c** What is an easy way to calculate the total number of segments after any cycle?

**d** If the number of cycles is $c$ and the number of segments is $s$, how can we represent this pattern mathematically?

**e** How many segments will there be after the 20th cycle?

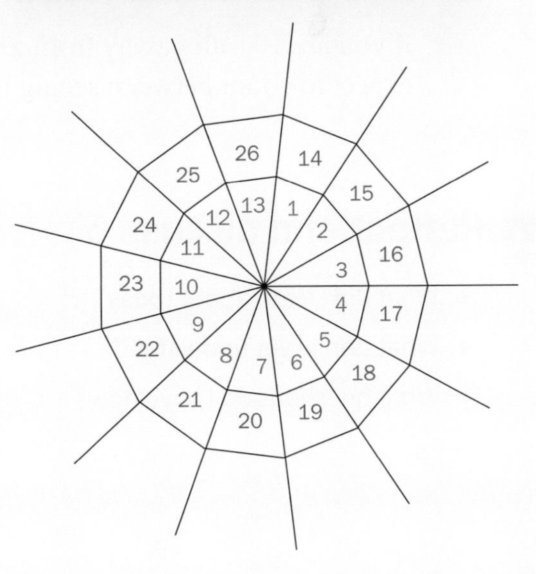


▶ Continued on next page

**3** As the orb spider adds more segments, the radius of the area where prey can be caught increases. The distance between any two rows is roughly 2 cm.

**a** If the number of rows is $n$, write a rule to determine the radius $r$ of the area where prey can be caught.

**b** Is this a linear pattern? Explain.

**c** If there are 11 rows, what is the radius of the web?

**d** If the radius of the web is 30 cm, how many rows are there?

ATL2

Pairs

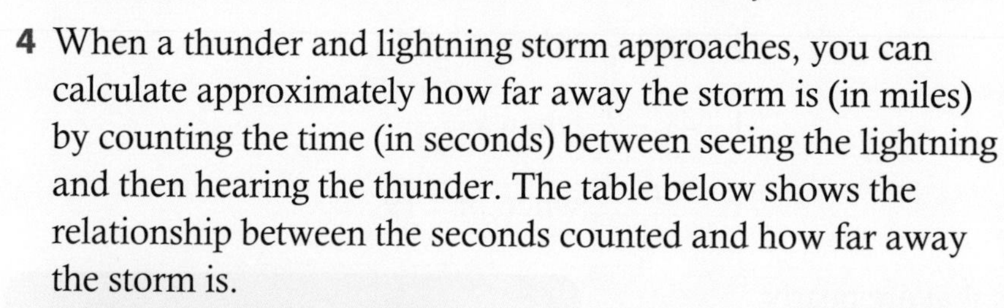

**4** When a thunder and lightning storm approaches, you can calculate approximately how far away the storm is (in miles) by counting the time (in seconds) between seeing the lightning and then hearing the thunder. The table below shows the relationship between the seconds counted and how far away the storm is.

| Seconds counted | 5 | 10 | 15 | 25 | 30 |
|---|---|---|---|---|---|
| Distance from storm (miles) | 1 | 2 | 3 | 5 | 6 |

**a** How can you calculate how far away the storm is after counting the seconds between seeing the lightning and hearing the thunder?

**b** If the time (in number of seconds counted) is represented by the letter $t$, write an expression that will calculate the distance $d$ of the storm in miles.

**c** If you are 10 miles away from a storm, how many seconds would you expect to count between seeing the lightning and hearing the thunder?

ATL1

## Reflect and discuss 6

- What did I learn about today?
- What don't I yet understand?
- What questions do I have now?

## Patterns as functions

Here is the Magic Math Machine.
A number (input) goes in one end, the machine applies a math rule to it, and a number comes out of the other end transformed (output). The machine will always apply the same rule to any number you put in.

Say you input 1 into the machine, and 5 comes out.
What could the rule be? There are two possibilities that involve only one operation:

- add 4 to the input
- multiply the input by 5.

Which rule is correct? Check by inputting another number . If you input 2, the number 10 comes out. Which rule works for both input numbers?

You can find the output by multiplying the input by 5.
This can be written as $5i$, where $i$ stands for *input*.
You can also write this as $o = 5i$ since the value of the output $o$ is equal to 5 times the value of the input $i$.

## Activity 4 - Function machines

Search online for 'Math Playground Function Machine' and select the game. Choose the Beginner level then click on Computer Decides Input and use the function machine to generate input and output numbers. When you think you have determined the rule for the function, type it into the calculator to see if you are correct.

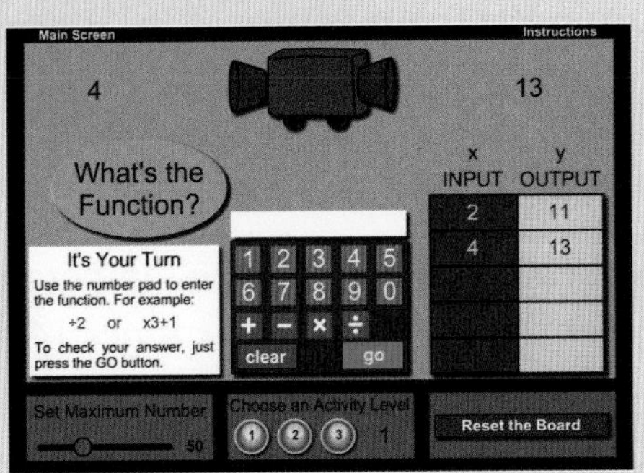

## Practice 5

**1** Some of the input and output values for a number of function machines are shown in the tables below.
For each machine, find the rule. Explain whether or not the pattern is linear and find any missing values.

**a**

| Input | Output |
|---|---|
| 1 | 4 |
| 2 | 8 |
| 3 | 12 |
| 4 | 16 |
| 100 | 400 |

**b**

| Input | Output |
|---|---|
| 1 | 2 |
| 2 | 4 |
| 3 | 6 |
| 4 | |
| 100 | |

**c**

| Input | Output |
|---|---|
| 1 | 6 |
| 2 | 7 |
| 3 | 8 |
| 4 | |
| 100 | |

**d**

| Input | Output |
|---|---|
| 1 | 11 |
| 2 | 12 |
| 3 | 13 |
| 4 | |
| 100 | |

**e**

| Input | Output |
|---|---|
| 1000 | 500 |
| 800 | 400 |
| 600 | 300 |
| 400 | |
| 50 | |

**f**

| Input | Output |
|---|---|
| 40 | 37 |
| 39 | 36 |
| 38 | 35 |
| 37 | |
| 11 | |

**g**

| Input | Output |
|---|---|
| 1 | 6 |
| 2 | |
| 3 | 18 |
| 4 | |
| 50 | 300 |

**h**

| Input | Output |
|---|---|
| 4 | 12 |
| 6 | |
| 10 | 18 |
| 15 | |
| 30 | 38 |

**i**

| Input | Output |
|---|---|
| 1 | 3 |
| 2 | 5 |
| 3 | 7 |
| 4 | |
| 100 | |

**j**

| Input | Output |
|---|---|
| 1 | 7 |
| 2 | 12 |
| 3 | 17 |
| 4 | |
| 50 | |

**k**

| Input | Output |
|---|---|
| 1 | 2 |
| 2 | 5 |
| 3 | 8 |
| 4 | |
| 100 | |

In parts **i**, **j**, and **k**, there are two operations involved.

**2** According to various sources and studies, you can estimate the outside temperature by counting the chirps of a certain cricket!

One study found that when the temperature is 70° F, the cricket chirps 30 times in each time period of 14 seconds. When the temperature is 80° F, the cricket chirps 40 times in the same period. And when the temperature is 85° F, the cricket chirps 45 times in the same period.

**a** Set up a table with two columns; one for the temperature in Fahrenheit and other for the number of chirps per 14 seconds.

**b** What is the relationship between the temperature and the number of chirps the cricket makes in the given time period? Write it as a mathematical rule.

**c** Is this a linear pattern? Justify your answer.

▶ Continued on next page

**3** The relationship between temperature and cricket chirps can also be studied using degrees Celsius, but it is more difficult to determine because it is a two-step process that involves division and addition. When the temperature is 19° C, the cricket chirps 45 times in a 25-second period. When the temperature is 21° C, the cricket chirps 51 times in the same period. And when the temperature is 24° C, the cricket chirps 60 times in the same period.

**a** Set up a table like you did in question **2**.

**b** What is the relationship between the temperature in degrees Celsius and the number of chirps the cricket makes in the given period? Write the relationship as a mathematical rule.

**c** Is this a linear pattern? Explain.

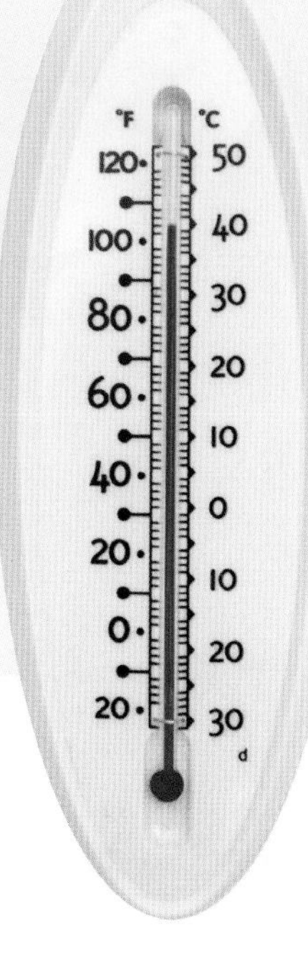


ATL1

## Reflect and discuss 7

- What did I learn about today?
- What don't I yet understand?
- What questions do I have now?

## Equations

In the previous section, you saw that you can write the rule for a function machine as, for example, $o = 5i$. This is referred to as an *equation*, since it relates two things that are equal. You may have already written statements like $2 + 3 = 5$, which is also an equation.

What if you were given the function rule $4n$ as well as the output 12? How would you find the input? This, too, can be represented as an equation:

$$4n = 12$$

An equation is a mathematical sentence where both sides are equal. There are many different kinds of equation, but they all have one thing in common: two sides are separated by an *equals* sign.

Some equations have a missing value or number that you have to find. You can write the missing number (or variable) as a question mark, box, letter, or any symbol at all. When you find the missing value, we say you have *solved the equation.*

For example:

$$3 + ? = 5$$

The question mark must have a value of 2 to make the equation true.

Pairs

## Activity 5 – Solving equations

Solve each equation, where each missing value is represented by a starfish image.

If $4 + \quad = 11$ then $\quad = $ ______

If $25 - 4 = \quad$ then $\quad = $ ______

If $5 \times \quad = 20$ then $\quad = $ ______

If $\quad - 4 = 9$ then $\quad = $ ______

If $5 + \quad = 14 - 6$ then $\quad = $ ______

If $6 \times 2 = 3 \times \quad$ then $\quad = $ ______

If $5 + 4 = \quad$ then $3 \times \quad = $ ______

If $\quad - 6 = 14$ then $4 \times \quad = $ ______

## Reflect and discuss 8

- What process did you use to solve the equations in Activity 5?
- How can you check to see if your answers are correct?

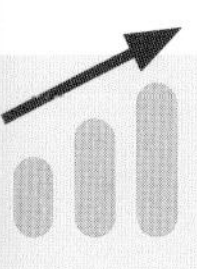

## Formative assessment centipedes vs millipedes

Centipedes and millipedes are often mistaken for one another. Both have segmented bodies and share a similar classification in biology (kingdom: animalia, phylum: arthropoda, subphylum: myriapoda).

One of the differences that you will explore is the relationship between their body segments and the number of legs they have. Centipedes have one pair of legs per body segment, while millipedes have two pairs.

### 1 Centipedes

Typically, a centipede has one pair of legs per body segment, not including the head. Here is a basic diagram of a centipede.

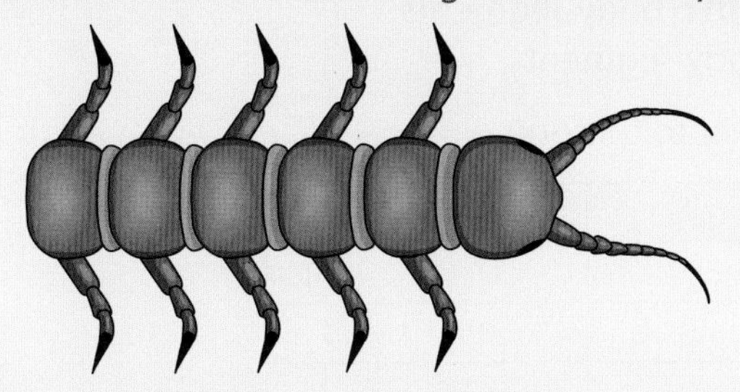

**Did you know...?**

The number of legs that centipedes have varies from 30 to over 300. As they **always** have an odd number of pairs of legs, there are no centipedes with exactly 100 legs!

**a** Draw a diagram of a centipede with 6 body segments. Draw another centipede with 7 body segments.

**b** Create and complete a table similar to this one.

| Number of body segments | Number of legs |
|---|---|
| 3 | |
| 4 | |
| 5 | |
| 6 | |
| 7 | |

**c** Explain the pattern in words. (Look at how many legs are added each time you add a body segment.) Is this a linear pattern? Explain.

**d** How can you calculate the number of legs if you know how many body segments there are? Write your rule as an algebraic equation.

**e** Using your rule, determine how many legs a centipede with 11 body segments would have.

**f** Using your rule, determine how many body segments there would be if the centipede had 58 legs.

**g** Explain the degree of accuracy of your answers in parts **e** and **f**. Describe whether or not your results in **e** and **f** make sense.

**h** What does the word *centipede* mean? How many body parts would a centipede have to have to be a 'true' centipede? Are there any centipedes with this many legs?

▶ Continued on next page

## 2 Millipedes

Typically, a millipede has two pairs of legs per body segment, not including the head or the last body segment. Here is a basic diagram of a millipede.

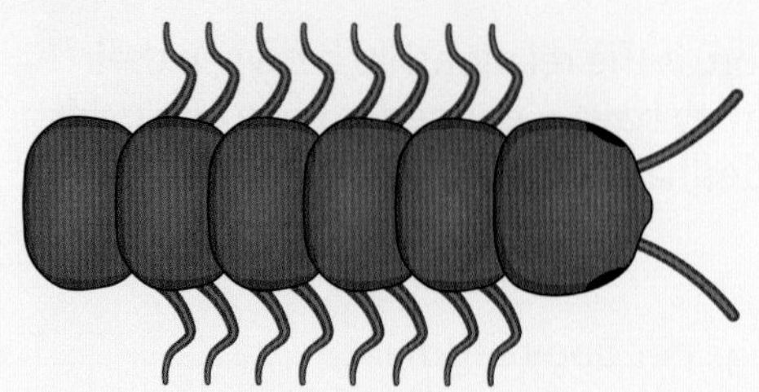

**a** Draw a diagram of a millipede with 6 body segments. Draw another millipede with 7 body segments.

**b** Create and complete a table similar to this one.

| Number of body segments | Number of legs |
|---|---|
| 3 | |
| 4 | |
| 5 | |
| 6 | |
| 7 | |

**c** Explain the pattern in words. Is this a linear pattern? Explain.

**d** How can you calculate the number of legs if you know how many body segments there are? Write your rule as an algebraic equation.

**e** Using your rule, determine how many legs a millipede with 10 body segments would have.

**f** Using your rule, determine how many body segments there would be if the millipede had 76 legs.

**g** Explain the degree of accuracy of your answers in parts **e** and **f**. Describe whether or not your results in **e** and **f** make sense.

**h** What does the word *millipede* mean? How many body segments would a millipede have to have to be a 'true' millipede? Are there any millipedes with this many legs?

Solving equations is one of the most important skills in mathematics. Many situations produce equations that will allow you to find the value of an unknown quantity. Equations, in fact, are some of the most powerful tools in mathematics for describing patterns and then making predictions.

## Example 3

**Q** Solve the following equations.

**a** $m + 7 = 10$ **b** $5k = 20$ **c** $\frac{p}{10} = 4$

**A** **a** $m + 7 = 10$

$m = 3$

If a number plus 7 equals 10, then that number is 3. Notice that this is the same thing as 10 – 7.

**b** $5k = 20$

$k = 4$

5 times a number is 20, thus the number is 4.

**c** $\frac{p}{10} = 4$

$p = 40$

Since $p$ divided by 10 is 4, then $p = 40$ (because $40 \div 10 = 4$).

With basic algebraic equations, you might be able to see at a glance what the missing number is, as in Example 3. If you cannot, it is good to have a process to follow that will work regardless of how difficult the algebraic equations are. To solve an algebraic equation, the end goal is to have the variable all on its own on one side of the equation and everything else on the other side.

To maintain balance, whatever operation you perform on one side of an equation must be performed on the other side as well.

For example, $2x + 10$ could be explained as 'I am thinking of a number, I double it, then add 10'. So, in order to figure out what the number is, you need to do the opposite: subtract 10 then divide by 2.

## Example 4

**Q** Solve the following equations.

**a** $5x + 13 = 28$ **b** $\frac{w}{3} - 7 = 12$

**A** **a** $5x + 13 = 28$

$5x + 13 - 13 = 28 - 13$

Subtract 13 from each side. The equation is still balanced.

$5x = 15$

$\frac{5x}{5} = \frac{15}{5}$

Divide each side by 5.

$\Rightarrow x = 3$

**b** $\frac{w}{3} - 7 = 12$

Add 7 to both sides.

$\frac{w}{3} - 7 + 7 = 12 + 7$

$\frac{w}{3} = 19$

Multiply both sides by 3.

$\frac{w}{3} \times 3 = 19 \times 3$

$w = 57$

## Practice 6

1 For each equation, determine which solution is correct.

**a** $x + 11 = 25$ $x = 14$ or $x = 36$

**b** $12y = 48$ $y = 36$ or $y = 4$

**c** $i + 34 = 52$ $i = 22$ or $i = 18$

**d** $f - 30 = 4$ $f = 26$ or $f = 34$

**e** $\frac{d}{100} = 2$ $d = 200$ or $d = 500$

**f** $\frac{m}{6} = 5$ $m = 30$ or $m = 65$

**g** $k - 17 = 18$ $k = 1$ or $k = 35$

**h** $3a = 75$ $a = 35$ or $a = 25$

ATL2

Pairs

2 Solve the following equations. Explain in words how you arrived at your answer.

**a** $q + 4 = 19$ **b** $9g = 72$ **c** $r - 12 = 20$

**d** $\frac{d}{2} = 25$ **e** $6c = 48$ **f** $y - 9 = 16$

**g** $\frac{z}{7} = 5$ **h** $u + 15 = 21$ **i** $11x = 88$

3 Which of these equations have the same solutions?

**a** $m + 9 = 11$ **b** $8s = 72$ **c** $V - 13 = 3$

**d** $\frac{h}{3} = 3$ **e** $b - 4 = 5$ **f** $\frac{e}{8} = 2$

**g** $10c = 20$ **h** $L + 18 = 27$ **i** $4r = 8$

**j** $U - 18 = 20$ **k** $\frac{q}{20} = 2$ **l** $Y + 30 = 32$

▶ Continued on next page

**4** Bees make a honeycomb in order to house their queen and to store honey. Worker bees can make a comb in as little as 7 days, or as long as 2 months, depending on the size. Honeycombs are made up of small hexagons (six-sided shapes) all packed together so there is no empty space.

A honeycomb is started with 10 hexagons on day 1, and a dozen hexagons are added on each subsequent day.

**a** Create a table for the number of hexagons in the honeycomb for the first 10 days.

**b** Write down an equation to determine the number of hexagons $h$ in the honeycomb after $d$ days.

**c** Use your equation to find how many hexagons will be in the honeycomb after 30 days.

**d** Use your equation to find how many days it will take before there are 240 hexagons.

**5** 'Frost flowers' form in the Arctic on very thin sea ice when the ice is at least 15° C warmer than the surrounding air. The number of frost flowers in a local area as time passes is represented in the table below.

| Time (minutes) | Number of frost flowers |
|---|---|
| 1 | 250 |
| 2 | 251 |
| 3 | 252 |
| 4 | 253 |
| 5 | 254 |
| 6 | 255 |

**a** Write down an equation to determine the number of frost flowers $f$ after $t$ minutes.

**b** Use your equation to find how many frost flowers there will be after two hours.

**c** Use your equation to find how long it will take before there are 1000 frost flowers.

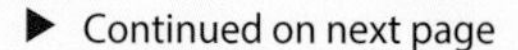

▶ Continued on next page

**6** The circumference of a tree trunk increases with age. The eucalyptus tree, indigenous to Australia, has a basic linear pattern connecting the circumference to the age of the tree. If someone wants to estimate the age of a eucalyptus tree, they just need to measure the circumference of the tree (at about shoulder height).

Look at the following table:

| Circumference (cm) | Age of the tree (in years) |
|---|---|
| 3.14 | 1 |
| 6.28 | 2 |
| 15.71 | 5 |
| 31.42 | 10 |

**a** Explain the pattern in words.

**b** What is an easy way to calculate the age of the tree?

**c** If the circumference is $c$ and the age of the tree is $a$, how can you represent this pattern as an algebraic equation?

**d** How old is a tree with a circumference of 62.83 cm?

**e** What would the circumference of a 50-year-old tree be?

ATL1

## Reflect and discuss 9

- What did I learn about today?
- What don't I yet understand?
- What questions do I have now?

# Unit summary

An ordered list of items (numbers, shapes, letters, etc.) is called a *sequence*. Items in a sequence are called *terms*.

A *pattern* is a set of items or numbers that follow a predictable rule. The *core* of a pattern is the part of the pattern that keeps repeating over and over.

In *algebraic patterns* you apply the same mathematical procedure each time. For example, you might add or multiply by the same number to determine the next term, or you might need a combination of different operations.

You can describe a pattern using words or mathematical symbols. Mathematicians call this *generalizing a pattern*. In order to generalize a pattern, you need to figure out the relationship between the terms.

When the difference between terms is always the same, it is called a *linear pattern*.

$-7 \quad -7 \quad -7 \quad -7$

$75 \quad 68 \quad 61 \quad 54 \quad 47$

The pattern rule describes how to find a term. The term can be based on the position of a term in the list or on knowing the term before it.

An algebraic expression consists of variables (typically letters), numbers and operations. When using algebraic letters, we do not use the multiplication symbol $\times$, to avoid confusion with the letter $x$. When we want to describe multiplication, we write an algebraic letter and a number together and put the number first with no operation symbol, for example $7h$ means '7 times h'.

An *equation* is a math sentence where both sides are equal. The two sides are separated by an equals sign.

Solving an equation involves performing the same operations on both sides of the equals sign until the variable is isolated. These operations are the opposite operations of those in the original equation.

This can often reveal an underlying pattern.

# Unit review

Launch additional digital resources for this chapter

Key to Unit review question levels:

Level 1–2 Level 3–4 Level 5–6 Level 7–8

**1** What are the next two terms in each of the following sequences? **Explain** your reasoning.

| | | | | | | | |
|---|---|---|---|---|---|---|---|
| **a** | 5 | 10 | 20 | 40 | 80 | ___ | ___ |
| **b** | 1 | 6 | 11 | 16 | 21 | ___ | ___ |
| **c** | 31 | 28 | 31 | 30 | 31 | ___ | ___ |
| **d** | 96 | 48 | 24 | 12 | 6 | ___ | ___ |
| **e** | 10 | 20 | 30 | 40 | 50 | ___ | ___ |

Not all sequences follow a mathematical pattern. In part **c**, look carefully at the numbers and think about what might come in groups of 31, 30, etc.

**2** **Identify** the core of the each pattern.

**a**

**b**

**c**

**3** Translate each of the following phrases into a mathematical expression. **Use** whatever letter or symbol you like for the original number.

**a** 28 divided by a number **b** A number decreased by 4

**c** 4 more than a number **d** 17 times a number

**e** A number added to 12 **f** A number multiplied by 13

**g** 6 less than a number **h** 3 more than twice a number

**4** Translate these mathematical expressions back into words.

**a** $4x$ **b** $\frac{63}{a}$ **c** $y - 15$ **d** $61 + z$ **e** $3x + 7$

**5** Substitute the given number into each expression to **find** the value of the expression.

**a** $x - 12$, when $x = 20$ **b** $15 - 3x$, when $x = 2$ **c** $\frac{x+4}{6}$, when $x = 20$

**6** **Solve** each equation.

**a** $x + 25 = 40$ **b** $7x = 56$ **c** $m - 10 = 11$ **d** $\frac{p}{8} = 3$

**e** $9b = 63$ **f** $w - 14 = 6$ **g** $h + 52 = 61$ **h** $z - 19 = 21$

**i** $\frac{d}{10} = 15$ **j** $12x = 96$ **k** $m + 25 = 44$ **l** $\frac{n}{7} = 7$

**m** $g - 15 = 23$ **n** $5y = 55$ **o** $a - 71 = 43$ **p** $36 - x = 12$

**7** **a** **Draw** the next two shapes in the sequence below.
**Explain** the pattern in words.

**b** Determine what the 10th term in the sequence would be, without drawing it. How do you know?

**c** Will the number 78 be in this sequence?
**Explain** why or why not.

**d** **Write down** a rule to find the next term of the sequence if you know the one before it.

**8** **a** **Draw** the next two shapes in the pattern and **explain** the pattern in words.

**b** Determine what the 20th term in the sequence would be, without drawing it. How do you know?

**c** Will the number 46 be in this sequence? **Explain** why or why not.

**d** **Write down** a rule to find the next term of the sequence if you know the one before it.

**9** a **Draw** the next two shapes in the pattern below and **explain** the pattern in words.

b Will the number 100 be in this sequence? **Explain** why or why not.

**10** On squared paper, **draw** a pattern for the first *five* terms in the sequence that begins 2, 6, 10, ….

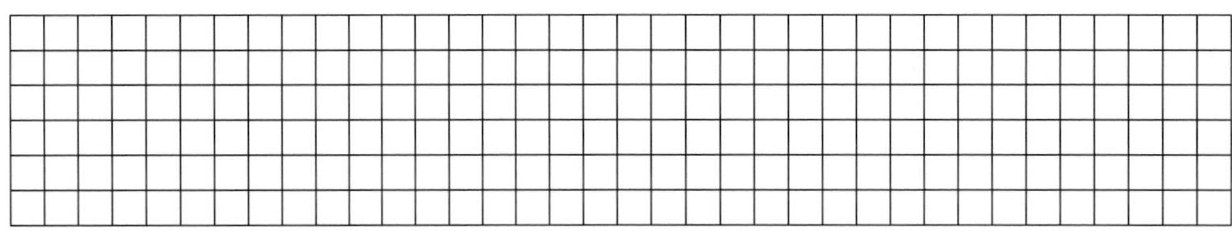

**11** **Solve** these equations.

a $2x + 10 = 16$

b $\frac{a}{4} = 7$

c $20 - 2g = 4$

d $\frac{4h}{11} + 23 = 15$

**12** Monarch butterflies are native to North America. They cannot survive the cold winter months of northern US and Canada so they migrate to southern California and Mexico. They use the Earth's magnetic field to help them travel an astonishing 3000 miles in some cases. On the first day they travel 75 miles. By the end of the second day they have travelled 150 miles. After one week, they have covered 525 miles.

a **Write down** an equation relating the number of days $d$ and the number of miles travelled $m$.

b How long does it take a Monarch butterfly to travel its entire journey of 3000 miles?

c If a Monarch butterfly has a lifespan of 6 to 8 months, what fraction of its life is spent in migration?

**13** **Use** the three equations below to determine the value of the snail, the dragonfly and the ladybug.

+ + = 60

+ + = 40

− = 5

Then **use** the equation below to check your answer.

+ + = 35

**14** Hawksbill turtles travel great distances to return to nesting sites to lay their eggs. A single female hawksbill can lay up to 200 eggs in one nest, called a *clutch*. In the first 3 seconds, 7 eggs are laid. After that, one new egg is laid in the clutch every second.

**a** Starting at 3 seconds, create a table for the number of eggs in the clutch during the first 10 seconds.

**b** **Write down** an equation to represent the relationship between the number of eggs $e$ and time $t$, when $t \geq 3$.

**c** **Use** your equation to find how long it will take for 200 eggs to be laid.

**d** **Use** your equation to find how long it will take for 153 eggs to be laid.

**15** Hawksbills, like most turtles, lay their eggs in a chamber which they dig out with their flippers. Newly hatched hawksbills are not strong enough to dig themselves out of the chamber so, as eggs hatch, they stimulate other eggs to hatch. The group of hatchlings can then dig themselves out together. The number of hatched eggs as a function of time is represented in the table.

| Time (seconds) | Number of hatched eggs |
|---|---|
| 1 | 2 |
| 3 | 6 |
| 10 | 20 |
| 15 | 30 |

**a** **Write down** an equation to represent the relationship between the number of hatched eggs $e$ and the elapsed time $t$.

**b** **Use** your equation to **find** how long it will take for all 200 eggs to hatch. Write your answer in minutes.

**c** How long will it take for 80 eggs to hatch?

# Summative assessment

In this task you will discover a famous sequence that occurs all around us. You will then research where you find this sequence and other special relationships that are based on the pattern of this sequence. On completion of this task, you can then decide if there is mathematical order in our natural world.

You will create a report presenting all your answers and submit this to your teacher.

## Part A – Discovering the pattern

**1** Here is the beginning of a sequence. Can you figure out what the next four numbers will be?

1, 1, 2, 3, 5, 8, . . .

**2** Explain the pattern in words.

**3** Check to see if your pattern is correct. Verify whether or not the following statements are true.

**a** 610 is in the sequence.

**b** 370 is not in the sequence.

**4** There are other interesting patterns within this sequence.

**a** Write down what you notice about:

**i** every 3rd number in the sequence

**ii** every 4th number in the sequence

**iii** every 5th number in the sequence.

**b** Find a pattern of your own within the sequence.

**c** From the sequence, take any three consecutive numbers (e.g. 1, 1, 2 or 3, 5, 8). Copy and complete this table to see a pattern.

| Pick any 3 consecutive numbers in the sequence. | Multiply the 1st and 3rd numbers together. | Multiply the middle number by itself. | Calculate the difference between the second and third columns. |
|---|---|---|---|
| | | | |
| | | | |
| | | | |
| | | | |

**d** Generalize this relationship with words or symbols.
Verify the relationship with two more examples from the sequence.

**5** This is a very famous sequence. What is it called? Why is it called that? When was it discovered?

## Part B – This pattern in nature

**1** The sequence in Part A was originally discovered while observing how generations of rabbits breed. Look at this scenario to see the pattern.

Suppose you start with a newly-born pair of rabbits: one male and one female. Rabbits are able to mate at the age of just one month, and pregnancy takes just one month. Thus, at the end of the second month the female can give birth to another pair of rabbits.

Suppose that none of your rabbits ever die and that each female always gives birth to one new male–female pair *every* month from the second month on.

At the end of the first month, the rabbits mate but there is still only one pair. At the end of the second month, the female gives birth to a new pair, so now there are two pairs of rabbits. At the end of the third month, the original female gives birth to a second pair, making three pairs in all. At the end of the fourth month, the original female has produced another new pair and the female born two months ago produces her first pair, making five pairs.

**a** Draw a diagram of this sequence for the first six months.

**b** How many rabbit pairs will there be after one year?

**2** This very famous pattern is often found in nature. Below right is a diagram of the sneezewort plant shown in the photo on the left.

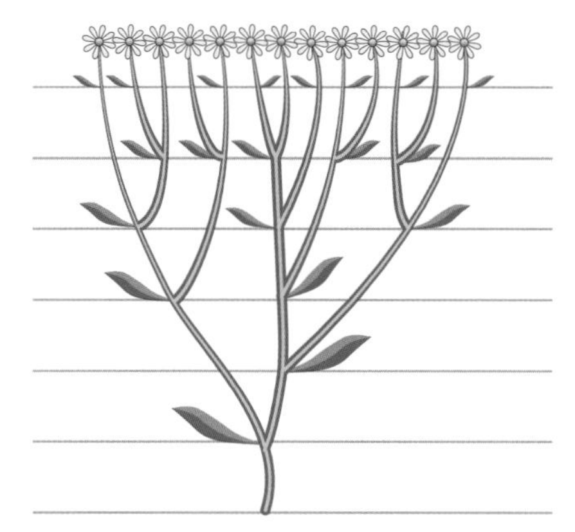

If we draw in horizontal lines to indicate the growth cycles of this plant, we can see how its stems relate to the famous sequence. Using the diagram, explain how this sequence can be found in the sneezewort plant.

**3** Now take a look at this pinecone.

**a** How many spirals turn left? (The yellow lines have been drawn to help you.)

**b** How many spirals turn right? (Try taking a pencil and tracing the spirals to help you count.)

**c** What do you notice about these spirals if you compare them to the famous sequence you discovered?

If available, your teacher might bring in real pinecones for you to analyze to see if all pinecone spirals form the same patterns.

**4** This interesting behavior is not only found in pinecones. Fruit, leaves, branches and petals can grow in spirals that have a connection to this famous sequence, too. Research five other examples in nature that have a connection to this famous sequence of numbers.

**a** Copy and paste a picture of each of these into your report and explain its connection to the numbers in the famous sequence. You may even find your own examples and take a photograph.

**b** Why do you think so many natural objects contain these numbers from the sequence? Research the answer and include the explanation in your report.

## Part C–Relationships between this famous sequence and a special ratio

| F | S | S ÷ F |
|---|---|---|
| 1 | 1 | 1 |
| 1 | 2 | 2 |
| 2 | 3 | 1.5 |
| 3 | 5 | |
| 5 | 8 | |
| 8 | | |
| | | |
| | | |

**1** Create a copy of the table here and add five more rows to it. Take your list of the sequence and complete columns F and S. Note that the sequence in column S starts with the second term of the original sequence. The last column is for writing down the calculation S ÷ F, which gives the ratio of successive terms.

**2** What do you notice about the number in the third column as you go down the table? This special number has a name. Research it and give a brief description of this ratio.

**3** Where is this ratio found in the natural world?

**4** What is the significance of this ratio?

## Reflect and discuss 10

Is there a mathematical order to our natural world?

# 4 Geometric constructions

The angle is one of the fundamental elements of geometry, and its applications in mathematics are endless. In this unit, you will study angles through the global context of personal and cultural expression. What would a study of angles look like in other contexts?

## Identities and relationships

### Health & well-being

When a patient is injected with medicine, the needle is inserted at a specific angle in order to reach the right spot. Injecting medicine in to the wrong layer may result in the vaccine or medicine not working.

Radiation therapy also relies on angles. Doctors need to adjust the angle at which radiation is administered so that the patient  does not receive a double dose in some spots. Too much radiation can cause problems, rather than solve them.

A new health concern, called "text neck", has been identified because of the angle of your neck while you use a device such as a smartphone. While the exact repercussions of this new behaviour have yet to be determined, it is assumed that, as they get older, some people will develop neck pain and other complications.

### Modernization and engineering

Airplane designs differ more than you might think. Some planes have a more "tail dragger" angle while others have a "tricycle undercarriage". Pilots have to be very aware of the angle at which they take off in order to avoid "tail strike" - when the tail of the plane rubs against the runway upon take-off.

In the past, buildings tended to have a shape that resembled a rectangular prism. Advances in engineering have allowed a more creative use of angles to give buildings a distinctive style, as well as increase safety.

# Geometric constructions

## Artistry and creativity

## KEY CONCEPT: FORM

## Related concept: Measurement

## Global context

In this unit you will learn about different angles and explore the relationships between them as you express your artistry and creativity and explore the global context of **personal and cultural expression**. From art to architecture, you will use mathematics to analyze and create artistic forms of expression, leaving you to answer whether art is more inspiration or calculation.

## Statement of Inquiry

Artistry and creativity are enhanced through an understanding of how measurement helps to define forms.

### Objectives

- Naming and classifying different geometric elements (point, ray, line, segment)
- Naming and classifying the different types of angle
- Constructing and measuring angles
- Solving problems using the various angle properties, including the angles in triangles
- Naming and classifying the different types of triangle

### Inquiry questions

**F** What is a line?

What can be measured?

**C** How do measurements help define different forms?

**D** Is art more inspiration or calculation?

**ATL1 Information literacy skills**

Use memory techniques to develop long-term memory

**ATL2 Transfer skills**

Make connections between subject groups and disciplines

## You should already know how to:

- write an equation with one variable
- solve single-step equations
- use a ratio to find an unknown quantity
- find a fraction of a number
- find a percentage of a number

# Introducing geometric constructions

Cubism was an art movement that began in the early 1900s. Pablo Picasso (1881–1973) was one of the most famous cubists, influencing the work of other artists for decades. Vasily Kandinsky (1866–1944) combined cubism with abstract art to create paintings like *Composition VIII*, shown below.

## Reflect and discuss 1

In a small group, answer the following questions.

- What do you see when you look at Kandinsky's painting?
- Why do you think this style of art is called *cubism*?
- Why do think some people say that mathematics influenced the cubist movement?

## Rays, line segments and lines

In mathematics, a *point* is a location and it is represented as a dot. Points are usually indicated with a capital letter, like the points $P$, $Q$ and $R$ here.

$P$ • $Q$ • $R$ •

Other basic elements of geometry are based on one or more points, as you will see in Activity 1.

## Activity 1 – Rays, line segments and lines

Here are some examples of rays, line segments and lines.

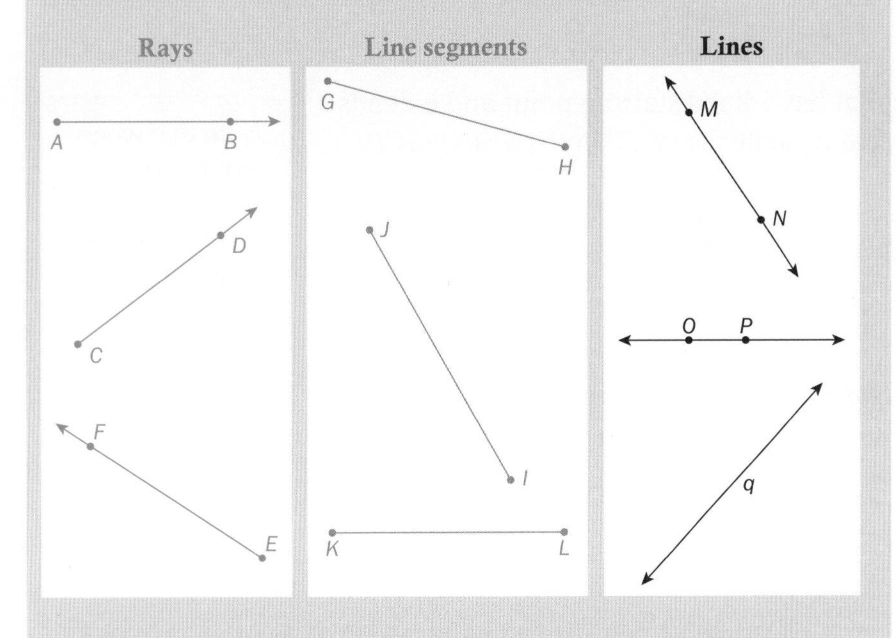


**1** What are the similarities between rays, line segments and lines?

**2** What are the differences between them?

**3** Based on your observations, write down a definition of:

**a** a ray **b** a line segment **c** a line.

## Reflect and discuss 2

In pairs, answer the following questions.

- Where do you see a point, line, line segment and/or ray in Kandinsky's *Composition VIII* painting, shown on page 124?
- For those elements that you could not find, how would Kandinsky's work have to be changed in order to see them? How do you think that would affect the overall impact of the work?

A *line* is a set of points that extends forever in two directions. Two points define a specific line, such as line $AB$ below. Using symbols, this is often written as $\overleftrightarrow{AB}$.

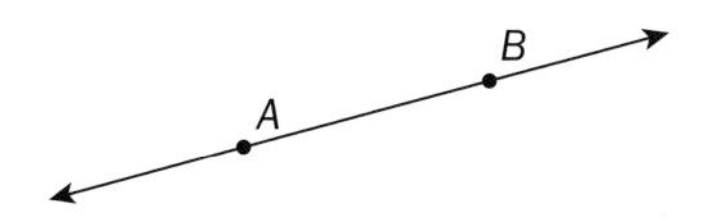

Sometimes, rather than using two points to name a line, lines are named using a single letter, such as line $m$ here.

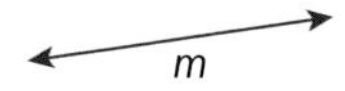


A *ray* is a part of a line that has a fixed starting point and extends forever in the other direction, such as ray $EF$ here (written as $\overrightarrow{EF}$).

Note that when writing a ray using letters, the first letter is the fixed starting point.

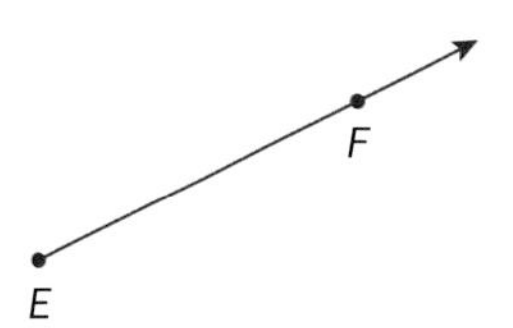


A *line segment* (or just *segment*) is a part of a line that has two fixed endpoints, such as segment $CD$ here (often written as $\overline{CD}$).

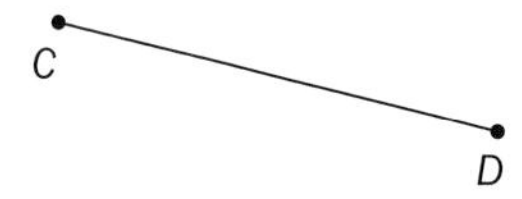


## Activity 2 – All around you

Your teacher will put you in a group to do this activity:

1 Work with your group to classify each of the following as an example of a point, ray, segment or line. Justify your choice.
- light shining from a flashlight
- the space bar on a computer keyboard
- the letter 'e' on a computer keyboard
- the edge of a picture frame
- the horizon
- the equator (or any other line of latitude)
- the tip of a needle

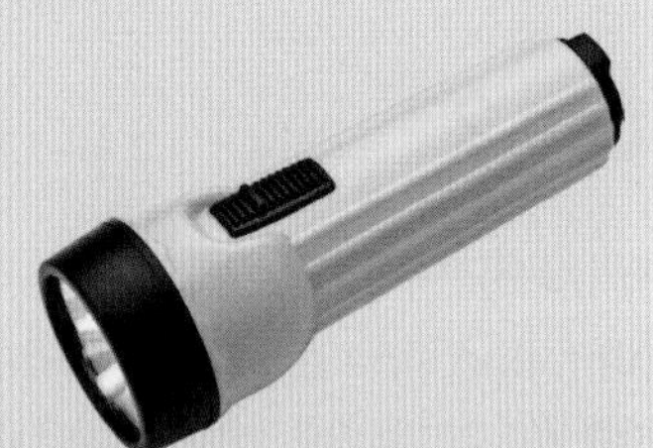

2 On your own, find two other examples each of points, rays, segments and lines that you see around you. Present them one at a time to your group. The other group members should try to classify each of your examples.

3 Select the best example of each found by your group and present them for the rest of the class to try to classify.

Because they extend forever, lines often intersect one another. However, *parallel lines* are lines that never meet. Two lines are said to be parallel if they never intersect each other at any point. Parallel lines are indicated using the notation *g*   *h*.

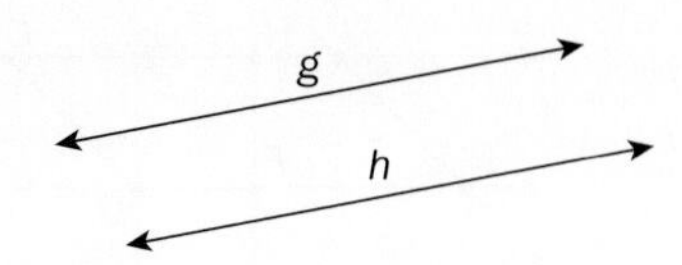


## Activity 3 – Parallel segments

1. In *Composition VIII*, shown on page 116, there appear to be many pairs of parallel segments. Find as many pairs of parallel segments as you can.
2. How can you tell if they are indeed parallel?
3. Describe how the distance between two lines can be used to determine whether or not they are parallel.
4. Can lines be described as parallel just because they *look* parallel? Explain.
5. Research parallel lines and the symbol mathematicians use to indicate lines that they are parallel.

## Practice 1

**1** Draw and label each of the following.

**a** $\overleftrightarrow{XY}$   **b** $\overline{KL}$   **c** $\overrightarrow{PQ}$   **d** point *A*   **e** line *n*   **f** $\overleftrightarrow{BC} \parallel \overleftrightarrow{FG}$

**2** Name each of the following.

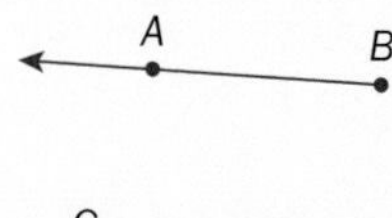


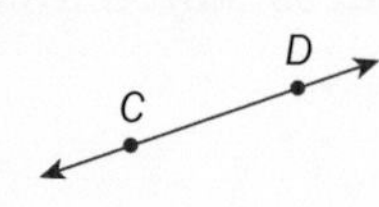

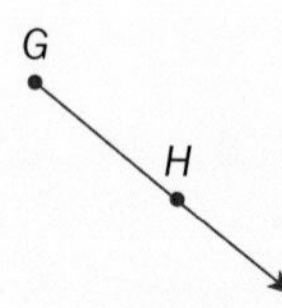

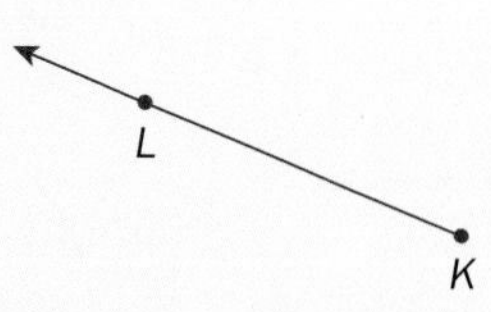


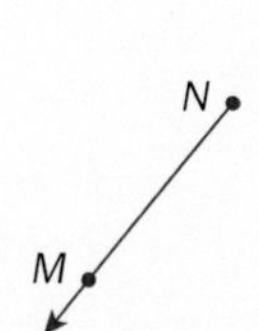

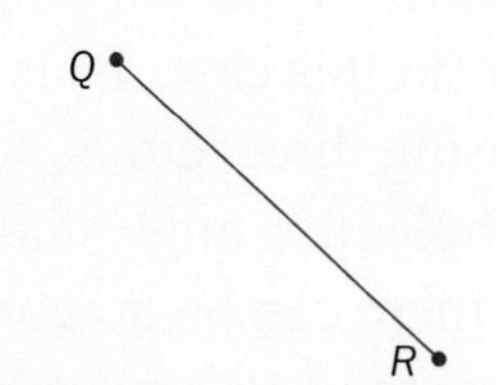


▶ Continued on next page

3 In the diagram on the right, find and name the following elements.

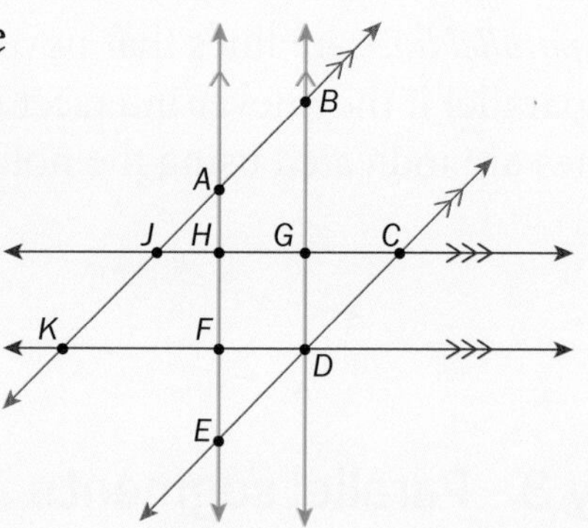


a 5 rays

b 6 points

c 4 lines

d 4 segments

e 2 sets of parallel lines

f 2 lines, and name each in two different ways

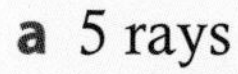


4 Making a two-dimensional piece of art look three-dimensional requires giving it some *depth*.

a In the painting on the right, which lines would be parallel in real life? How do you know?

b What technique does the artist use on the parallel lines to give the work depth?

## Angles

Two rays with a common endpoint form what is called an *angle*.

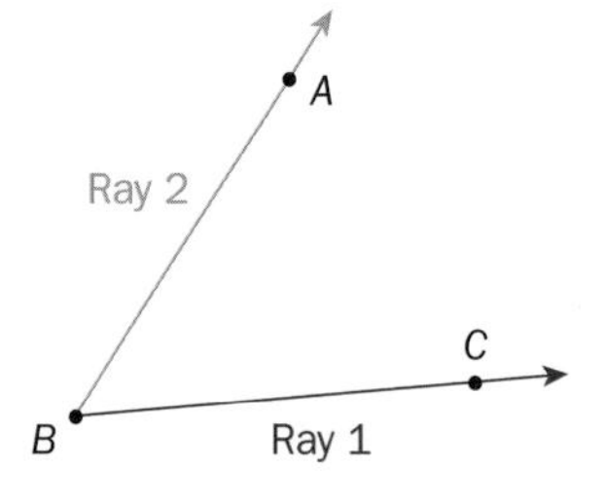


The rays are the *sides* or *arms* of the angle, while the common endpoint (in this case, $B$) is called the *vertex*. An angle is usually named using three letters, with the vertex in the middle. For example, this angle could be called $\angle ABC$ or $\angle CBA$. Because there can be no confusion with other angles, this angle could also be called $\angle B$.

The size of an angle depends on how far apart the rays are from each other. There are several different units for measuring the space between the arms. One unit of angle measurement is *degrees*, which can be measured with a *protractor*. A circle is defined as having 360 degrees, so one degree is one 360th of the way around any circle. 360 degrees can also be written as 360°.

## Did you know...?

A circle is defined as having 360 degrees. One theory suggests that 360 was chosen because it is close to the number of days in a year. Ancient astronomers saw that the sun made its orbit once every year, or roughly once every 360 days. They believed that the sun moved one degree every day.

This is how to measure the size of an angle using a protractor.

1. Place the middle of the protractor directly on top of the vertex of the angle.
2. Line up one of the sides of the angle with the 'zero line' of the protractor.
3. Use the scale that starts at zero. This could be the inside or outside scale.
4. Read the degrees at the point where the other side of the angle crosses the number scale.

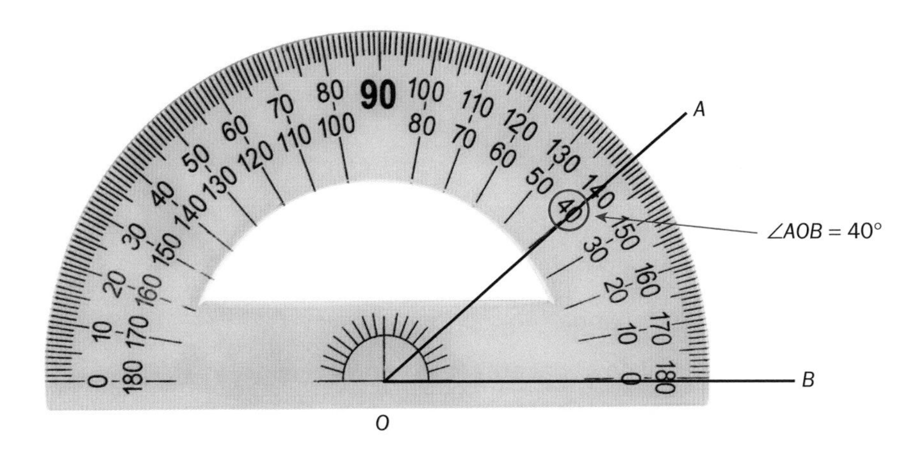

## Activity 4 – Classifying angles

Angles can be classified as *acute*, *obtuse*, *right* or *straight*. Use a protractor to measure each of these angles and then answer the questions that follow.

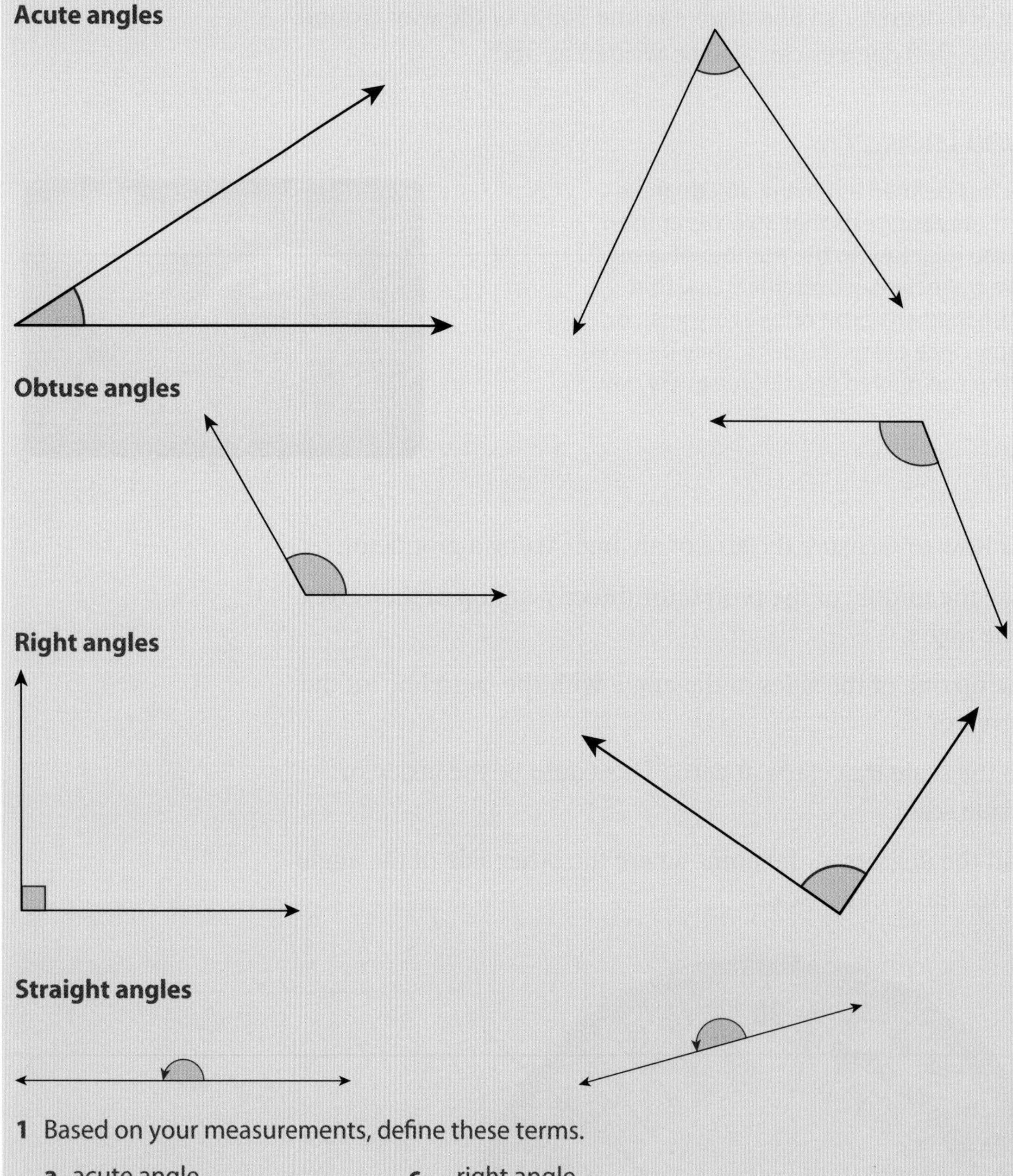


1 Based on your measurements, define these terms.

  **a** acute angle
  **b** obtuse angle
  **c** right angle
  **d** straight angle

2 Can an angle measure more than 180°? Explain. What are these angles called?

3 Write down the different terms used to classify an angle as its measure increases from 0 to 180 degrees. Indicate the measurement at which each classification changes.

## Reflect and discuss 3

ATL2

In his painting *Portrait of Wilhelm Uhde*, Pablo Picasso painted the author and art collector using a cubist style.

- In the painting, identify three examples of acute, obtuse, right and straight angles.
- How do you think the use of angles changed the portrait? What do the angles contribute to the portrait?
- Do acute angles and obtuse angles serve different purposes in the portrait? Explain.

## Activity 5 – Walkabout

Pairs

1. In pairs, look around your classroom, school or school grounds for at least three examples each of acute, obtuse and right angles in architecture. Take a photo of each.
2. Share your photos one at a time with another pair of students to see if they can name the types of angles in your pictures.
3. What types of angles did you see the most? What types did you see the least?
4. Do different types of angles have different purposes in architecture? Explain.

Two lines that meet to form a right angle are called *perpendicular lines*. In diagrams, a right angle is indicated using the square symbol:

Perpendicular lines are indicated using the notation $\overleftrightarrow{AB} \perp \overleftrightarrow{MN}$.

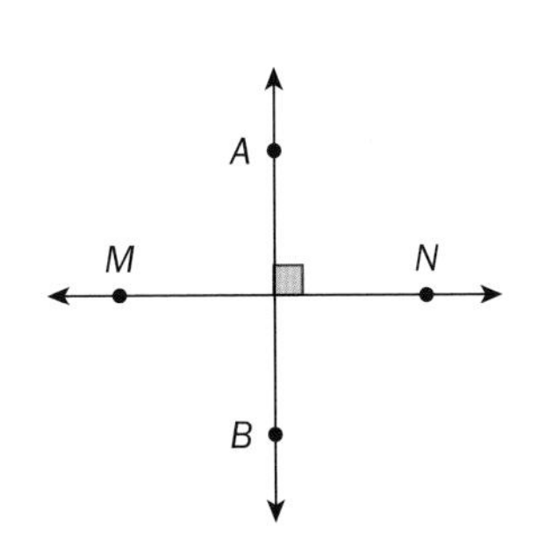

## Reflect and discuss 4

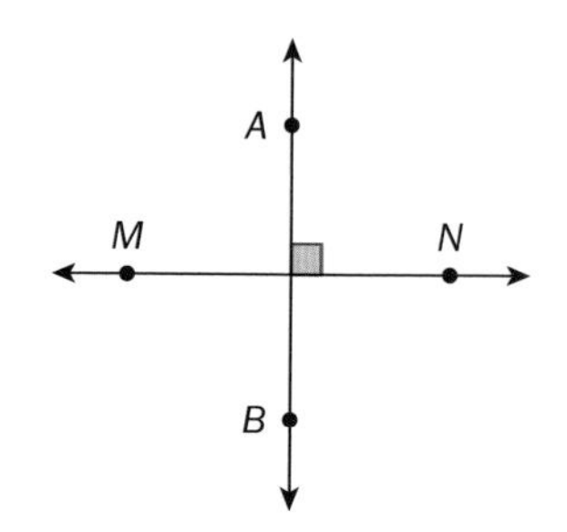


- State the measure of the other angles in the diagram here. Justify your answers.
- What is the sum of all the angles at a point? Justify your answer.
- What is the sum of all the angles on a line? Justify your answer?

If you are drawing a sketch of a right angle, be sure to indicate it using the appropriate symbol, especially if the angle doesn't quite look like 90°. Drawing more precise angles with a specific measure requires the use of special tools. While you have already seen how useful they are for measuring angles, protractors can also be used to draw angles.

### Example 1

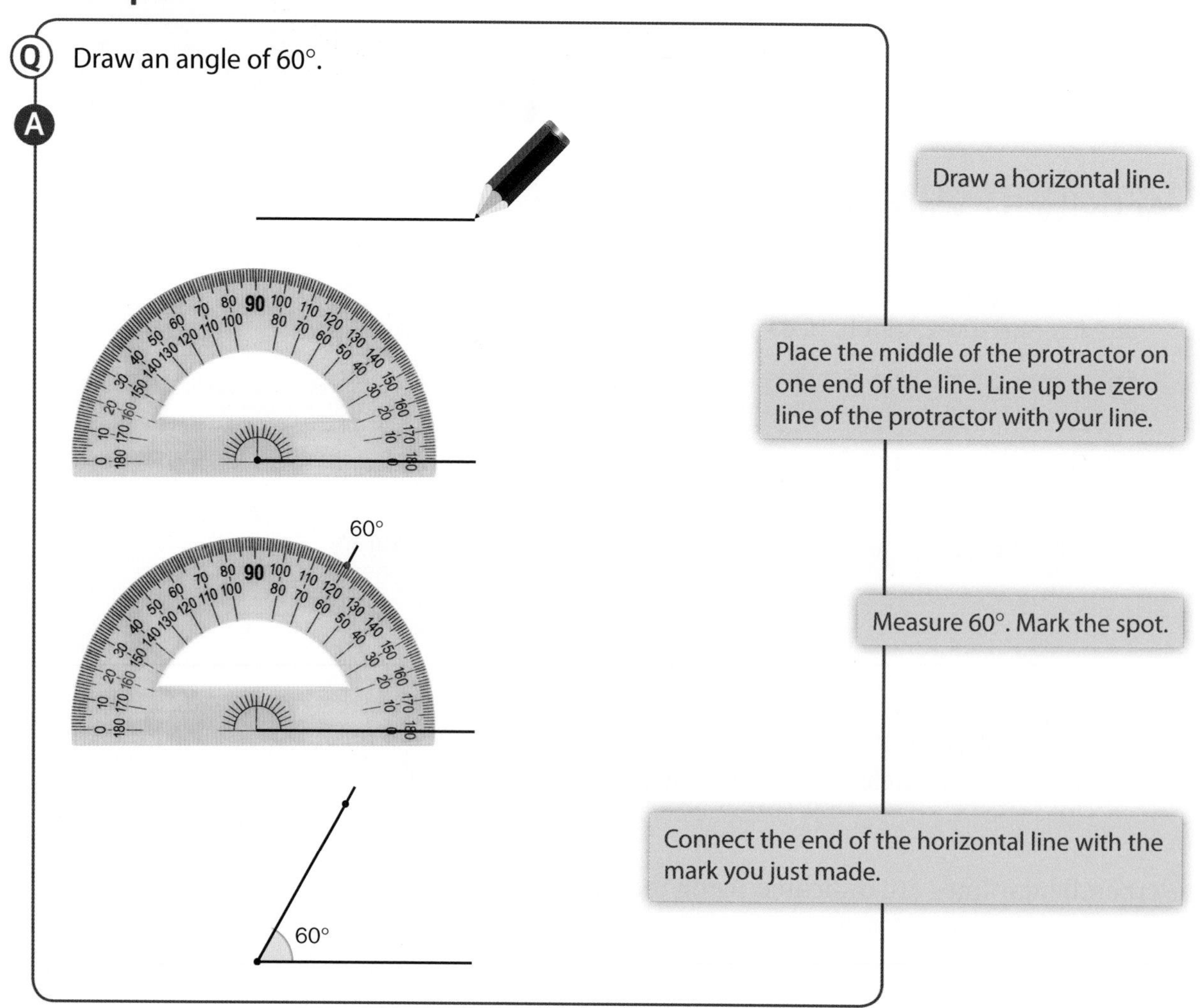

## Practice 2

**1** Identify each of these angles as acute, right, obtuse or straight. Justify your classification.

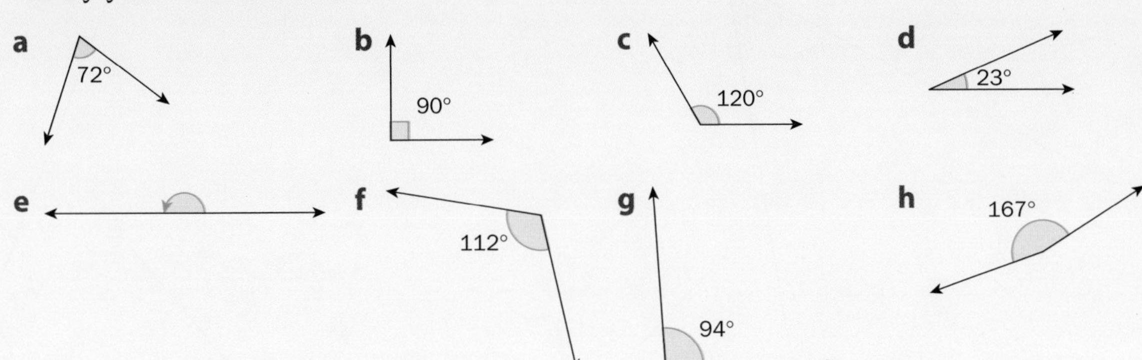


**2** State the measure of these angles and classify them.

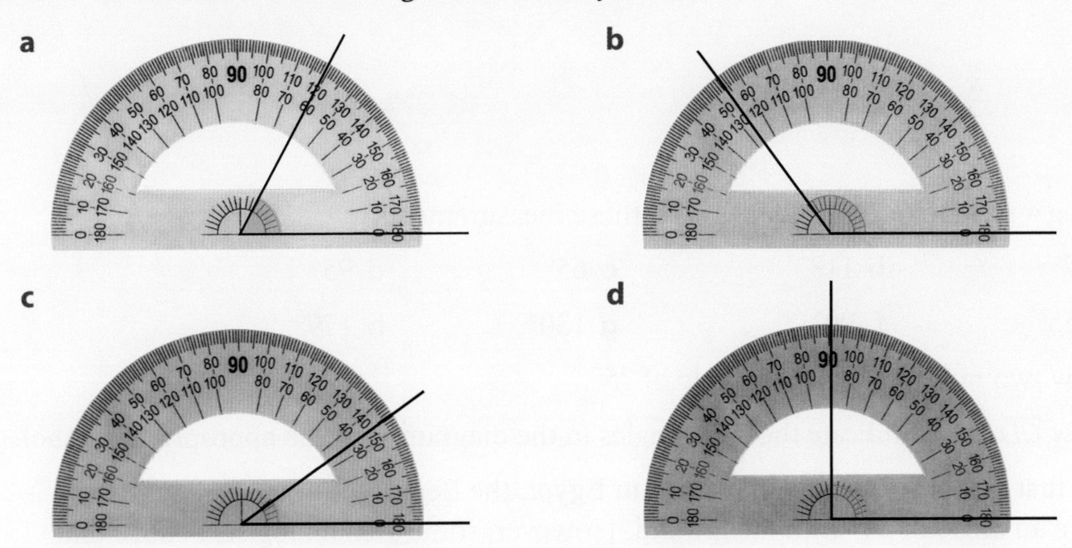


**3** Using a protractor, measure and classify each angle.

▶ Continued on next page

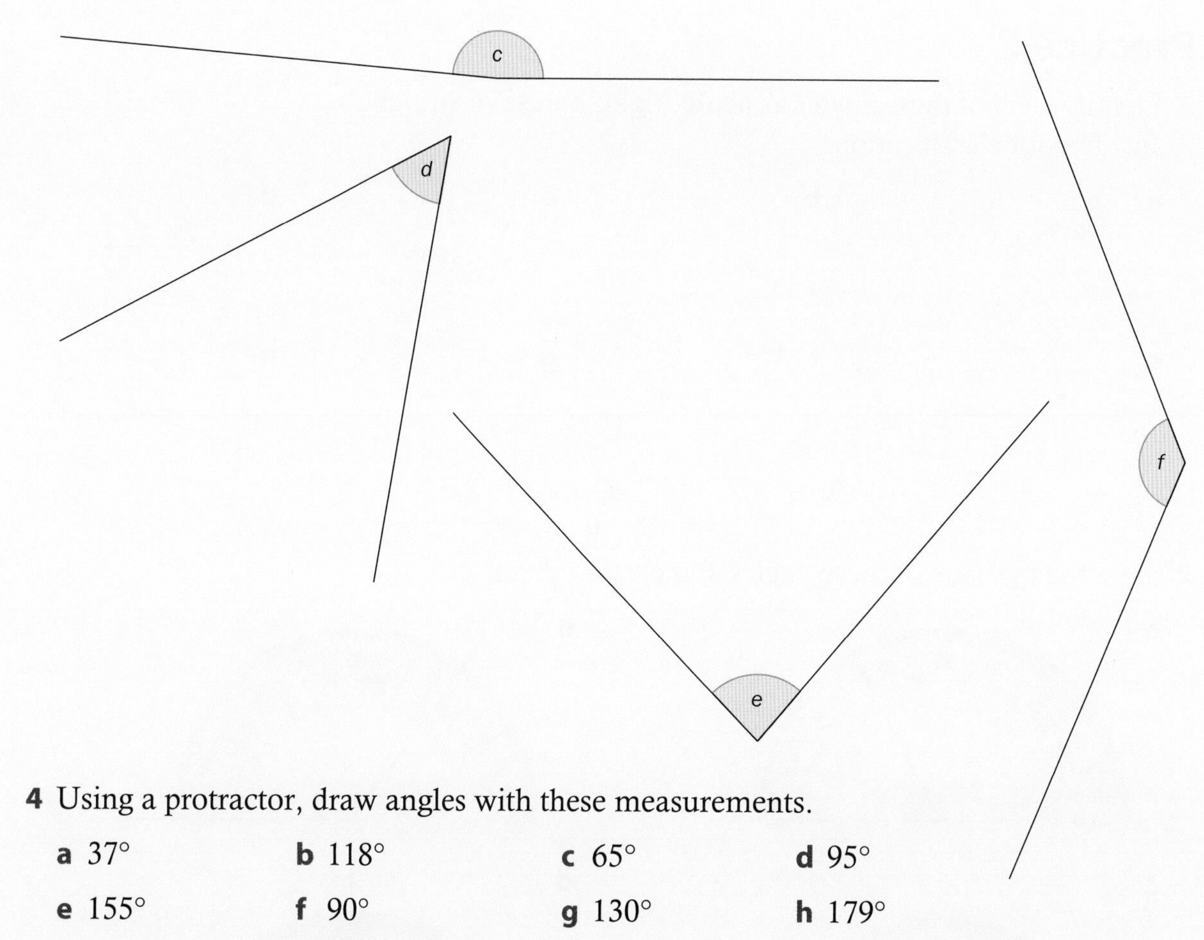


**4** Using a protractor, draw angles with these measurements.

**a** 37° **b** 118° **c** 65° **d** 95°

**e** 155° **f** 90° **g** 130° **h** 179°

**5** Draw two rays that make an angle of 75°.

**6** Draw $\overline{CD} \perp \overline{EF}$. Indicate the right angles in the diagram with the appropriate symbol.

**7** The first smooth-sided pyramid built in Egypt, the Bent Pyramid, has sides that make an angle of 54° with the ground. However, some archaeologists believe that the structure may have been too steep to be stable, because the angle was changed partway through construction. The top section of the pyramid has an angle that is 80% of the angle of the lower section (rounded to the nearest degree). Use a protractor to draw a sketch of the front face of the Bent Pyramid.

▶ Continued on next page

**8** Emerald Plaza is a group of five office towers in San Diego, California, that has a geometric design. The architect designed the slope of the roofs of the buildings so that their angle (measured in degrees) is the same as the latitude of San Diego (33°).

A side view of each tower is a trapezoid, like the one on the right. A scale diagram of one tower has a width of 4 cm and a height of 20 cm at its lowest point. Use a protractor and a ruler to draw a side view of the tower. Be sure to mark the 33° angle clearly as well as the height of the longer side.

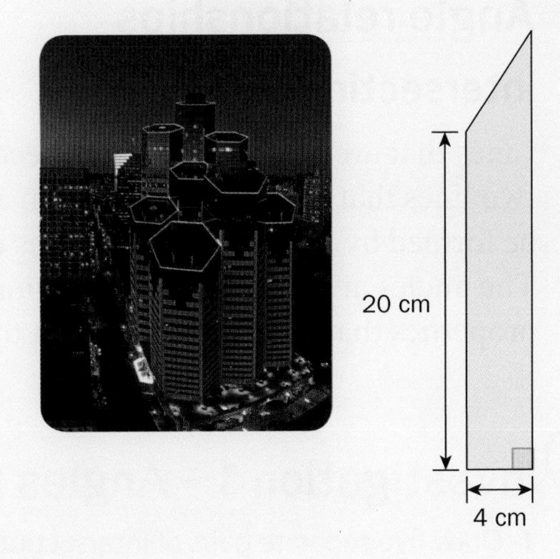


**9** **a** Draw two line segments that intersect and create an angle of 50°.

**b** What are the other angles that are created by these two line segments?

**c** How does this demonstrate that the sum of the angles at a point is 360°?

**d** How does this demonstrate that the sum of the angles on a line is 180°?

## Activity 6 – Acting out

ATL1

**1** A strategy that can help you remember information is to use movement. In pairs, find a way to use your arms and/or hands to act out the following vocabulary from this unit:

Point  Segment  Ray
Acute angle  Line  Obtuse angle
Right angle  Parallel lines
Angle  Perpendicular lines

**2** In small groups, or as a whole class, play a game like 'Simon Says' using these movements.

Point  Segment  Ray
Acute angle  Obtuse angle
Parallel lines

# Angle relationships

## Intersecting lines

Lines that are parallel never intersect each other. However, two lines that do intersect will form angles. Angles, then, can be formed by intersecting rays, lines or even line segments. The angles at the point of intersection of two lines have properties that you will discover in the next investigation.

## Investigation 1 - Angles formed by lines

1 Draw five separate pairs of intersecting lines.

2 Starting anywhere and moving clockwise, label the angles in each pair $a$, $b$, $c$ and $d$, as in this example.

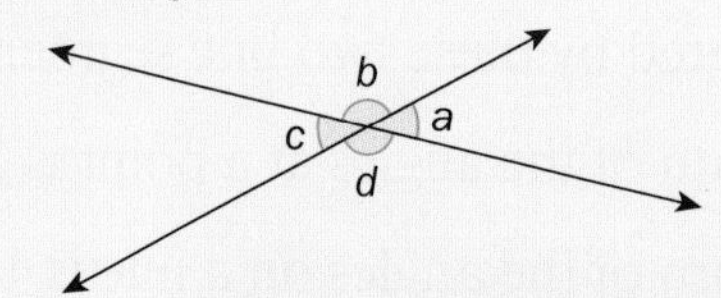


3 Measure the four angles in each pair and write them down in a table like the one here.

| Measure of $\angle a$ (degrees) | Measure of $\angle b$ (degrees) | Measure of $\angle c$ (degrees) | Measure of $\angle d$ (degrees) |
|---|---|---|---|
| | | | |
| | | | |
| | | | |
| | | | |
| | | | |

This investigation could also be done using dynamic geometry software.

4 What do you notice about the measures of the angles? Look for as many patterns as you can find, including angles that have the same measure and angles that have a particular sum.

5 Write what you notice as generalized rules.

6 Verify your rules by drawing two more pairs of intersecting lines and measuring the angles formed.

7 Research what the angles in each of these cases are called.

8 Write down the names you found and define them. Draw diagrams to support your definitions.

## Reflect and discuss 5

- Explain why certain pairs of angles should add up to 180°.
- Is it possible for more than two angles to add up to 180°? Use a diagram to explain your answer.
- How does this investigation support the property that angles at a point sum to 360°?

## Supplementary angles

Two angles that add up to 180° are called *supplementary* angles. One angle is the supplement of the other. In this diagram, $a$ and $b$ are supplementary angles.

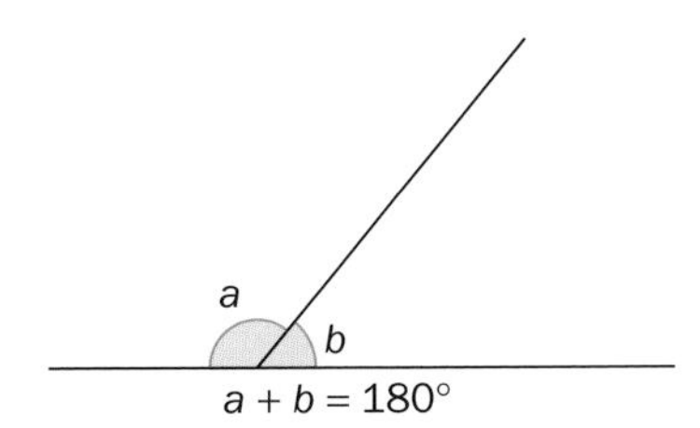


## Complementary angles

Two angles that add up to 90° are called *complementary* angles. One angle is the complement of the other. In this diagram, $a$ is the complement of $b$, and $b$ is the complement of $a$.

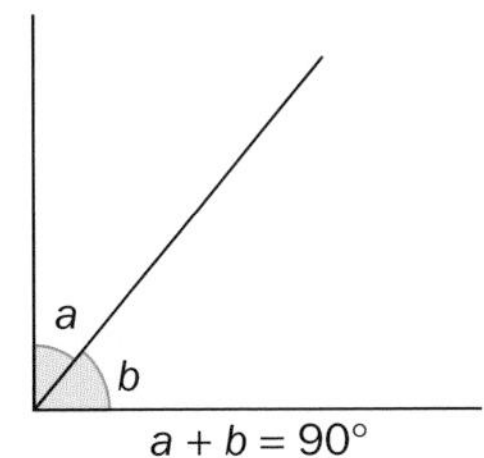


## Example 2

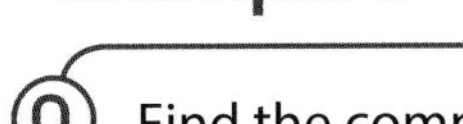

**Q** Find the complement and supplement of a 48° angle by first writing an equation and then solving it.

**A** $a + 48° = 90°$

Complementary angles add to 90°. Call the missing angle $a$.

$a + 48° - 48° = 90° - 48°$
$a = 42°$

Subtract 48° from each side.

The complement of 48° is 42°.

$b + 48° = 180°$

Supplementary angles add to 180°. Call the missing angle $b$.

$b + 48° - 48° = 180° - 48°$
$b = 132°$

Subtract 48° from each side.

The supplement of 48° is 132°.

## Reflect and discuss 6

- Does every angle have a complement? Explain.
- Does every angle have a supplement? Explain.

## Practice 3

**1** In each diagram,

**a** state two pairs of vertically opposite angles

**b** state two pairs of supplementary angles.

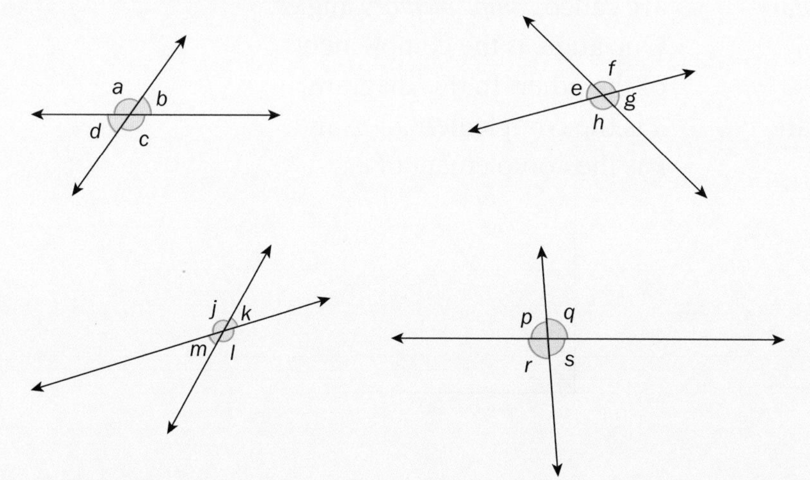


Vertically opposite angles are formed by intersecting lines.

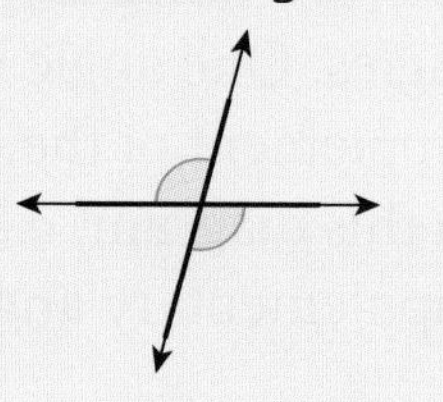

**2** In the diagram below, state all of the pairs of vertical angles and all of the pairs of supplementary angles.

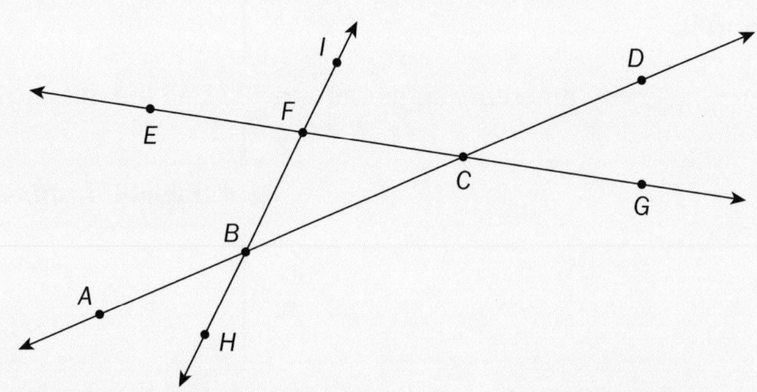


**3** Determine the complement and the supplement of each angle, if it exists. In each case, write an equation and solve it.

**a** $35^\circ$ **b** $11^\circ$ **c** $84^\circ$ **d** $9^\circ$ **e** $52^\circ$

▶ Continued on next page

**4** Name the missing angles and determine their measure. Justify each answer.

**a**

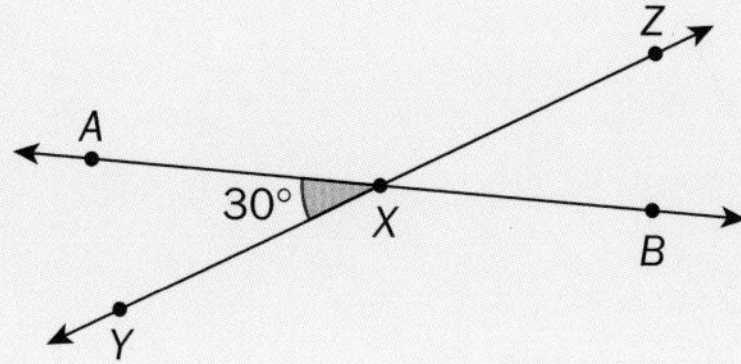


**b**

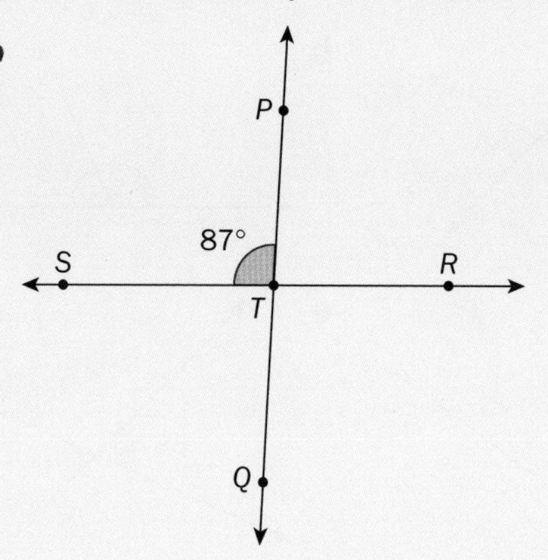


**c**

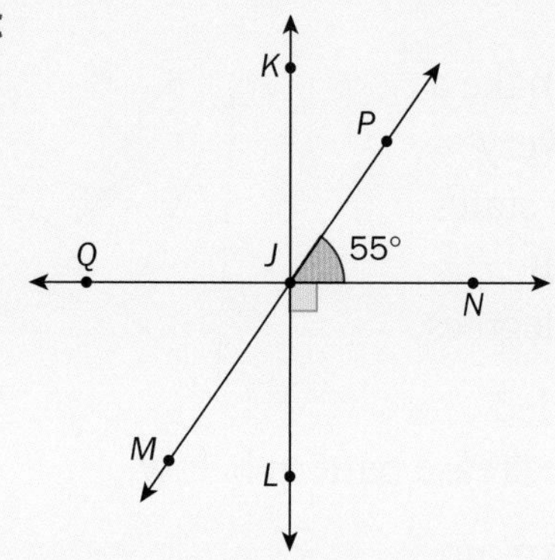


**d**

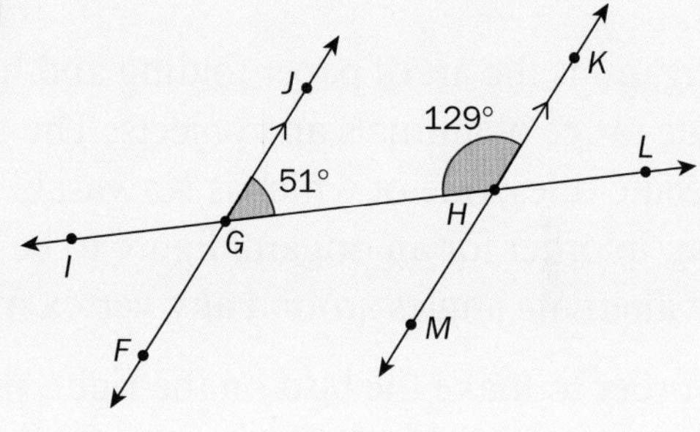


**e**

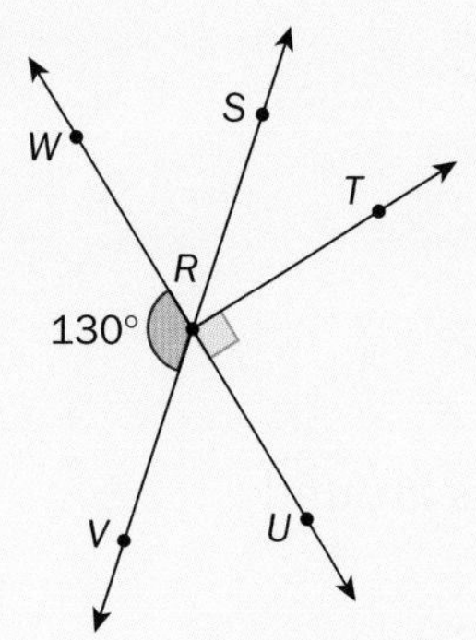


**f**

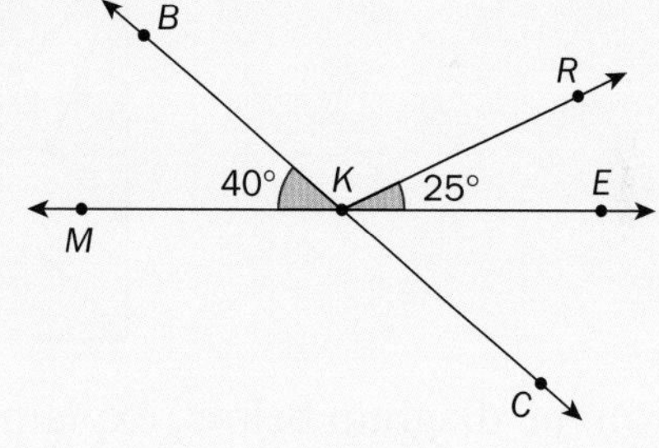


**g**

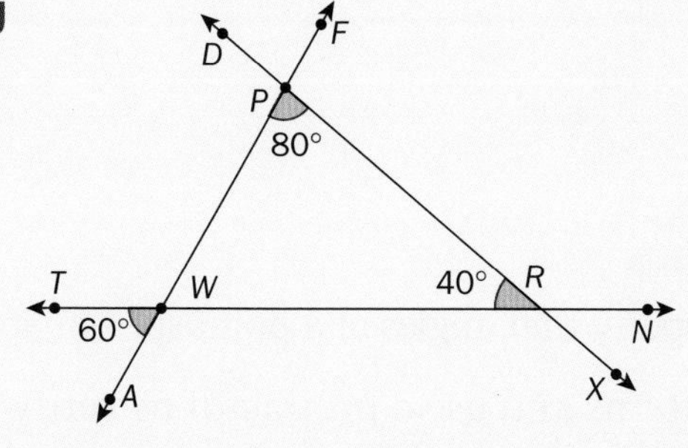


**h**

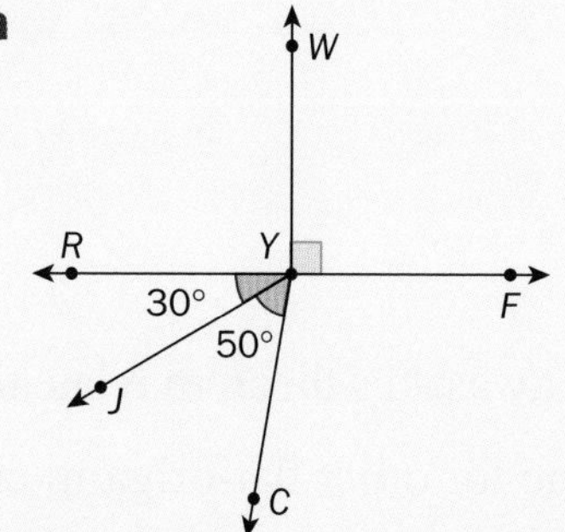


▶ Continued on next page

**5** For each diagram, write down an equation and solve it to determine the value of $x$.

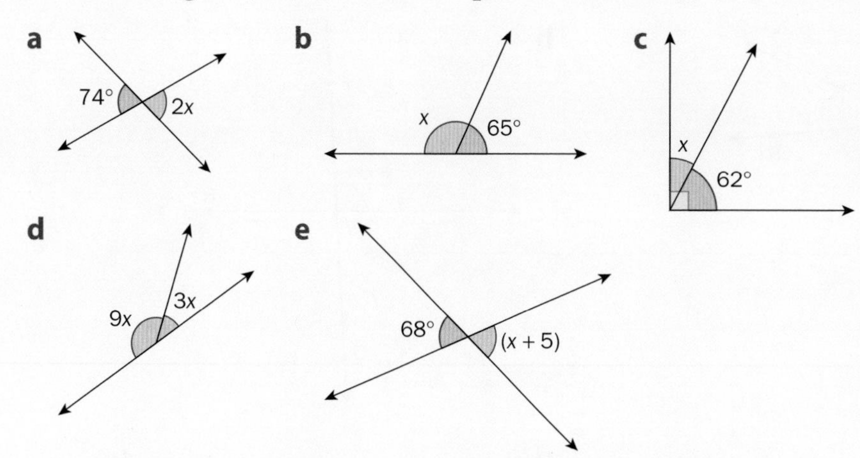


**6** Origami is the art of paper folding and has been used to make a wide range of animals and objects. The art form follows very specific rules, one of which is Kawasaki's theorem which states that, in order for an origami figure to be able to be laid flat, the alternate angles around any vertex must sum to 180 degrees.

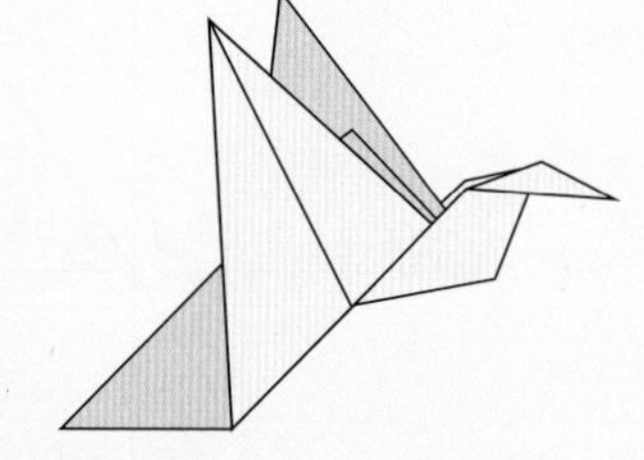

In order to make the bird on the right, paper must be folded many times. When the paper is unfolded, it produces the crease pattern below.

**a** Are there any vertical angles in the crease pattern? Explain.

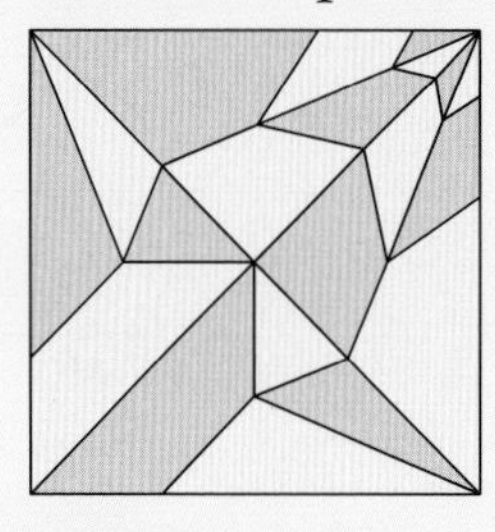

**b** Based on the diagram below, explain why the alternate angles around a vertex must add to 180 degrees.

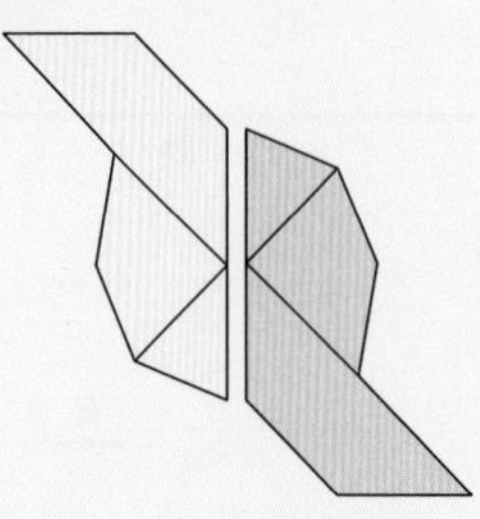

**c** How does Kawasaki's theorem relate to the property that angles at a point sum to 360°?

**d** Search online for other flat-origami crease patterns and use a protractor to verify Kawasaki's theorem.

## Did you know...?

Flat origami isn't just an art form. It has also been used to solve problems, like how to pack large solar panels for travel on a rocket, how to flatten an air bag so it can be put into a car, or even how to insert a heart stent in a patient. All of these require items that need to be made smaller until they are needed. There is an interesting talk available online about how origami has been used to solve some of these real-life problems. You can find it here: https://www.ted.com/talks/robert_lang_folds_way_new_origami/transcript

## Parallel lines and transversals

Parallel lines never meet, so they don't form any angles with each other. However, if a third line, a transversal, intersects them, angles are formed with interesting properties.

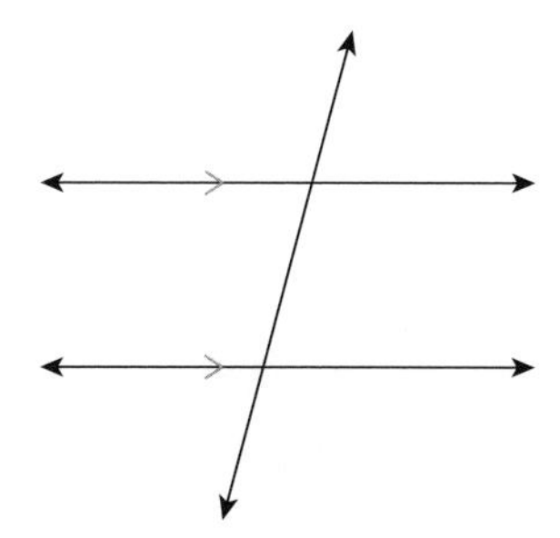

Investigation 2 will help you determine these properties.

## Investigation 2 – Parallel lines and transversals

You can perform the following investigation using paper and pencil, or with dynamic geometry software.

1 Draw two parallel line segments. You can use both edges of a ruler, existing lines on a piece of paper or any other method. Draw a third line segment (the transversal) that intersects both the other segments.

2 Using a protractor, measure one of the angles formed by the parallel line segments and the transversal.

3 Calculate the measures of as many remaining angles as you can. If you get stuck, measure another angle and then use it to calculate the ones you do not yet know.

4 Measure those angles that you calculated to verify your results.

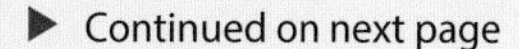
▶ Continued on next page

5 Which pairs of angles have equal measures? Find as many pairs as you can. (Hint: There should be three different kinds of pairs.)

6 Verify your results by drawing two more pairs of parallel lines cut by a transversal and measuring the angles formed.

7 Research the names of the different equivalent pairs.

## Reflect and discuss 7

ATL1

- The different pairs of equivalent angles resemble some letters of the alphabet. Classify the angle pairs based on the letter of the alphabet that they make. Refer to the three diagrams below.
- How do the angles in this activity validate the claim that angles at a point add up to 360°?

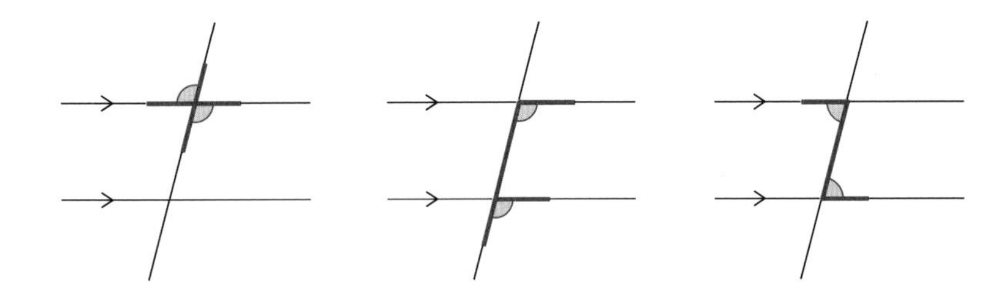

## Activity 7 - Angles

ATL2

What made cubism so different from other artistic styles at the time was that it no longer conformed to the typical rules for art. Cubist artists didn't feel they needed to make their paintings look like the 'real world'; not in shape, color or even space. Cubist painters redefined reality and suddenly were able to present several sides of an object all at once.

Juan Gris' *Portrait of Picasso*, shown here, uses angles and line segments in the classic cubist style.

1 Find 3 pairs of alternate interior angles.

2 Find 3 pairs of corresponding angles.

3 Find 3 pairs of vertically opposite angles.

4 Give an example of how angles and line segments in the painting allow the viewer to see different sides of the same object/person.

## Practice 4

**1** Match each term with the appropriate description.

| | |
|---|---|
| Alternate interior angles | are angles that add up to 180°. |
| Vertically opposite angles | form F-angles. |
| Complementary angles | form X-angles. |
| Supplementary angles | form Z-angles. |
| Corresponding angles | are angles that add up to 90°. |

**2** In this diagram, name three pairs each of vertical angles, alternate interior angles and corresponding angles.

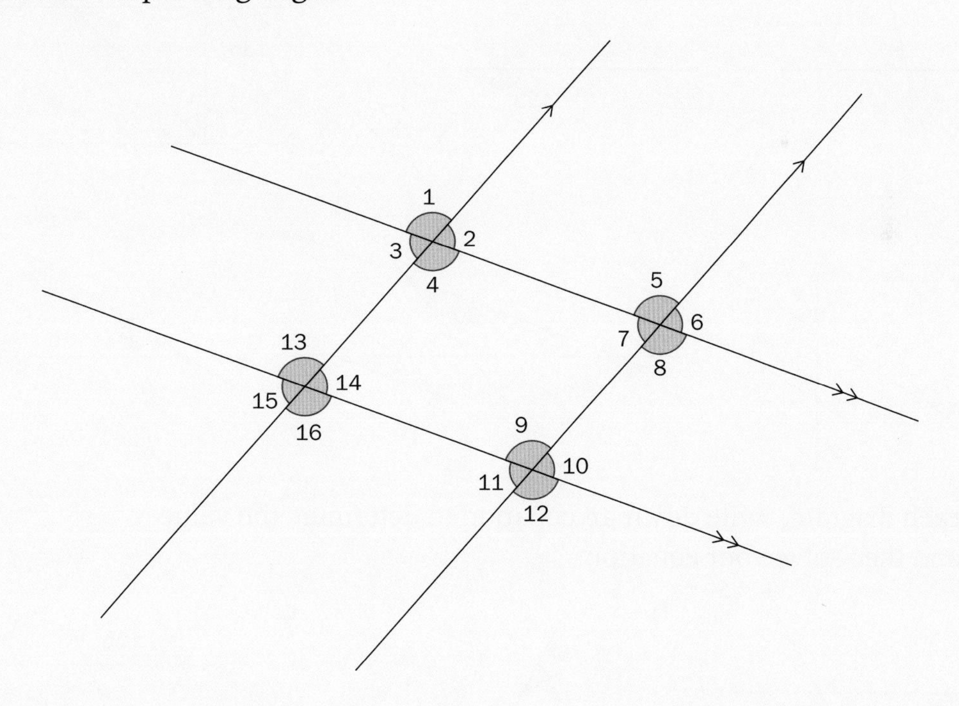


▶ Continued on next page

**3** In this diagram, identify all the angles which are equal in measure to:

**a** $\angle c$ **b** $\angle h$.

In question **3**, be sure to justify your answers.

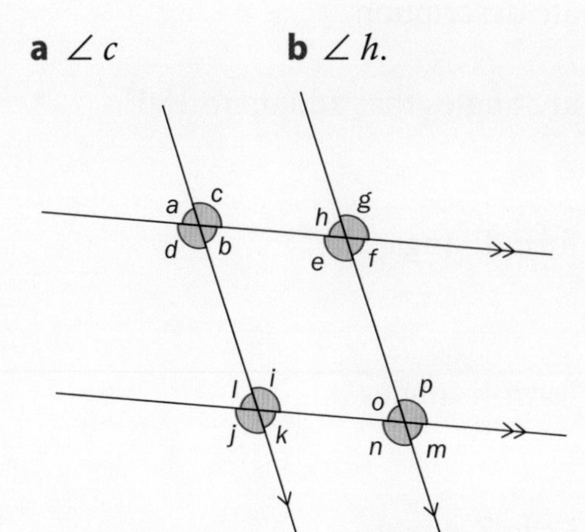


**4** Determine the measure of the indicated angles.

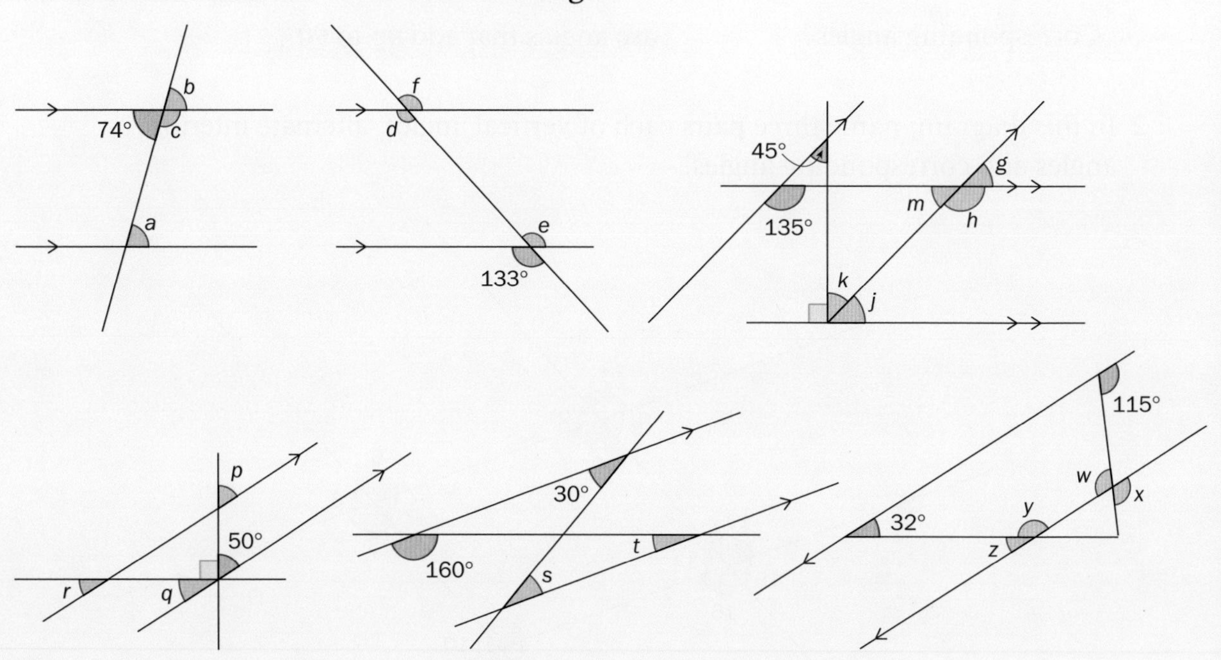


**5** For each diagram, write down an equation to determine the value of $x$ and then solve your equation.

**a**

120°
2x

**b**

35°
x

**c**

115°
(x + 5)

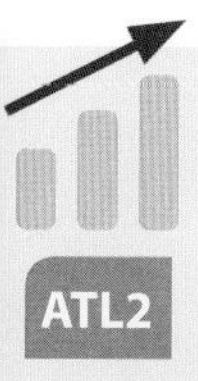

ATL2

## Formative assessment – Lakota beadwork

criterion C

The Lakota are a tribe of indigenous people that live in the northern plains of North America. The patterns they use in textiles and beadwork represent their tribal identity as well as family values. The Lakota Star, pictured here, is an example of such a pattern.

1 Download and print a copy of the Lakota Star (not necessarily in colour).

2 On your copy, clearly indicate the geometric elements listed below. Be sure to name them appropriately and indicate any required points on the diagram. Try to use the same points as often as possible so as to limit the number of markings on the copy. For the acute and obtuse angles, use a protractor to find their measure.

- 4 points
- 2 pairs of supplementary angles
- 4 rays
- 2 pairs of complementary angles
- 4 lines
- 4 pairs of vertically opposite angles
- 4 line segments
- 2 pairs of corresponding angles
- 4 acute angles
- 2 pairs of alternate interior angles
- 2 pairs of parallel lines
- 4 right angles
- 4 obtuse angles

3 Justify your decisions for each of the pairs of angles. Use a diagram to support your explanations.

## Triangles

A triangle is a shape with three angles and three sides. Triangles can be classified by the lengths of their sides or by the measures of their angles. In Activity 8, you will discover and define these classifications.

## Activity 8 – Classifying triangles

A triangle can be classified as an *acute* triangle, an *obtuse* triangle or a *right-angled* triangle (also sometimes called a right triangle). Look at the triangles below and answer the questions that follow.

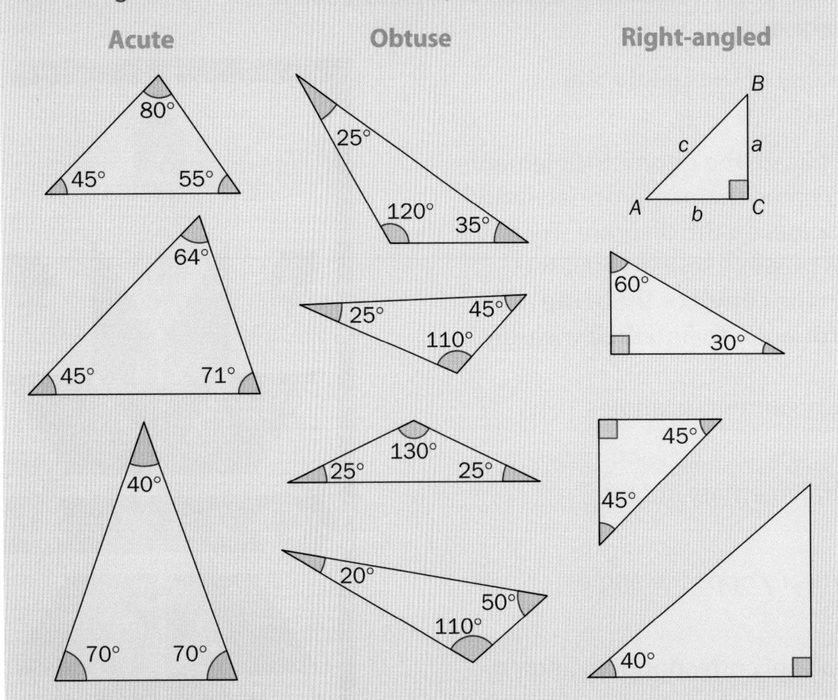


**1** How would you define each of the following?

**a** an *acute* triangle **b** an *obtuse* triangle **c** a *right-angled* triangle

**2** Can a triangle be two of those classifications at the same time? Explain.

## Reflect and discuss 8

Mosaics can be created using small tiles or pieces of glass. This same idea can also be used in print art, like the flower in the diagram below.

- Where do you see the following elements?
  - acute triangles
  - obtuse triangles
  - right-angled triangles
- What assumptions did you make in finding these elements?
- What effect is produced by having these kinds of different triangles next to each other? Explain.

Regardless of how a triangle is classified, it shares certain properties with every triangle. You will investigate one of these properties in the next investigation.

## Investigation 3 – Interior angles of a triangle

Criteria B, C

You can perform the following investigation using paper and pencil or with dynamic geometry software.

LINK For example, go to mathsisfun.com/geometry/protractor-using.html, and scroll down to 'Have a Go Yourself'

1 Using a ruler, draw six different triangles. Use the angles listed in the table below for the first three triangles. Draw the other three triangles as you wish.

2 Use a protractor to measure the unknown angle(s) in each triangle. Write down your measurements in a table like the one below.

| Triangle | 1st angle | 2nd angle | 3rd angle |
|---|---|---|---|
| 1 | 40° | 80° | |
| 2 | 90° | 60° | |
| 3 | 50° | 70° | |
| 4 | | | |
| 5 | | | |
| 6 | | | |

3 What pattern do you see related to the measures of the angles in each triangle?

4 Draw a few other triangles and test your theory for other cases.

5 Write down what you found as a general rule.

ATL1

6 Cut out one of the triangles and fold the angles inward so that the vertices all touch each other but don't overlap. What does that show?

7 Cut out another triangle and rip off the corners. Explain how you could demonstrate your rule with these corners.

## Reflect and discuss 9

- Can a right-angled triangle also be an obtuse triangle? Explain.
- An *exterior* angle of a triangle is an angle outside of the triangle, as shown here. How does the measure of this angle relate to the measures of the two interior angles (*A* and *B*) that are not next to it?

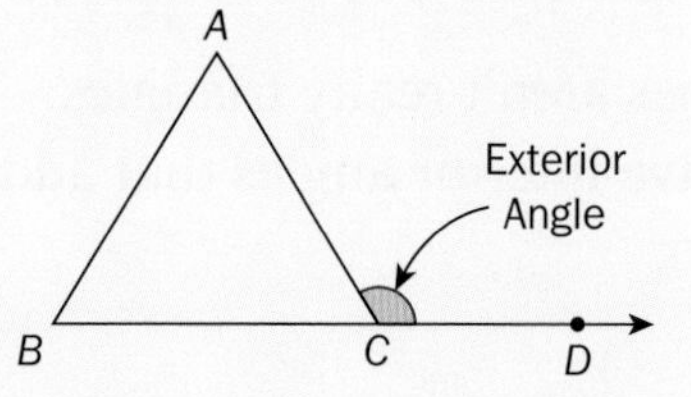


- How does the sum of the angles inside a triangle relate to the sum of the angles inside a rectangle? Explain.

The sum of the angles inside any triangle (often referred to as the *interior angles*) is always 180°. Knowing this allows you to calculate the measures of unknown angles.

## Example 3

**Q** Determine the measure of the unknown angles.

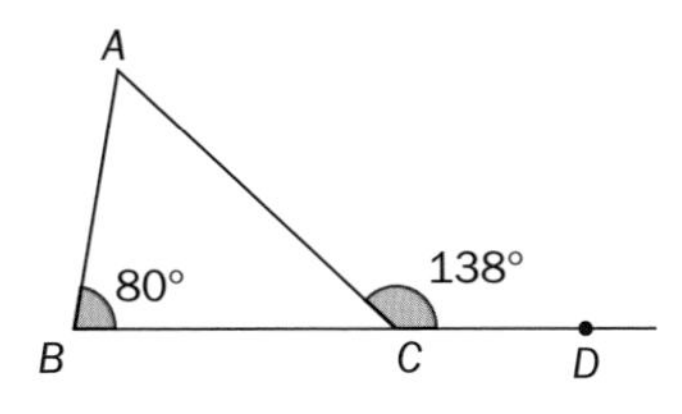


**A** $\angle BCA + \angle DCA = 180°$

Angles on a line add up to 180°.

$\angle BCA + 138° = 180°$
$\angle BCA + 138° - 138° = 180° - 138°$
$\angle BCA = 42°$

$\angle DCA = 138°$.

$\angle BAC + \angle ACB + \angle CBA = 180°$
$\angle BAC + 42° + 80° = 180°$
$\angle BAC + 122° = 180°$
$\angle BAC + 122° - 122° = 180° - 122°$
$\angle BAC = 58°$

The sum of the interior angles of a triangle is 180°. If you know the measures of two of the angles of a triangle, you can calculate the measure of the third.

## Practice 5

**1** Art can be created using geometric designs, such as the horse head design here.

**a** Find examples of acute, obtuse and right triangles in the image.

**b** What effect do the triangles give to the piece?

**c** Some of these shapes aren't really triangles. Do these shapes have interior angles that add up to 180°? Explain.

▶ Continued on next page

**2** Calculate the measure of each missing angle and justify your answer with a reason.

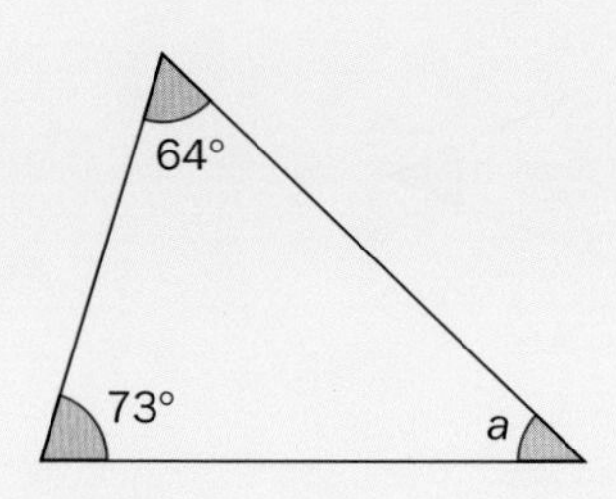


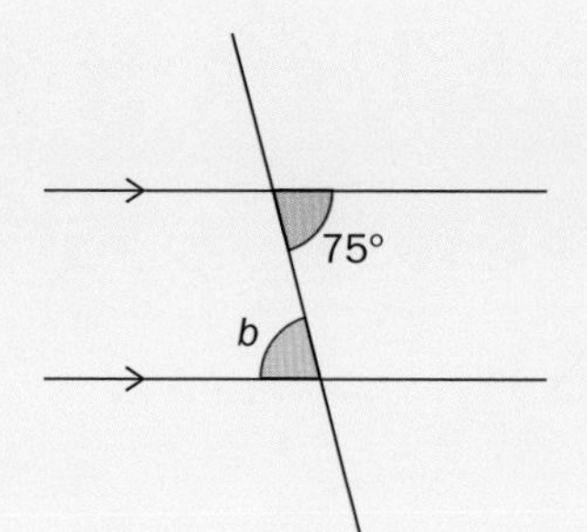


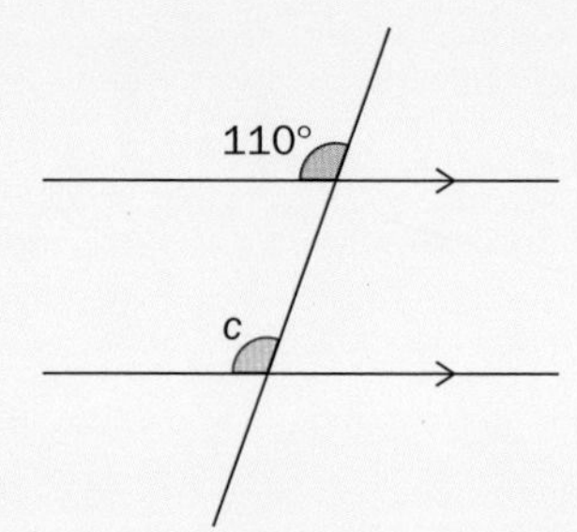


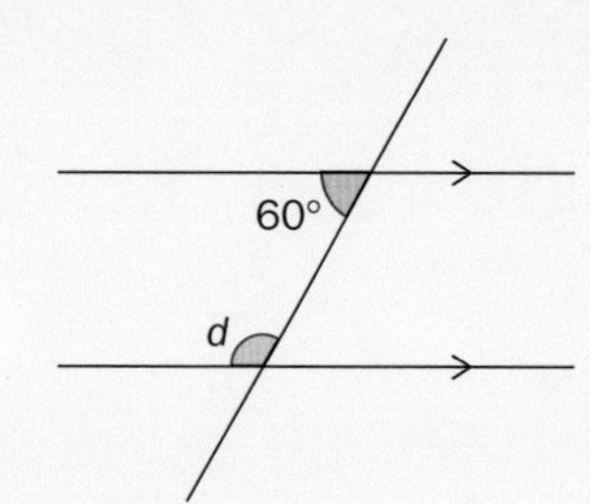


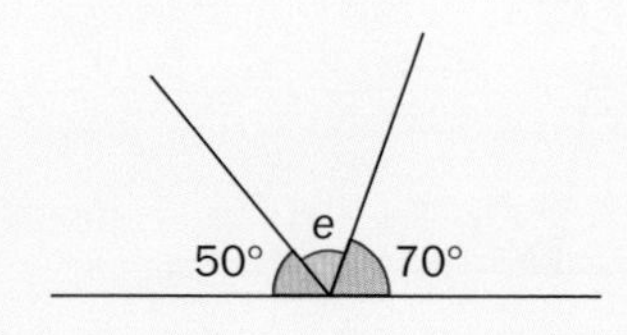


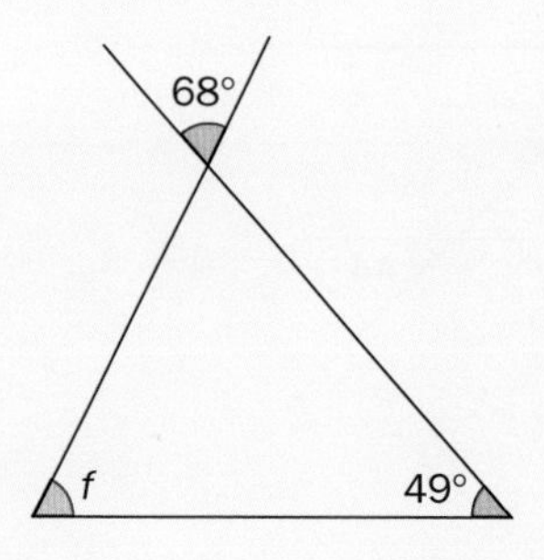


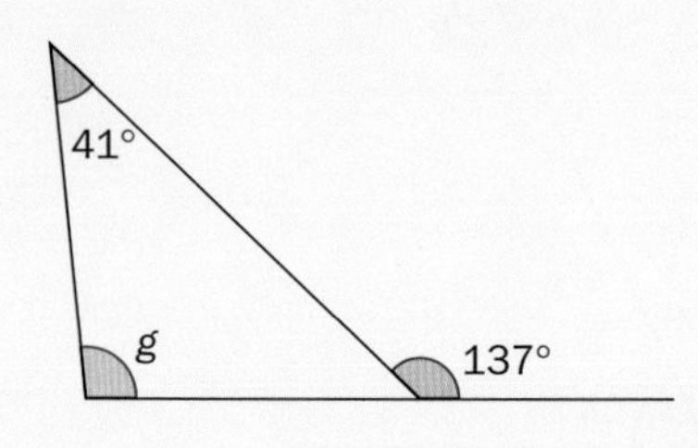


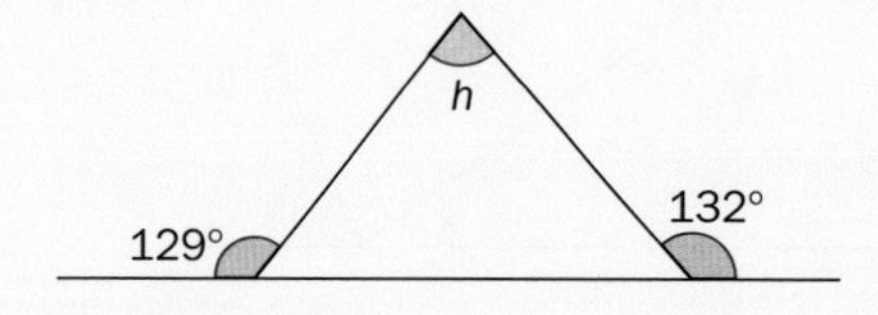


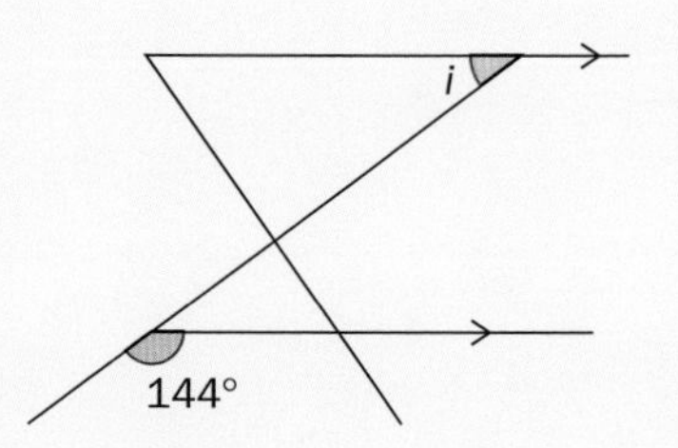


▶ Continued on next page

**3** Determine the measure of the indicated angles.

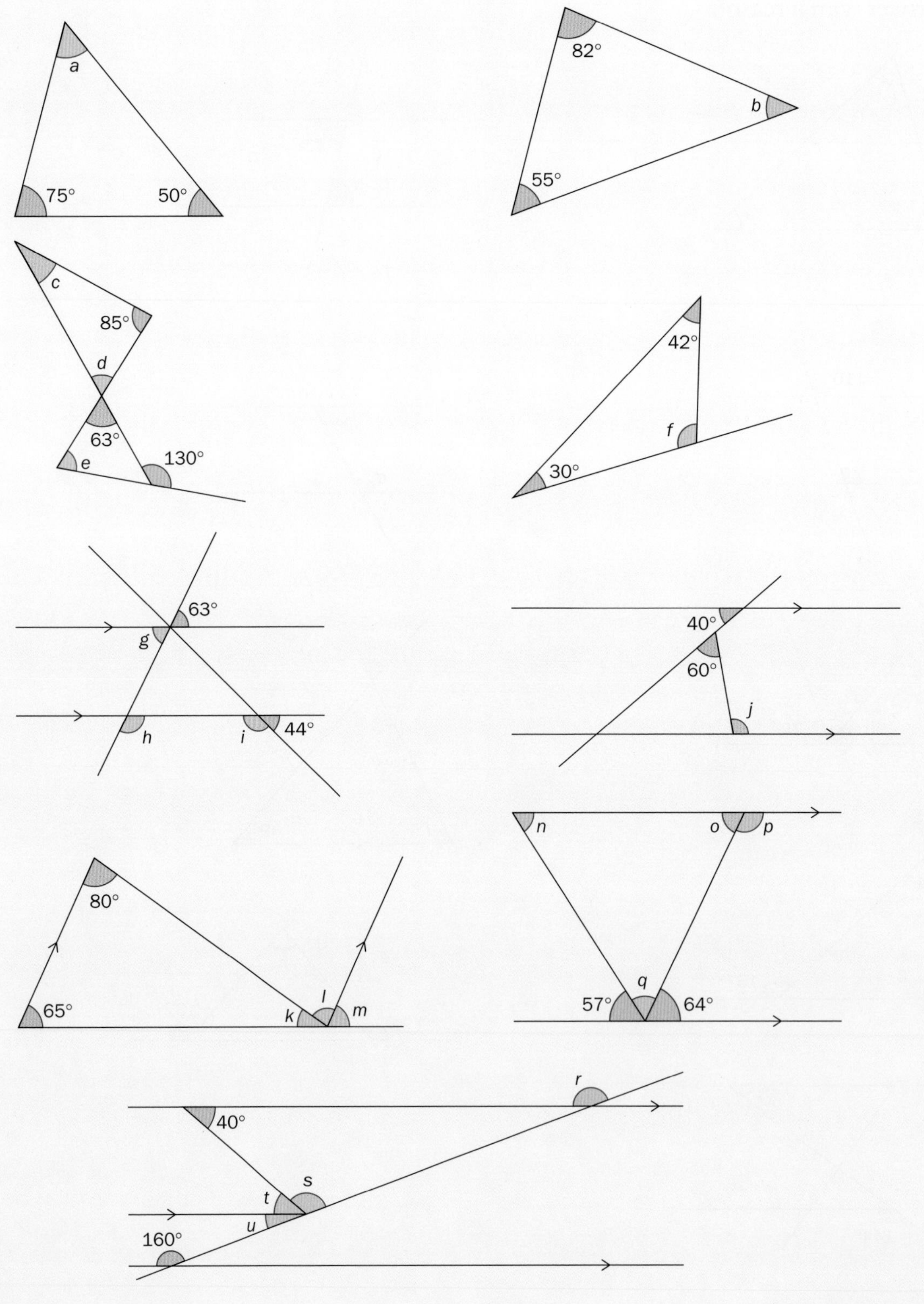


▶ Continued on next page

**4** In this diagram, show how you can use the relationship between the measure of angles $\angle c$, $\angle d$ and $\angle e$ to prove that the interior angles of a triangle sum to 180°.

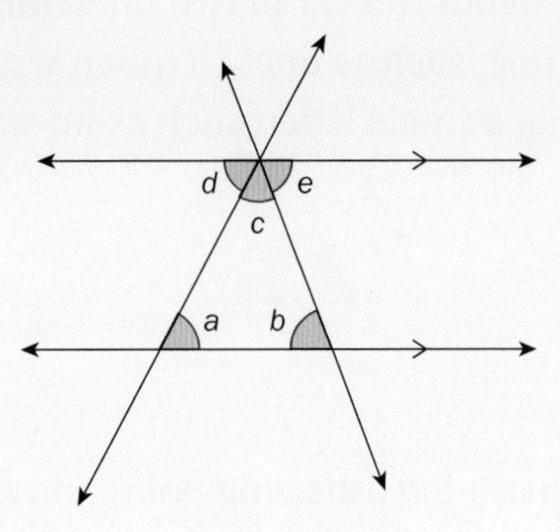


ATL2

**5** Apart from Mathematics and Art, describe subject areas where you have had to work with angles or triangles. How has knowledge in one subject or discipline impacted what you have learned or been able to do in another subject?

Musuem of Modern Art, Bonn, Germany

# Unit summary

A *point* is a location and is represented as a dot.

A *line* is a set of points that extends forever in two directions. Two points define a specific line, such as line *AB* (often written as $\overleftrightarrow{AB}$). Lines can also be named using a single letter, such as line *m*.

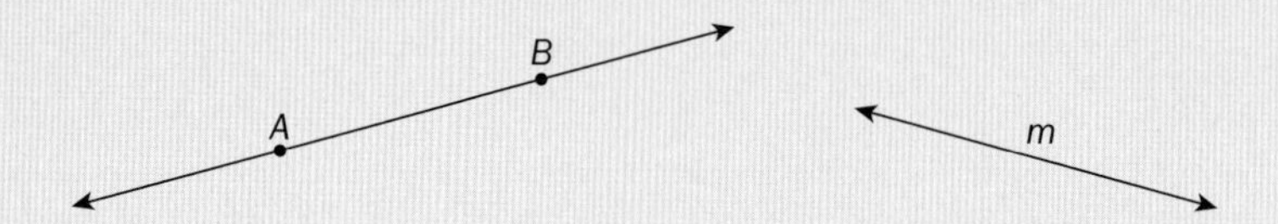


A *ray* is a part of a line that has a fixed starting point and extends forever in the other direction, such as ray *EF* (often written as $\overrightarrow{EF}$).

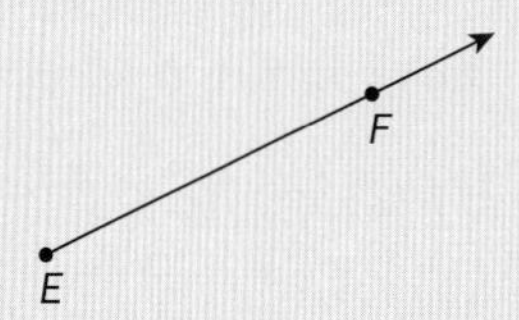


A *line segment* (or *segment*) is a part of a line that has two fixed endpoints, such as segment *CD* here (often written as $\overline{CD}$).

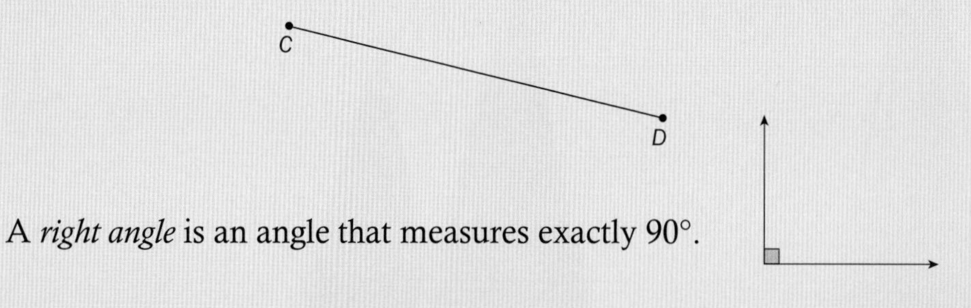


A *right angle* is an angle that measures exactly 90°.

An *acute angle* measures less than 90°.

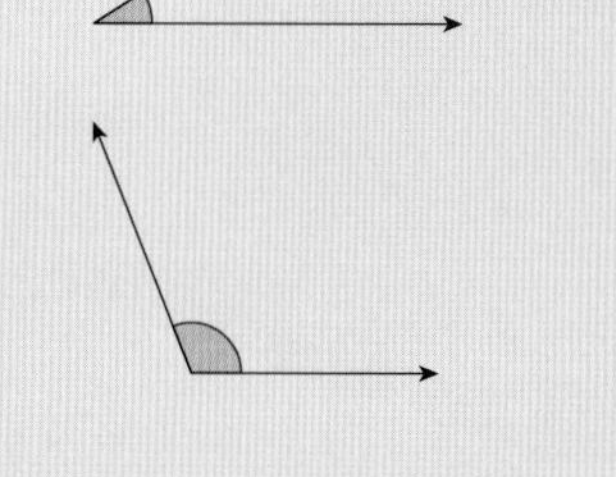

An *obtuse angle* measures more than 90 degrees.

A *straight angle* measures 180°.

## Supplementary angles

Two angles that add up to 180° are called *supplementary* angles. One angle is the supplement of the other. In this diagram, $a$ and $b$ are supplementary angles.

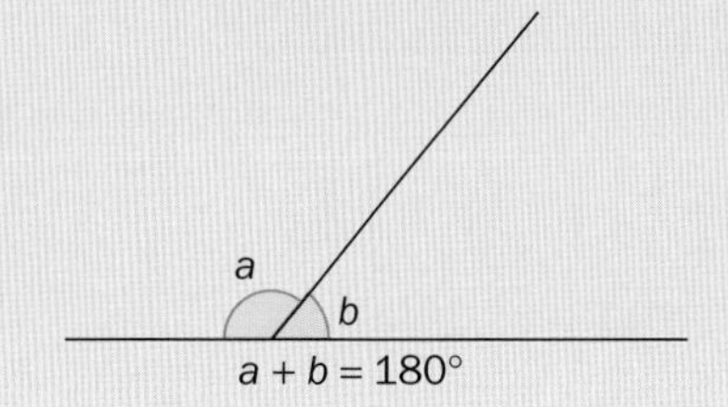


## Complementary angles

Two angles that add up to 90° are called *complementary* angles. One angle is the complement of the other. In this diagram, $a$ is the complement of $b$, and $b$ is the complement of $a$.

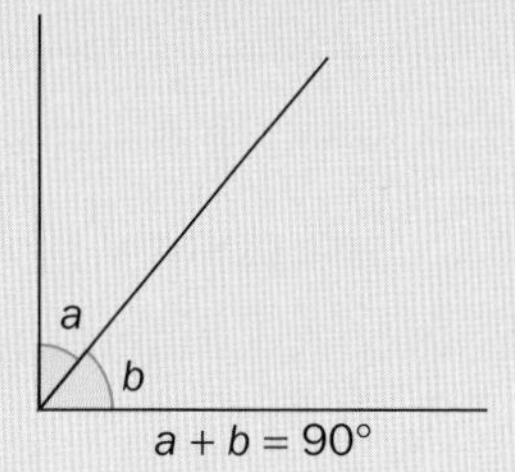


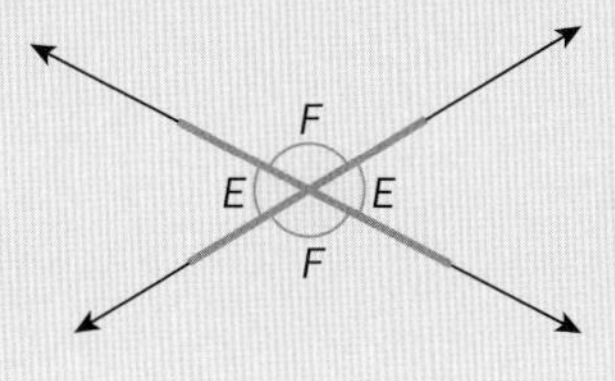


Intersecting lines form two pairs of *vertically opposite* angles. These angles (also called *vertical* angles) have equal measures and form an X.

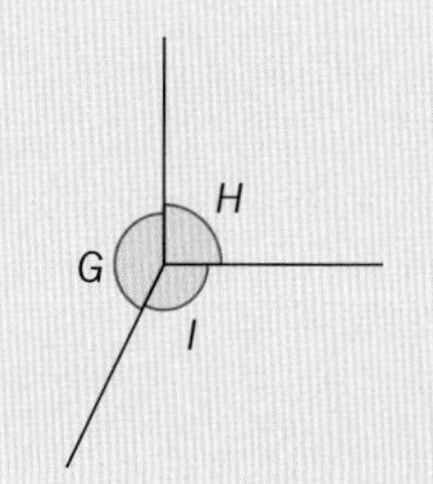


Angles at a point add up to 360°. In this diagram, $\angle G + \angle H + \angle I = 360°$.

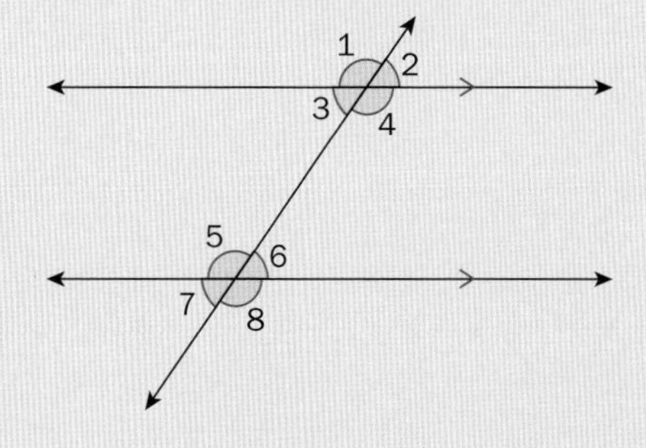


Parallel lines cut by a transversal create pairs of angles that have equal measures.

Corresponding angles make an F (for example, $\angle 4$ and $\angle 8$).

Alternate interior angles form a Z (for example, $\angle 3$ and $\angle 6$).

The sum of the interior angles of a triangle is always 180°.

Parallel lines are indicated using the notation $g \parallel h$.

Perpendicular lines are indicated using the notation $\overleftrightarrow{AB} \perp \overleftrightarrow{MN}$.

# Unit review

Launch additional digital resources for this chapter

Key to Unit review question levels:

Level 1–2 Level 3–4 Level 5–6 Level 7–8

1 **Draw** an example for each of the following.

**a** $\overleftrightarrow{EF}$ **b** $\overline{XY}$ **c** $\overrightarrow{RS}$ **d** point $T$ **e** $\overline{AD} \perp \overline{JM}$ **f** $\overleftrightarrow{TV} \parallel \overleftrightarrow{YZ}$

2 **Write down** expressions for each diagram, using mathematical notation.

**a**

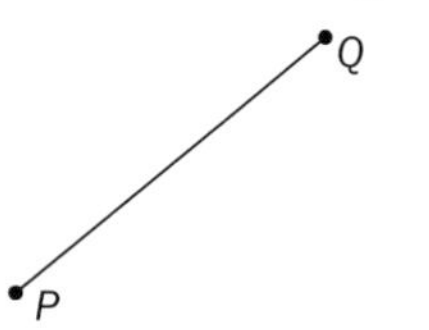


**b**

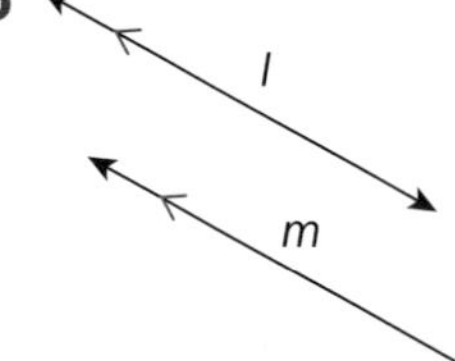


**c**

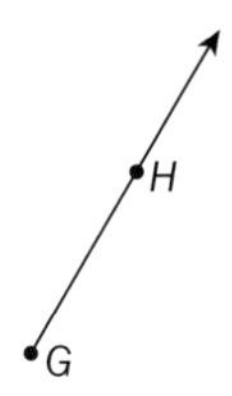


**d**

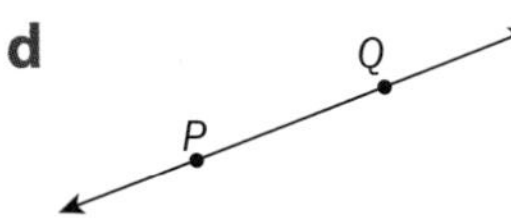


**e**

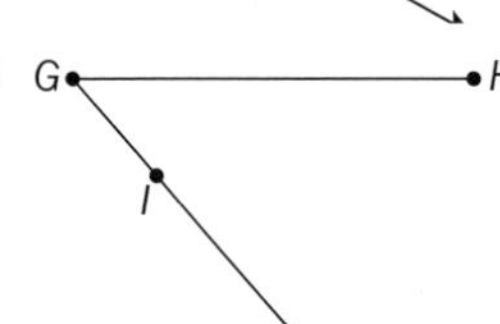


**f**

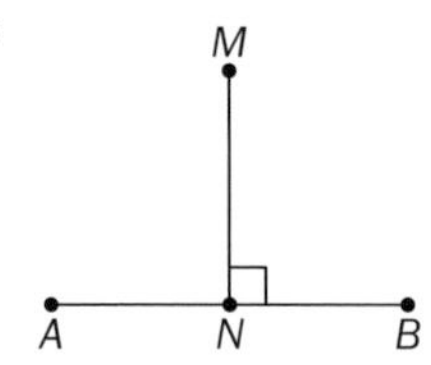


3 Using a protractor, **measure** and classify each angle.

**a**

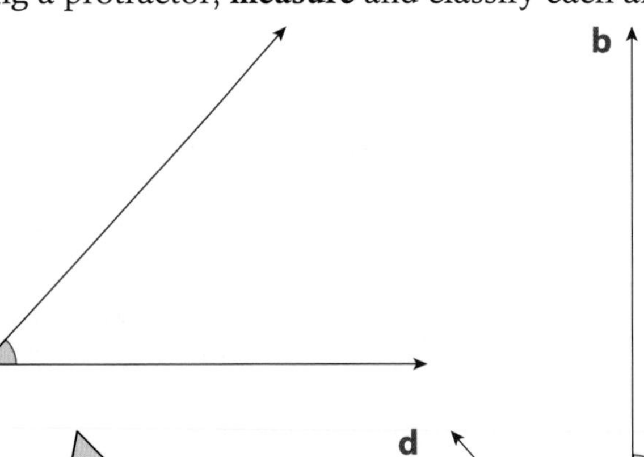

**b**

**c**

**d**

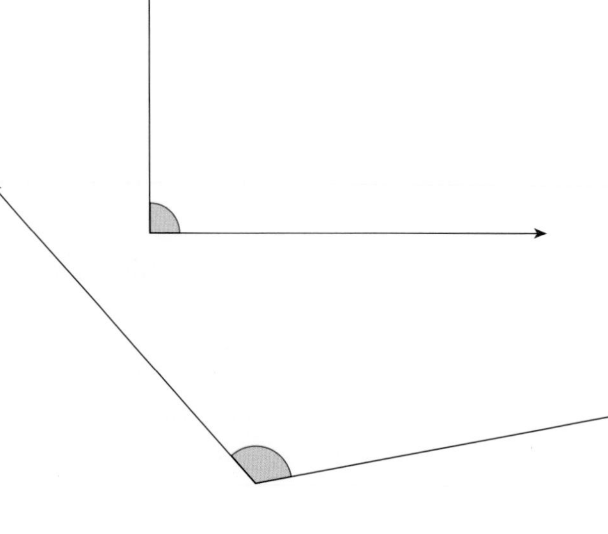

**4** Using a protractor, **draw** the following angles.

**a** 152° **b** 68° **c** 34° **d** 119° **e** 90° **f** 14°

**5** Find the complement and supplement of each angle in question **4**, if it exists.

**6** For each diagram, **write down** an equation and **solve** it to determine the value of $x$.

**a**

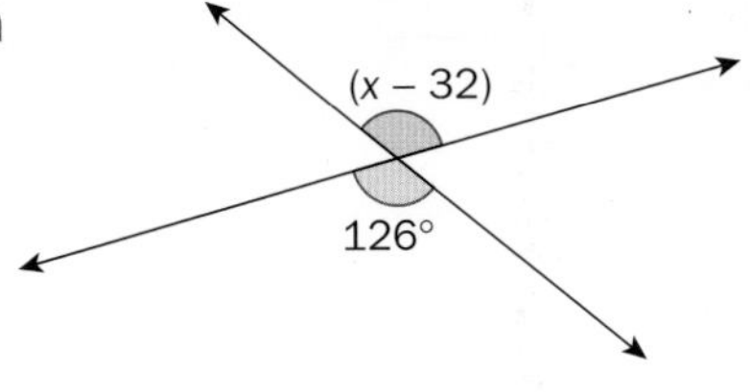


**b**

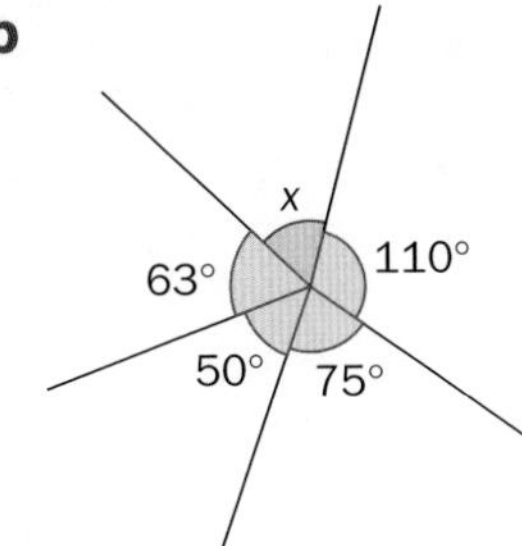


**c**

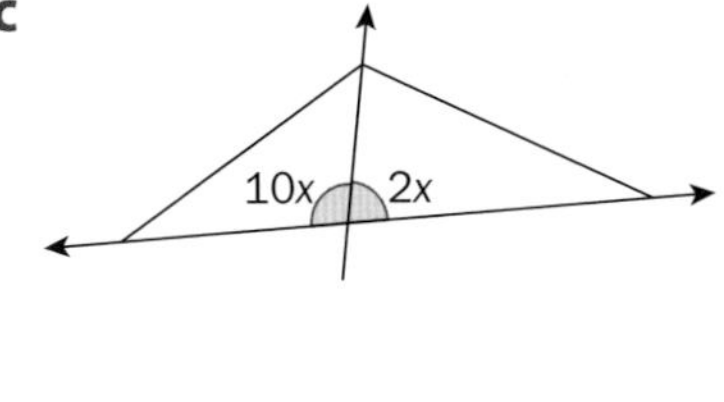


**d**

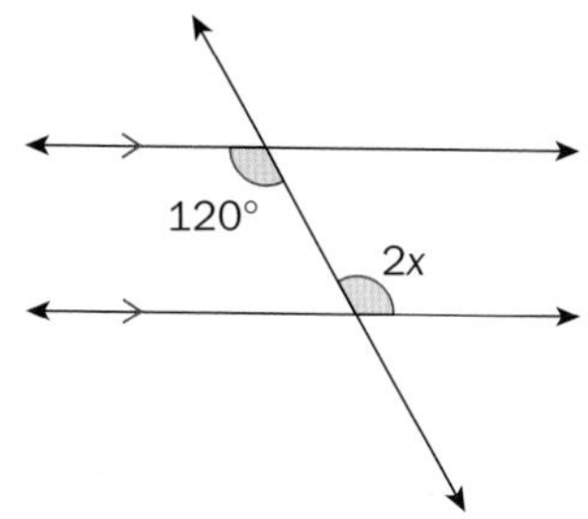


**e**

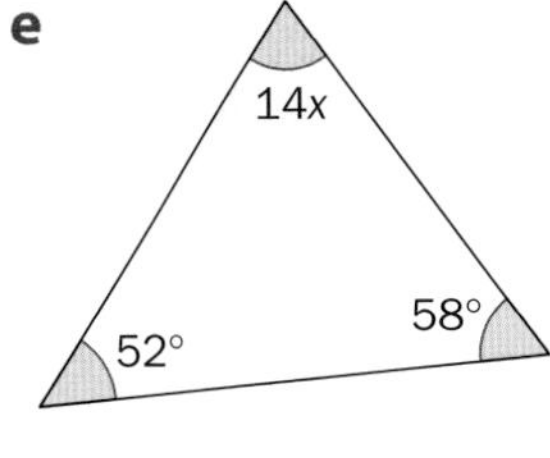


**7** **Show** that the two acute angles in a right-angled triangle must be complementary.

**8** Find the measures of the indicated angles. **Justify** your answers.

**a**

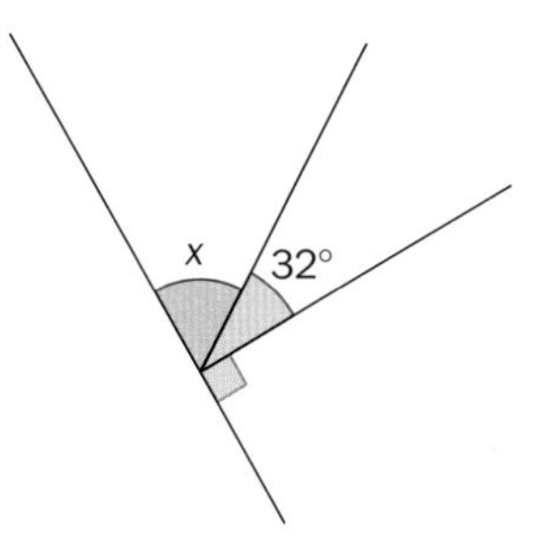


**b**

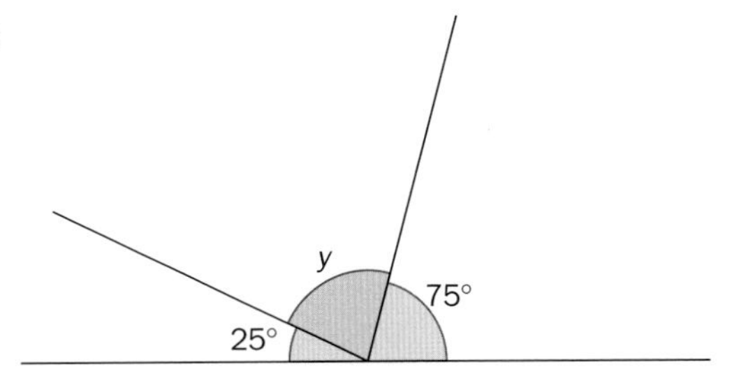


**c**

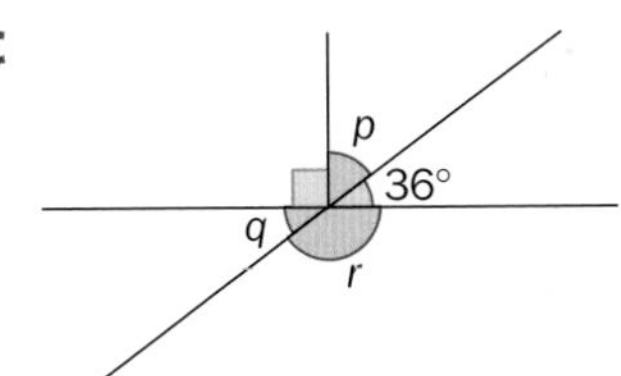


**d**

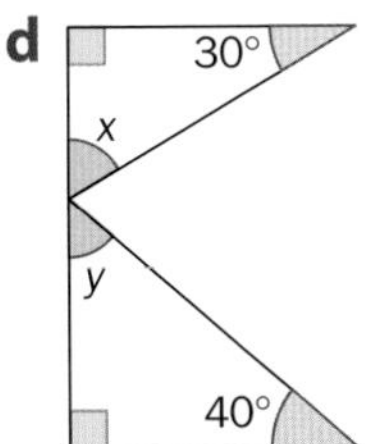

**9** For each diagram, **write down** an equation and **solve** it to determine the value of $x$.

**a**

**b**

**10** Find the value of each indicated angle.

**a**

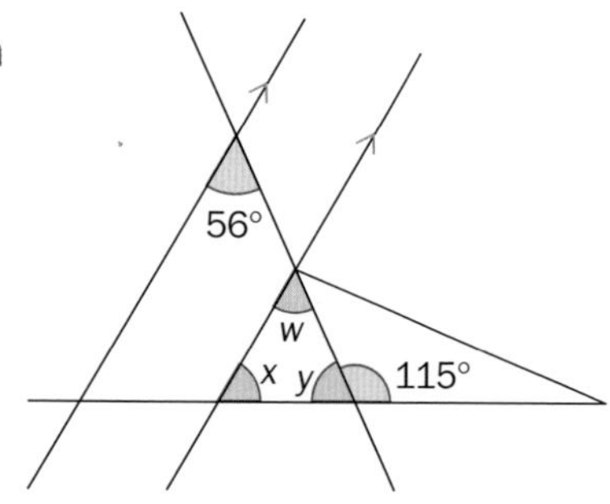


**b**

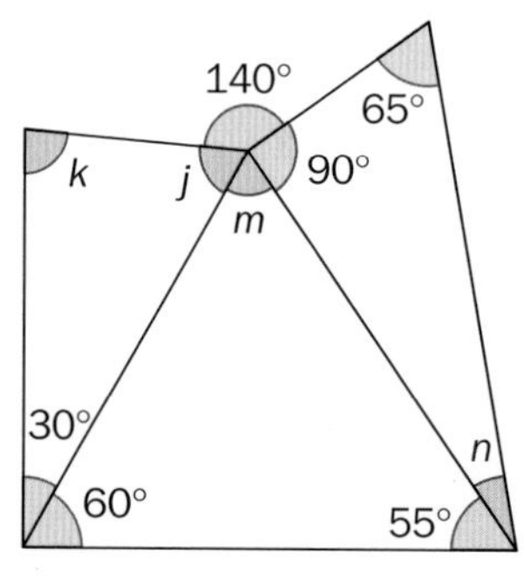


**c**

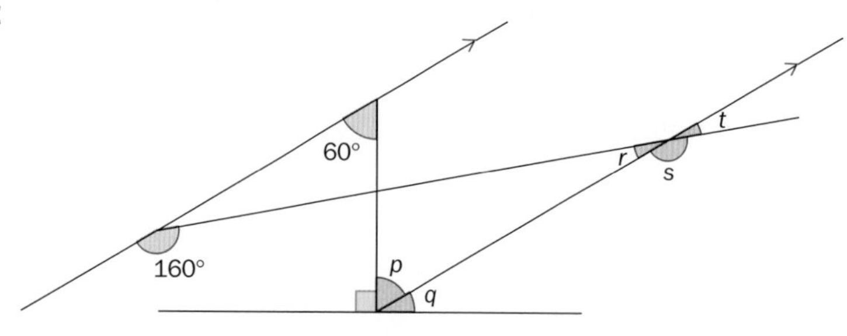


**d**

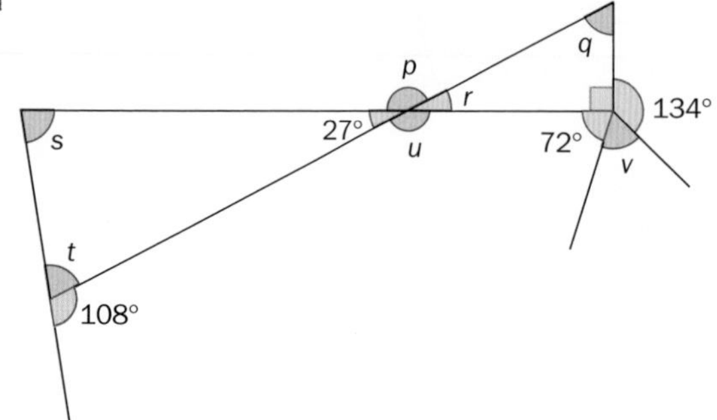

**e**

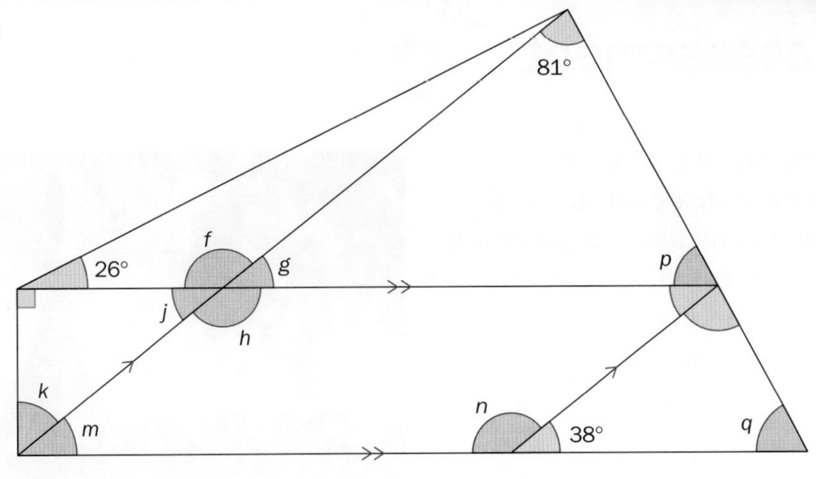


**f**

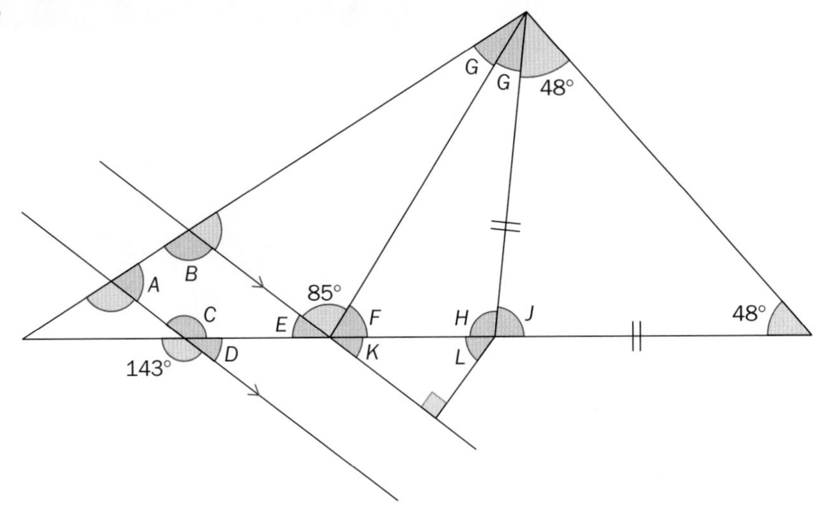


In part **f**, notice that the two angles marked *G* are equal, as are the two side lengths marked with ticks.

ATL2

**11** In Science class, during a study of mirrors, Svetlana noticed that the diagrams showing how light reflects off of a curved mirror used many of the same elements that she found in her Mathematics class.

**a** Looking at the diagram here, **identify** as many elements from this unit as you can. These can include basic elements like rays and lines as well as all the types of angle that you have seen so far.

**b** How does mathematics help you understand what is represented in the image?

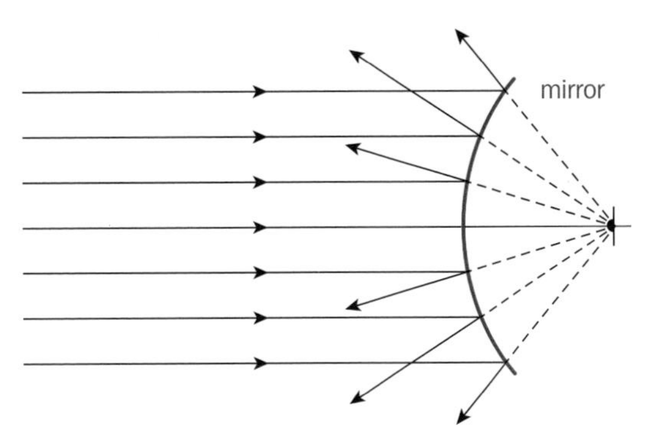

## Summative assessment

criterion C

You have seen in this unit how mathematics can be used to analyze and even create artistic works. From cubism to origami to architecture, geometry can be both functional and beautiful. Your task is to create your own work of art that includes the elements that you have seen throughout the unit.

You may create a painting, drawing, sculpture, mosaic or any other piece of art. You will have a 'draft', 'sketch' or 'blueprint' that will demonstrate the required mathematics, as well a 'final product', which will be the art piece, a scale model or a full-color image (for example a painting or mosaic).

The following elements must be clearly indicated in your draft and listed on a separate sheet of paper, including proper mathematical notation and names. These elements must also be visible in your final product, but without the labels.

- 2 rays
- 2 segments
- 2 lines (not including the ones that are parallel)
- 2 sets of parallel lines
- 2 acute angles (not including the ones in the triangles)
- 2 obtuse angles (not including the ones in the triangles)
- 2 right angles (not including the ones in the triangles)
- 2 clear examples of supplementary angles
- 2 clear examples of more than two angles on a line that add to 180°
- 2 clear examples of complementary angles
- 2 right-angled triangles
- 2 acute triangles
- 2 obtuse triangles
- the measures of all of the angles in the above triangles, clearly showing the sum within each triangle

You must give your final product a title and write a brief description of what the artwork represents. Within the description, you must explain how at least three of the geometric elements relate to what is being represented in the artwork.

# 5 Fractions

In this unit the study of fractions will help you to better understand daily interactions with our fellow human beings. Dividing things– such as blocks of land or grains of wheat – using fractions is a great way to demonstrate similarities and differences. You could use fractions to better understand issues of fairness and sustainability.

## Fairness and development

### Sharing finite resources with other people and with other living things

In the year 1750, the global population was approximately 800 million people. It is currently over 7.4 billion. How many of those people live in Asia? How many in Europe? How much of the planet does Asia occupy?

An understanding of fractions can help us to analyze and describe the global population that lives in each continent. We can then look at how human population growth has affected the environment and the population of other living things. Do we share the planet we call "home" equitably?

At the same time, humans use far more natural resources than any other living things. How we use our resources and how fairly that is done is easily expressed and analyzed with fractions. Determining whether we all have access to our fair share of the planet's resources is an application of fractions well worth investigating.

## Globalization and sustainability

### Urban planning, strategy and infrastructure

You may not have thought about it, but how was the city in which you live constructed? Does it follow a clear pattern? Some cities are radial cities where everything seems to converge at one point. Fractions can be used to plan the city and its infrastructure and then to analyze each section, ensuring that the city does not become too crowded.

Other cities have been built using a strategy that looks more like a grid. As with radial cities, fractions can be an invaluable tool to help plan and build these communities, but they can also be used to analyze trends and issues within the grid. For example, by what fraction have temperatures increased due to the urban heat island effect?

# 5 Fractions

## Human connections

## KEY CONCEPT: LOGIC

## Related concepts: Quantity and Simplification

## Global context

In this unit you will use fractions to explore the global context of **identities and relationships**. Humans engage in many different activities which allow us to connect with one another, within families, communities and cultures. Learning how to represent fractions and perform operations with them will help you uncover some of these activities, including social media, music and even food!

## Statement of Inquiry

Using logic to simplify and manipulate quantities can help us explore human connections within families, communities and cultures.

### Objectives

- Representing and comparing fractions in different forms
- Simplifying different forms of fractions
- Using the four number operations (addition, subtraction, multiplication and division) with fractions
- Applying mathematical strategies to solve problems involving fractions

### Inquiry questions

**F** What does it mean to simplify?

What operations can you perform on quantities?

**C** How is logic used to manipulate quantities?

**D** What promotes human connections?

What hinders human connections?

**ATL1 Critical thinking skills**

Draw reasonable conclusions and generalizations

**ATL2 Communication skills**

Give and receive meaningful feedback

## You should already know how to:

- express a fraction as a decimal
- express a basic fraction in its simplest terms
- change an improper fraction to a mixed number
- change a mixed number to an improper fraction
- change a fraction to a decimal and a percentage

For practice with estimating fractions and decimals graphically, go to brainpop.com and search for 'Battleship Numberline'.

# Introducing fractions

Family trees are used to represent many generations within a family and to trace relatives as far back as possible. Family trees can help you discover historical figures in your family history or even the cultural background that is a part of your heritage.

In recent years, it has become possible for people to try to trace their ethnic ancestry using something as simple as a saliva sample. The DNA in the saliva is analyzed and the genetic ethnicity is predicted. The results can lead to new family connections or improve research so you can add more branches to your family tree.

A typical ancestry report might look like this.

| Country | Approximate amount |
|---|---|
| Algeria | $\frac{1}{20}$ |
| Morocco | $\frac{1}{100}$ |
| Benin | $\frac{1}{25}$ |
| Cameroon | $\frac{7}{100}$ |
| Spain | $\frac{7}{50}$ |
| Portugal | $\frac{1}{4}$ |
| Italy | $\frac{6}{25}$ |
| England | $\frac{1}{5}$ |

## Reflect and discuss 1

- What do the fractions in the table mean? What does each numerator and denominator mean in this report?
- What would you expect the fractions to add up to? Explain.
- How could you tell which country has the highest amount? Explain.
- How could this kind of information lead to more human connections in a family?

# Representing and comparing fractions

## Representing fractions

A fraction represents a part of a whole. It can be represented using a wide range of objects. For example, the fraction $\frac{3}{4}$ can be represented by each of these pictorial representations. The fraction $\frac{3}{4}$ can also be represented by this whole group of representations. Can you see how?

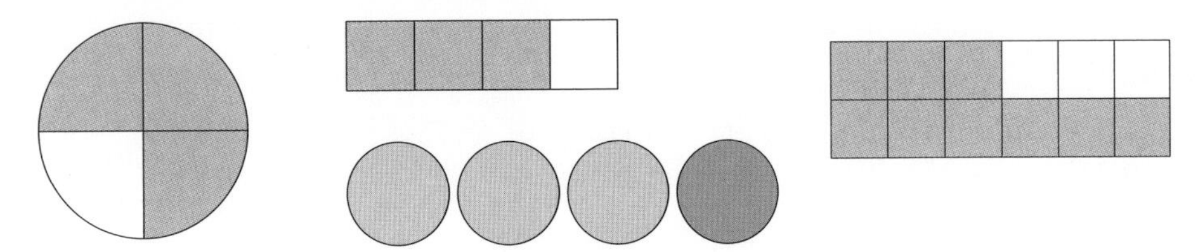

The way the fraction is represented depends on what 'the whole' is.

### Activity 1 – Tangram fractions

You will need a tangram set to complete this activity. If you do not have one, your teacher will provide you with a paper set to cut out.

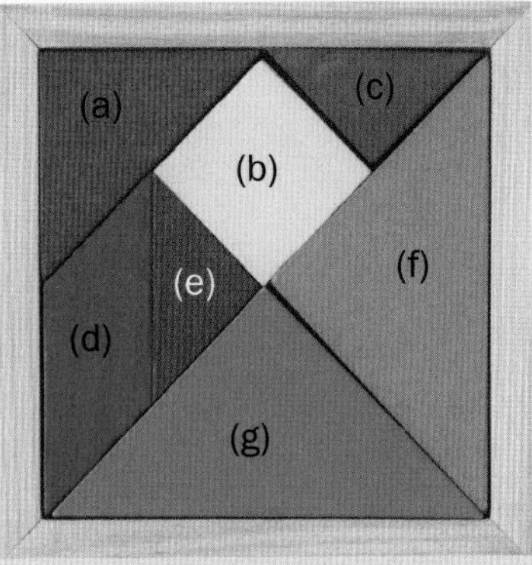


Working with a tangram set, answer the following questions.

**1** What fraction of each shape is (c), the smallest triangle? Show your working. Include diagrams if they are helpful.

| Shape | Number of (c) triangles that would fit inside | What fraction of this shape is (c)? |
|---|---|---|
| Largest triangle (f) or (g) | | |
| Medium triangle (a) | | |
| Parallelogram (d) | | |
| Square (b) | | |

Triangles (c) and (e) are both the same size, so step **1** could also be answered in terms of triangle (e).

▶ Continued on next page

2 What fraction of triangle (f) is each smaller shape listed in the table below? Show your working.

| Shape | Fraction of triangle (f) |
|---|---|
| Medium triangle (a) | |
| Parallelogram (d) | |
| Square (b) | |

3 What fraction of the whole tangram set is represented by each color? Show your working.

4 Which different shapes have the same area? Explain how you know this.

5 Show two examples of how different colored pieces can be added together to equal the area of a different colored piece.

ATL2

Pairs

6 With a partner, look at each other's answers to step **5**. Take turns giving feedback by first asking questions to clarify anything that isn't quite clear. Then, give at least one positive comment. Follow that with a suggestion for improvement, if appropriate.

Triangles (f) and (g) are both the same size, so step **2** could also be answered in terms of triangle (g).

Hold on to the tangram pieces, as you will be doing more with them during this unit.

WEB LINK

Try the Sushi Fractions activity on the Mr Nussbaum website at mrnussbaum.com/sushi-fractions. Divide the sushi between tuna, avocado and shrimp to match the fractions you are given.

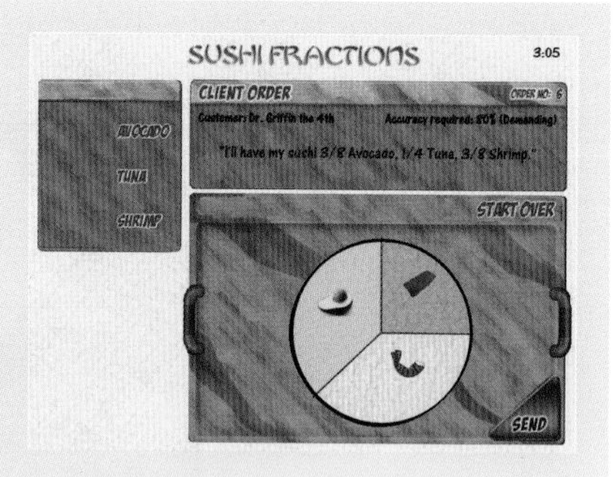


## Activity 2 – Money diagram

1 Create a visual representation showing how 1 base currency unit in your country (for example: 1 US dollar or 1 euro coin) can be broken down into all of the smaller currency denominations. You can be creative with how you set up the visual, as long as you include the face value of the coins and notes, and what fraction of the base unit they represent.

ATL2

Pairs

2 With a partner, look at each other's work from step **1**. Take turns giving feedback by first asking questions to clarify anything that isn't quite clear. Then, give at least one positive comment. Follow that with a suggestion for improvement, if appropriate.

▶ Continued on next page

3 Do the actual sizes of the coins and notes reflect the fraction represented by their monetary values? Explain by giving specific examples.

4 Select two other currencies from around the world and repeat step **1** with each currency. What are the similarities and the differences to your home currency?

5 Determine five different ways you could pay someone 1 base unit of your currency.

## Comparing fractions

We can compare fractions to find out which fraction represents more of the whole.

### Reflect and discuss 2

- Explain what the numerator and denominator represent in a fraction.
- Explain two different methods you could use to compare the fractions $\frac{5}{11}$ and $\frac{9}{20}$.
- Explain what happens to the value of a fraction if the denominator stays the same and the numerator increases.
- Explain what happens to the value of a fraction if the numerator stays the same and the denominator increases.

It is easiest to compare fractions that are out of the same whole. For example, it is easy to tell that $\frac{5}{18} < \frac{11}{18}$ because the denominators are the same, and 5 is less than 11.
When comparing fractions that are *not* out of the same whole, one method is to convert the fractions so that they have the same denominator.

### Example 1

Compare the fractions $\frac{4}{5}$ and $\frac{7}{8}$ using either $<$ or $>$.

$\frac{4}{5} = \frac{4 \times 8}{5 \times 8} = \frac{32}{40}$

$\frac{7}{8} = \frac{7 \times 5}{8 \times 5} = \frac{35}{40}$

Therefore, $\frac{7}{8} > \frac{4}{5}$ since $\frac{35}{40} > \frac{32}{40}$.

The LCM of 5 and 8 is 40.
Make each fraction out of 40.

Sometimes, you will be able to compare fractions by using other fractions as benchmarks. If you can compare them to the same fraction, then you may be able to compare them to each other.

### Example 2

Q Compare the fractions $\frac{5}{12}$ and $\frac{11}{20}$ using either > or <.

A $\frac{6}{12} = \frac{1}{2}$ so $\frac{5}{12} < \frac{1}{2}$

$\frac{10}{20} = \frac{1}{2}$ so $\frac{11}{20} > \frac{1}{2}$

Therefore, $\frac{5}{12} < \frac{11}{20}$

Compare both fractions to $\frac{1}{2}$.

## Practice 1

**1** At the beginning of every New Year, Noë's family has a tradition where they gather around the kitchen table with a bowl full of grapes. Along with his mother, father and sister, each family member makes 12 wishes and eats one grape per wish. They take turns, making a wish and eating a grape each round, until they have completed 12 rounds.

**a** How many grapes will be eaten in all? Show your working.

**b** At what point will half of the grapes be eaten? Justify your answer.

**c** At what point will $\frac{1}{4}$ of the grapes be eaten? Justify your answer.

**d** At what point will $\frac{9}{16}$ of the grapes be eaten? Justify your answer.

**e** After seven rounds, what fraction of the grapes have been eaten? Find your answer in two different ways.

**f** When only two grapes are left, what fraction of the grapes have been eaten? Justify your answer.

**g** Discuss any rituals that your family has to celebrate or anticipate the New Year. How do these traditions develop connections between family members?

▶ Continued on next page

**2** Compare each pair of fractions using > or <. Explain your reasoning.

**a** $\frac{5}{6}, \frac{8}{9}$ **b** $\frac{12}{25}, \frac{5}{9}$ **c** $\frac{3}{7}, \frac{21}{50}$ **d** $\frac{5}{24}, \frac{2}{5}$ **e** $\frac{3}{20}, \frac{7}{24}$

**3** According to a recent study, $\frac{1}{4}$ of the people surveyed use their phones primarily to communicate with others using social networks, since they prefer to socialize online more than in person. Of the same group of people surveyed, $\frac{1}{3}$ use their phones primarily for sending texts, and $\frac{7}{25}$ use them primarily for making and receiving calls.

**a** Order the modes of socializing (social networks, texting, calling) from lowest to highest.

**b** Despite so many people having smartphones, global voice usage declined for the first time ever in 2013. Why do you think this happened? What effect will it have on human connections? Explain.

**c** Many people say that, ironically, smartphones are eliminating the need for people to speak to each other. What do you think of this theory? Explain.

**4** Arrange the fractions $\frac{2}{11}, \frac{11}{40}, \frac{1}{4}, \frac{1}{5}, \frac{3}{9}, \frac{8}{33}$ from least to greatest. Show all of your working.

**5** You can also convert fractions into decimal form to compare them. Using the symbols < or >, compare these fractions after representing them in decimal form.

**a** $\frac{2}{3}, \frac{5}{7}$ **b** $\frac{3}{5}, \frac{7}{12}$ **c** $\frac{9}{10}, \frac{5}{6}$ **d** $\frac{2}{7}, \frac{3}{8}$ **e** $\frac{8}{13}, \frac{5}{8}$

**f** $\frac{4}{9}, \frac{6}{13}$ **g** $\frac{5}{11}, \frac{10}{19}$ **h** $\frac{1}{4}, \frac{2}{9}$ **i** $\frac{3}{4}, \frac{6}{7}$ **j** $\frac{10}{13}, \frac{7}{11}$

**6** Which is easier to use when comparing numbers: decimal form or fractional form? Justify your choice.

▶ Continued on next page

**7** Chinese Lunar New Year celebrates family, health and prosperity. It is the holiday which involves the most travel in the world, as it is the time when whole families come together.

**a** Approximately $\frac{1}{6}$ of the world's population celebrates the Chinese New Year in many countries around the world. How many people is this?

> Research the current world population to answer part **a**.

**b** Chinese New Year, also known as Spring Festival, is celebrated over 15 days. What fraction of the calendar year is the duration of this festival?

**c** The Chinese zodiac follows a 12-year cycle, with each year represented by one of the 12 animals from the Chinese zodiac. Your birth year corresponds to one of these 12 animals. If average life expectancy is approximately 80 years, approximately what fraction of a person's lifetime will be spent in the years of their Chinese zodiac sign?

Many of China's people work in big cities and travel back to their home town for Spring Festival. For some people, it is the only time in the year when they will visit their families. The modes of transportation taken by these people break down as follows:

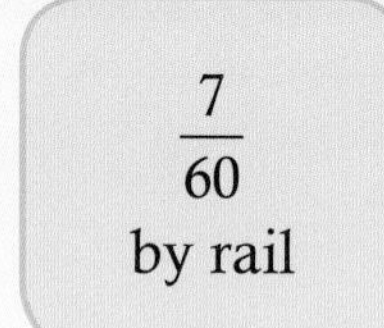

| $\frac{7}{60}$ by rail | $\frac{3}{200}$ by boat | $\frac{1}{50}$ by airplane | $\frac{5}{6}$ by road |
|---|---|---|---|

**d** Rank the modes of transportation from most to least popular.

**e** If there were approximately 3 billion trips altogether, how many trips were taken by each mode of transportation?

**8** Look back at the ancestry report on page 154 of this unit.

**a** What is 'the whole' in these fractions?

**b** How is it possible that the denominators are all different if they relate to the same whole?

**c** Order the countries from lowest to highest by finding a common denominator.

**d** Order the countries from lowest to highest by first converting them to percentages.

Go to nrich.maths.org and search for the 'Chocolate' activity. Work with a partner on this challenge. Try to record your working in a systematic way so you can keep track of your ideas and compare the different methods that you try.

## Activity 3 – Fraction card game

This is a game for two to four players. Use cards Ace–9 from a standard deck of playing cards, with the Aces counting as ones. Six cards are dealt face up and placed so that all players can see them at the same time. The object is for each person to use two of the cards to make a fraction that is as close as possible to one half.

Your teacher will set the time limit for a round. The person whose fraction is closest to one half (and who can explain why) wins that round, and then the next round begins. If two or more students identify the same fraction, or ones that are equally close to one half, then it is a tie.

You can also play a variation where every player is also dealt their own card, which they keep to themselves until they reveal their fraction.

# Operations with fractions

As with whole numbers, fractions can be added, subtracted, multiplied and divided. In this section you will discover the process for performing each of these.

## Multiplying fractions

In the following investigation, you will determine an algorithm for multiplying fractions.

**algorithm:** a process or set of rules to be followed

## Investigation 1 – Multiplying fractions

You will use a visual method of multiplying fractions to help you determine the algorithm. You will start with the example given and then create five more multiplication questions of your own.

Example: Multiply $\frac{1}{4}$ by $\frac{3}{5}$

1 Draw a rectangle on a plain piece of paper.

2 Draw lines to divide the rectangle into 4 equal columns. Make sure they are of equal width.

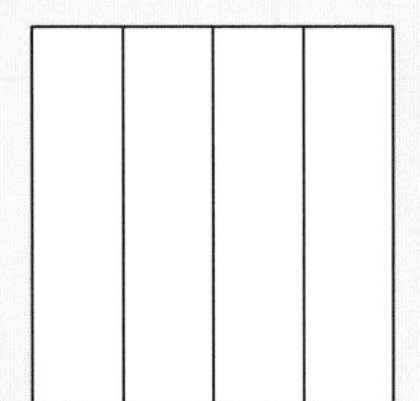

3 Shade one column using a highlighter. This represents the fraction $\frac{1}{4}$.

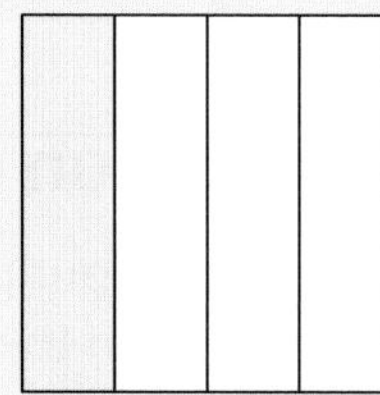

4 Now divide the rectangle into 5 equal rows.

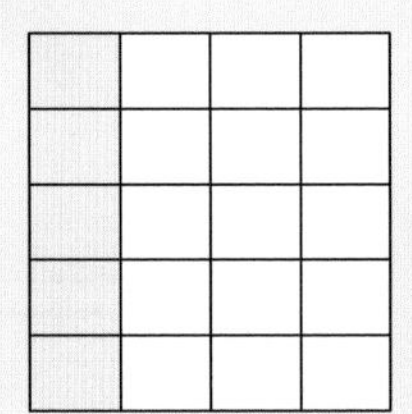

5 Shade three rows using a highlighter of a different color. The cells shaded with this second color represent the fraction $\frac{3}{5}$.

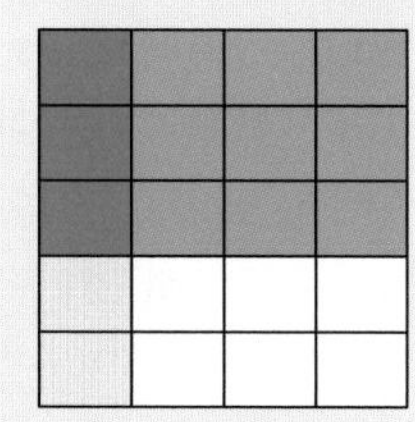

6 State the fraction of the whole rectangle that is highlighted with both colors. In this case, it is $\frac{3}{20}$.

7 Create a table like this one. Add five more rows for steps **8** to **10**.

| Fraction | Fraction | Number of cells shaded with both colors | Total number of cells | Product of the two fractions |
|---|---|---|---|---|
| $\frac{1}{4}$ | $\frac{3}{5}$ | 3 | 20 | $\frac{3}{20}$ |

8 Create five more examples by drawing five new rectangles and choosing your own proper simplified fractions to multiply.

9 Record the information in your table.

10 What patterns do you see? Be sure to look at the numerator and denominator in each original fraction and their relationship to the numerator and denominator in the product.

ATL1

11 Write down the algorithm for multiplying fractions.

12 Verify your algorithm using two more examples.

▶ Continued on next page

13 Go to pbslearningmedia.org and search for 'Area Models for Multiplication of Fractions'. Use the area model there to verify your algorithm for three more cases, changing the fractions to whatever you want.

ATL2

Pairs

14 With a partner, look at each other's algorithms. Take turns giving feedback by first asking questions to clarify anything in the work that isn't quite clear. Then, give at least one positive comment. Follow that with a suggestion for improvement, if appropriate.

## Reflect and discuss 3

- Why does this algorithm work? Explain.
- What is the benefit of finding an algorithm? Explain.
- Is it easier to understand how this algorithm works by looking at a visual form or a numeric form? Explain your choice.

In the method that you discovered in Investigation 1, you multiplied the numerators together, multiplied the denominators together, and then reduced or simplified the answer.

There is another method, in which you do the reducing/simplifying at the beginning and then multiply the numerators together and the denominators together. Both methods are shown in Example 3.

## Example 3

Multiply $\frac{2}{5} \times \frac{10}{11}$ by: **a** multiplying first **b** simplifying first.

**A** **a** $\frac{2}{5} \times \frac{10}{11} = \frac{20}{55}$

Multiply first. Remember, multiply across.

$= \frac{20 \div 5}{55 \div 5}$

Simplify last. Both numerator and denominator can be divided by 5.

$= \frac{4}{11}$

**b** $\frac{2}{5} \times \frac{10}{11} = \frac{2}{{}_1\not{5}} \times \frac{{}^2\not{10}}{11}$

Simplify first. One of the numerators and one of the denominators can be divided by the same factor, 5.

$= \frac{2}{1} \times \frac{2}{11} = \frac{4}{11}$

## Reflect and discuss 4

- In Example 3, what number was used to simplify the fractions in each method? What does this say about simplification when multiplying fractions?
- How do you know when you can simplify first?
- What is the advantage of simplifying first?
- What would you need to do with mixed numbers before multiplying them? Explain.

## Practice 2

**1** What multiplication is being represented in each diagram? Write down the question and the simplified answer.

**a** 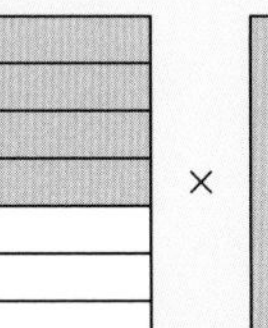 × 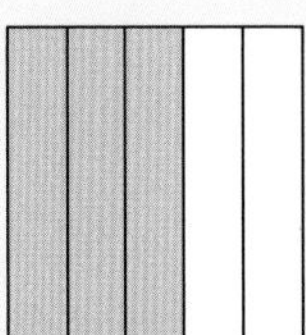 = 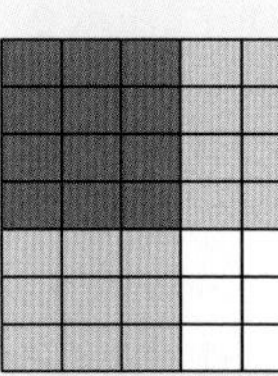

**b** 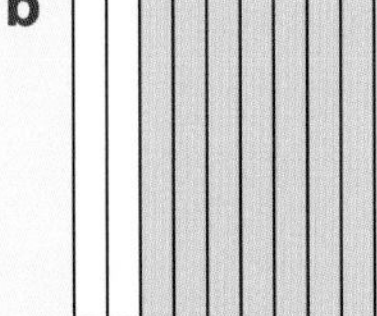 × = 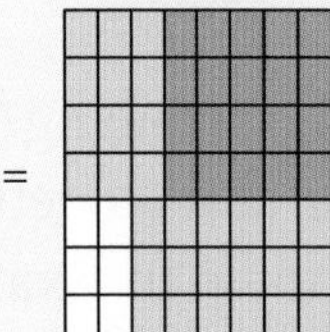

**c** 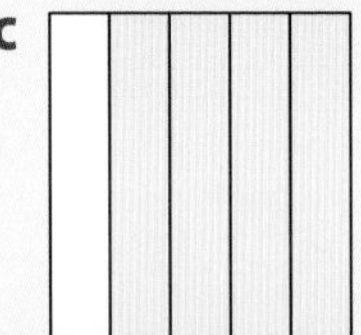 × = 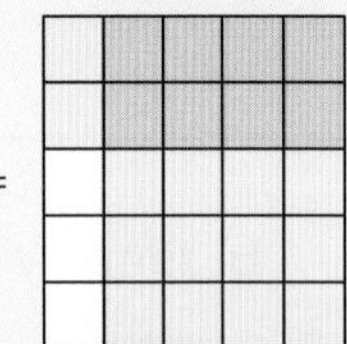

**d** 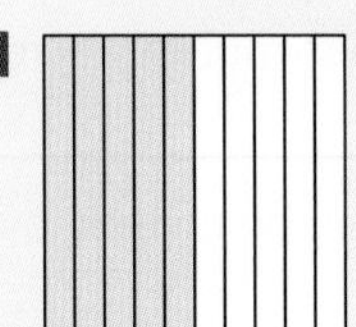 × = 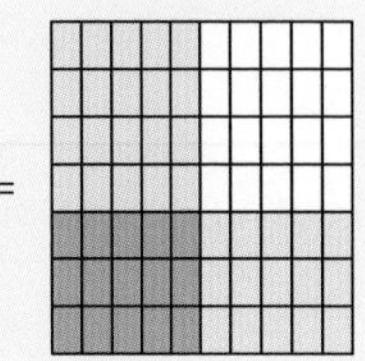

**2** Multiply these fractions first, then simplify your answers.

**a** $\frac{1}{4} \times \frac{7}{9}$ **b** $\frac{4}{5} \times \frac{15}{16}$ **c** $\frac{7}{11} \times \frac{6}{42}$

**d** $2\frac{4}{5} \times \frac{5}{8}$ **e** $3\frac{1}{3} \times \frac{6}{8}$ **f** $5\frac{1}{3} \times \frac{1}{4}$

▶ Continued on next page

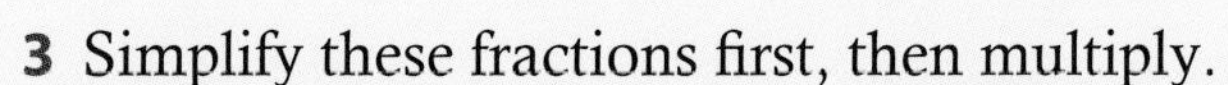

**3** Simplify these fractions first, then multiply.

**a** $\frac{5}{9} \times \frac{3}{10}$ **b** $\frac{3}{4} \times \frac{10}{27}$ **c** $\frac{6}{7} \times \frac{5}{12}$

**d** $\frac{2}{15} \times \frac{6}{14}$ **e** $\frac{4}{11} \times \frac{3}{16}$ **f** $\frac{8}{9} \times \frac{15}{16}$

**4** Multiply these fractions using the method of your choice.

**a** $\frac{2}{10} \times \frac{5}{6}$ **b** $\frac{4}{9} \times \frac{27}{32}$ **c** $\frac{1}{3} \times \frac{6}{7}$

**d** $\frac{5}{8} \times \frac{4}{15}$ **e** $\frac{4}{12} \times \frac{3}{5}$ **f** $\frac{2}{3} \times \frac{9}{14}$

**5** According to a survey at the time of writing, approximately $\frac{21}{25}$ of American teenagers own a cell phone.

**a** Of these teenagers, $\frac{1}{3}$ of them use messaging apps.
What fraction of American teenagers does this represent?

**b** Of the teenage cell phone owners, $\frac{1}{10}$ of them hardly ever use their phone to make or receive calls.
What fraction of American teens does this represent?

**c** Of the teenage cell phone owners, $\frac{5}{7}$ of them say they use one or more social network sites.
What fraction of American teens does this represent?

**d** How has the use of social network sites and messaging apps affected connections among American teens?
Explain your answer.

## Did you know...?

In the United States and in the Philippines, they are called *cell phones*. The name comes from the cell-like structure of the networks in which they operate. In Italy they are called *telefoninos*, which means 'little phones'. In the United Kingdom, India, New Zealand, and other countries they are referred to as *mobiles*. In Germany they say *das Handy*, and in Indonesia you will hear people say 'ha-pay' which is how to pronounce *hp*, the abbreviation for *hand phone*.

## Dividing fractions

Dividing fractions can be explained using several different models.

### Investigation 2 – Dividing whole numbers by fractions

WEB LINK

1 Go to the Grade 6 section of Mathgames.com and click on 'Divide by Fractions with Models' in the Division section.

2 Use the visual models to help you solve each question. Record your answers to all the questions in a table like the one below.

For example, if the question is $2 \div \frac{1}{2} = 4$ the first row shows how to enter the numbers.

| Dividend | Divisor | Quotient |
|---|---|---|
| 2 | $\frac{1}{2}$ | 4 |
| | | |
| | | |
| | | |
| | | |

**Light Blue - Dark Blue**

Stage: 2 ★★

Investigate the successive areas of light blue.

What fraction of the total area of the square does the area of light blue take up in each case?

Can you work out what the next two diagrams would look like?

3 What patterns do you see? Be sure to look at individual numbers within the fractions and their relationship to the numbers in the solution.

4 In your own words, explain this relationship.

ATL1 5 Determine the algorithm for dividing whole numbers by fractions.

6 Verify your algorithm using two more examples.

ATL2 7 With a partner, look at each other's algorithm. Take turns giving feedback by first asking questions to clarify anything that isn't quite clear. Then, give at least one positive comment. Follow that with a suggestion for improvement, if appropriate.

Also try your hand at the Light Blue – Dark Blue investigation at nrich.maths.org/2105. It is an investigation involving fractions and patterns.

The process in Investigation 2 seems to work when dividing whole numbers by fractions. Will this algorithm work when dividing one fraction by another?

When you first learned to divide, you often asked yourself the question, 'How many times does some number go into some other number?'

For example, 21 ÷ 7 meant: 'How many times does 7 go into 21?' The answer is of course 3.

In the investigation, you were determining how many times a fraction, say $\frac{1}{3}$, goes into 4. In Investigation 3 you will follow this same process where both numbers are fractions, using fraction bars to help you.

## Investigation 3 – Dividing fractions by fractions

ATL1

Use fraction bars to find the answer to each question. Make a table like the one below and fill in the rows using these fraction divisions:

$\frac{1}{4} \div \frac{1}{8}$ $\frac{2}{3} \div \frac{1}{6}$ $\frac{1}{2} \div \frac{1}{4}$ $\frac{5}{6} \div \frac{1}{12}$ $\frac{1}{2} \div \frac{1}{6}$ $\frac{5}{12} \div \frac{1}{6}$ $\frac{4}{5} \div \frac{3}{10}$

| Division | In words | Pictorial representation (Fraction bar) | Answer |
|---|---|---|---|
| $\frac{1}{2} \div \frac{1}{6}$ | How many times does $\frac{1}{6}$ go into $\frac{1}{2}$? | $\frac{1}{2}$ / $\frac{1}{6}$ $\frac{1}{6}$ $\frac{1}{6}$ | $\frac{1}{2} \div \frac{1}{6} = 3$ |
| | | | |
| | | | |
| | | | |
| | | | |
| | | | |

1 Does the algorithm you found in Investigation 2 work here? Explain using two more examples.

2 When does the use of fraction bars for dividing fractions become difficult? Explain.

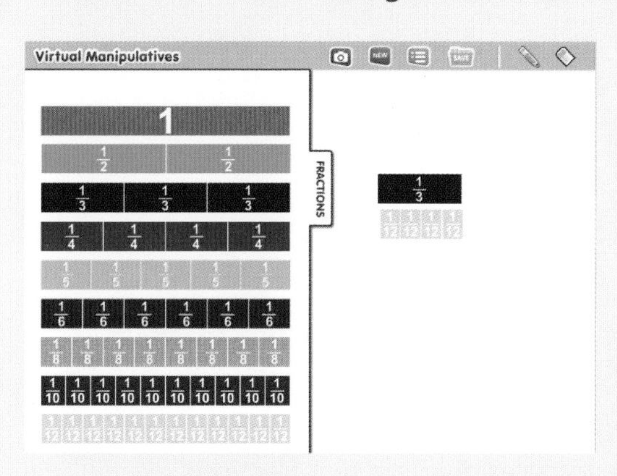


You can experiment with the Virtual Manipulatives at: abcya.com/fraction_percent_decimal_tiles.htm

When asked to divide two fractions, invert the second fraction (the dividend) and multiply the first fraction (the divisor) by it. This is called *multiplying by the reciprocal.*

## Reflect and discuss 5

- Explain why it is important to have an algorithm to divide fractions and not just rely on fraction bars or diagrams.
- What would you do if you were asked to perform a division that involved one or more mixed numbers?
- Does the algorithm you've discovered work for divisions like $12 \div 3$? Explain.

Pairs

## Activity 4 – Another way of looking at it

In pairs, tackle each of the following statements and questions. Be sure to explain your responses clearly before moving on to the next question.

1 'The easiest number to divide by is 1.' Explain.

2 What fraction can you multiply $\frac{2}{3}$ by to obtain 1? How about $\frac{3}{4}$? What about $\frac{5}{7}$? Research what the fraction that has this property is called in mathematics.

3 Instead of seeing two fractions being divided as $\frac{2}{3} \div \frac{5}{6}$, think of it more like this: $\frac{\frac{2}{3}}{\frac{5}{6}}$.

What number would you need to multiply the denominator by to make the denominator equal to 1?

4 Remember, when you multiply the denominator by a number, you must also multiply the numerator by the same number:

$$\frac{\frac{2}{3} \times \frac{6}{5}}{\frac{5}{6} \times \frac{6}{5}}$$

What does that give you? How does that relate to the process you discovered in Investigations 2 and 3?

5 Explain in your own words why multiplying by the reciprocal of the divisor accomplishes the same thing as dividing two fractions.

## Example 4

 Divide and simplify if possible: $\frac{2}{5} \div \frac{12}{7}$

 $\frac{2}{5} \div \frac{12}{7} = \frac{2}{5} \times \frac{7}{12}$ Invert the dividend (the second fraction) and multiply.

$= \frac{1}{5} \times \frac{7}{6}$ Simplify whenever possible, unless instructed otherwise.

$= \frac{7}{30}$

Once you have inverted the fraction and it is set up like a multiplication question, you may find it easier to simplify first before multiplying.

## Practice 3

**1** Find the reciprocal of each fraction.

**a** $\frac{2}{9}$ **b** $\frac{3}{5}$ **c** $\frac{10}{21}$

**d** $\frac{1}{8}$ **e** 4 **f** $\frac{4}{11}$

**2** Solve these fraction division problems. Be sure your answers are simplified.

**a** $\frac{3}{11} \div \frac{9}{22}$ **b** $\frac{24}{5} \div \frac{9}{20}$ **c** $\frac{1}{4} \div \frac{7}{16}$

**d** $\frac{7}{12} \div 14$ **e** $2\frac{4}{5} \div \frac{7}{10}$ **f** $\frac{3}{8} \div 2\frac{1}{12}$

**3** Explain in words the difference between finding $5 \times \frac{1}{2}$ and finding $5 \div \frac{1}{2}$. Calculate both answers and compare.

**4** Athletic events have always drawn humans together, whether set on the world stage, like the Olympics, or locally, like a little league hockey match. Cheering for the same team, country or individual can promote a sense of community and belonging that increases human connections.

In 2016, at one of the early rounds in Olympic soccer, only $\frac{3}{5}$ of the seats were occupied. Of that number, $\frac{1}{4}$ were cheering for the visiting team.

**a** What fraction of the seats were occupied by people cheering for the visiting team?

**b** What fraction of the seats were occupied by people cheering for the home team?

**c** Tickets for half of the unoccupied seats were given away at halftime. What fraction of the total number of seats does this represent?

▶ Continued on next page

**5** Students are voting for their school president, which has produced a lot of interactions among students and candidates. In order to win, a candidate must receive at least two-thirds of the votes in each of grades 4 to 6. If there are 30 students in grade 4, 24 students in grade 5, and 21 students in grade 6, what is the minimum number of votes the winning candidate needs from each grade?

**6** Scandinavian Midsummer is a celebration of the longest day of the year. In Finland, families go to their favorite lake and build a bonfire on the shore. They light the bonfire at midnight and then dance beside it. In Sweden, people often consume a 'Midsommer' cake.

**a** Suppose $\frac{3}{5}$ cup of flour is used in each 'Midsommer' cake and the bakery used $8\frac{4}{10}$ cups of flour yesterday. How many cakes did the bakery make?

**b** If $\frac{2}{3}$ cup of sugar is used in each cake, how many cakes can be made with $5\frac{1}{3}$ cups of sugar?

**c** A baker had $\frac{7}{8}$ of a tin of corn starch left and used $\frac{1}{4}$ of it. What fraction of the tin was used?

**7** A survey shows that an estimated $\frac{2}{3}$ of all teenagers own a smartphone. Of those teens, $\frac{3}{4}$ said they have had a dispute with a friend via text.

**a** What fraction of all teens have had a dispute with a friend via text?

**b** What are some issues with having a dispute via text that are not issues when having the same discussion face-to-face?

## Formative assessment – Peppermint bark

criteria C, D

Peppermint bark is a minty chocolate treat that has become very popular in some countries, where it is made and served at Christmas time. When families get together, or when you visit another family around this holiday, peppermint bark is often given as a gift. It consists of peppermint candy pieces with layers of white and milk or dark chocolate. Sometimes peppermint essence and cream are added for texture and flavor.

Suppose you want to make peppermint bark to give to your family and friends.

1 Decide on the number of people you will be making the bark for (must be more than 10 people).

2 You will have to determine how to cut the bark so that you will know how many people each batch will provide for.

One batch of the peppermint bark measures 24 cm by 30 cm.

- You want all the pieces to be equally-sized triangles for the best possible presentation.
- The triangles can be whatever size you wish, but each gift portion should be at least 50 cm$^2$.
- Draw a rectangle layout of how you will cut the bark.
- Calculate the dimensions and area of a single triangular piece.
- Determine the number of pieces in each gift and calculate the total area of bark in each gift.
- Determine how many batches you will need to make.

3 Once you have determined the number of batches you will need for all your friends and family, calculate the ingredient totals you will need, using a copy of the table below.

| Ingredient | Imperial measurement for one batch | Number of batches | Ingredients needed for all of your batches |
|---|---|---|---|
| White chocolate | $1\frac{3}{4}$ cup | | |
| Hard peppermint candies or candy canes | $\frac{1}{2}$ cup | | |
| Dark chocolate | $\frac{7}{8}$ cup | | |
| Peppermint extract | $\frac{3}{4}$ teaspoon | | |

▶ Continued on next page

4 You typically buy chocolate in ounces (oz.) or grams, instead of cups. Research how to convert between these units and convert the total number of cups you need into ounces or grams.

5 A half cup of peppermint candies is approximately 6 candy canes or 30 hard peppermint candies. How many of each would you have to buy?

ATL2

6 With a partner, look at each other's work for this formative task. Take turns giving feedback by first asking questions to clarify anything that isn't quite clear. Then, give at least one positive comment. Follow that with a suggestion for improvement, if appropriate.

## Adding and subtracting fractions

### Activity 5 – Tea time

Drinking tea has become customary in many parts of the world. In Turkey, offering and drinking tea with someone is regarded as a sign of friendship, and tea is often consumed during business transactions. Tea gardens in Turkey are places for social gatherings, full of music, games, and lively conversation.

In Japan, the tea ceremony ('Cha-no-yu'), has been practiced for centuries as a way to nourish the body, mind and soul with invited guests. Japanese tea gardens are peaceful, calm places where these ceremonies are performed.

In England, where 'tea time' or 'afternoon tea' is a regular event, the goal is for people to take a break, eat a small snack and spend time with friends and family. With each of these traditions, the tea is served in very different types of cups, as seen here.

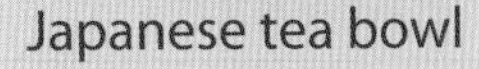

Japanese tea bowl

English tea cup

Turkish tea glass

1 What do you notice about the shapes of these drinking vessels?

2 Which one do you think holds the most tea? Explain your reasoning.

3 Edwin invites his friends and family over for his own version of a tea ceremony where he has blended elements from a variety of ceremonies. Edwin brews several different kinds of tea and uses 3 Japanese tea bowls, 6 English tea cups and 6 Turkish tea glasses. Is it fair to say that, during the ceremony, he and his friends and family drink 15 cups of tea? Explain.

▶ Continued on next page

In reality, two Turkish tea glasses fill one English tea cup; three of them fill a Japanese tea bowl.

4 How many English tea cups will fill a Japanese bowl? Show your work.

5 Show that the liquid from one Japanese tea bowl would fill exactly one English tea cup and one Turkish glass.

6 If the tea that Edwin and his friends drank was measured in Turkish glasses, how many did they drink? Show your working.

7 If the tea that Edwin and his friends drank was measured in English cups, how many did they drink? Show your working.

8 If the tea that Edwin and his friends drank was measured in Japanese bowls, how many did they drink? Show your working.

For the people in Western Sahara, offering tea to someone is considered a sign of generosity, something not to be refused. The guest will often drink three cups in one sitting, each one having its own taste and symbolism.

Similar to the drinking vessels that were all different sizes, fractions can be added and subtracted only if they are out of the same whole. When fractions have *common denominators*, their numerators can be added together to find out what part of the whole the total represents.

## Investigation 4 – Adding fractions

Adding fractions can also be demonstrated with an area model.

1 Copy this table and add four more rows to it.

| Addition | Area model of addition | Area model of answer | Summary |
|---|---|---|---|
| $\frac{1}{5}+\frac{2}{5}$ | + | | $\frac{1}{5}+\frac{2}{5}=\frac{3}{5}$ |

In the four additional rows, enter these fraction additions.

$\frac{2}{7}+\frac{4}{7}$ $\quad$ $\frac{2}{10}+\frac{3}{10}$ $\quad$ $\frac{1}{8}+\frac{5}{8}$ $\quad$ $\frac{5}{6}+\frac{1}{6}$

Fill in the rest of the table, and simplify your answers where appropriate.

2 What makes it possible to add the fractions in the table? Explain.

▶ Continued on next page

**3** When the denominators are not the same, the area models can be adjusted so that they are.

Copy this table and add four more rows to it.

| Addition | Area model of addition | Adjusted area model | Summary |
|---|---|---|---|
| $\frac{1}{3}+\frac{3}{4}$ | | | $\frac{1}{3}+\frac{3}{4}=\frac{4}{12}+\frac{9}{12}$ $=\frac{13}{12}$ *or* $1\frac{1}{12}$ |

In the four additional rows, enter these fraction additions.

$\frac{2}{3}+\frac{1}{5}$ $\frac{1}{2}+\frac{1}{6}$ $\frac{5}{6}+\frac{1}{3}$ $\frac{3}{5}+\frac{3}{4}$

Fill in the rest of the table, and simplify your answers where appropriate.

**4** What enabled you to add the fractions in the table?

ATL1 **5** Write down the algorithm for adding any two fractions.

**6** Verify your algorithm for two more examples.

## Reflect and discuss 6

- What do you think the algorithm will be for subtracting fractions? Explain.
- How do you think you will add mixed numbers, such as $1\frac{1}{2}+3\frac{3}{4}$?

Go to visualfractions.com and select 'Unlike Fractions With Circles 'from the ADD menu. Use your algorithm to complete the additions. Then select the 'Strict With Circles' option from the ADD menu to try adding some mixed numbers. You can then explore the 'Easy' and 'Strict' options from the SUBTRACT menu.

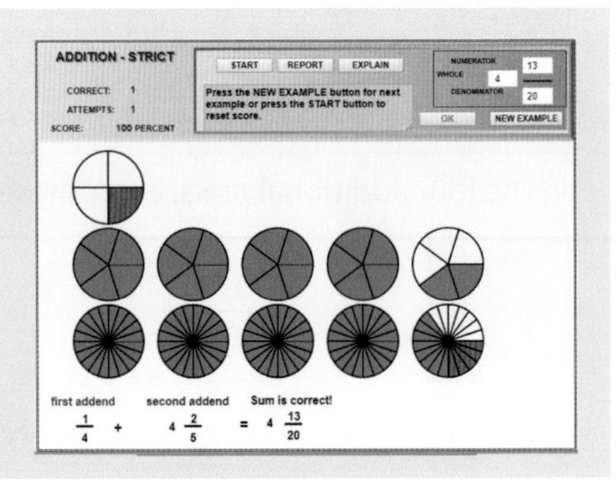


When adding or subtracting fractions, first rewrite them so that they have the same denominator. These fractions are equivalent to the original ones and they are said to now have a *common denominator*.

## Example 5

**Q** **a** Add: $\frac{5}{6}+\frac{2}{5}$

**b** Subtract: $\frac{2}{3}-\frac{1}{4}$

**A** **a** $\frac{5}{6}+\frac{2}{5}=\frac{25}{30}+\frac{12}{30}$

$=\frac{25+12}{30}=\frac{37}{30}$

$=1\frac{7}{30}$

Rewrite using a common denominator.

Your teacher will tell you if you should express your answer as a mixed number or if you can leave it as an improper faction in lowest terms.

**b** $\frac{2}{3}-\frac{1}{4}=\frac{8}{12}-\frac{3}{12}$

$=\frac{8-3}{12}=\frac{5}{12}$

Rewrite using a common denominator.

## Activity 6 – Revisiting the tangram

1 What fraction of the whole does each piece of the tangram represent?

2 Using the tangram you have kept from earlier in this unit, create four pictures that use the following fractions of the whole area.

**a** $\frac{5}{8}$ **b** $\frac{7}{8}$ **c** $\frac{9}{16}$ **d** $\frac{15}{16}$

3 Draw each picture, showing the breakdown of how the pieces fit together. Write down the fraction addition of the pieces.

4 Now create your own picture using some of the pieces. Draw the picture, showing the breakdown of how the pieces fit together. Write down the fraction addition of the pieces.

## Reflect and discuss 7

- Explain why you cannot simply add numerators and denominators when adding fractions. For example, explain why $\frac{1}{2}+\frac{1}{3}$ cannot be equal to $\frac{2}{5}$.
- Explain how the algorithms for adding/subtracting fractions are different than the one for multiplying fractions.

## Activity 7 – Fraction card game

In groups of three, you will have two players and a judge. From a standard deck of playing cards, discard the picture cards, leaving the cards one (Ace) to ten. The judge deals four cards, face up, for both players to see. The players have 30 seconds to come up with an addition of two fractions using the four numbers. The player whose sum is closest to 1 is the winner. Once a player has submitted their fraction addition, the second player cannot submit the same fractions, but will have another 30 seconds to try to come up with a better combination. The judge will verify the winner. Play the best of three rounds and then rotate jobs so that each student plays two games and is judge once.

ATL2

When you are the judge, look at one player's answer and give constructive feedback by first asking questions to clarify anything that isn't quite clear. Then, give at least one positive comment followed by a suggestion for improvement, if appropriate. Repeat for the other player.

The judge can have a calculator on hand to verify answers.

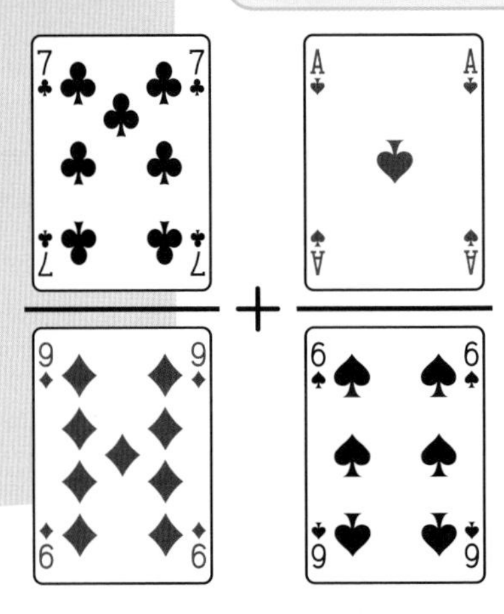

## Practice 4

**1** With each operation, simplify your answer and write as a mixed number if appropriate.

**a** $\frac{5}{7} - \frac{2}{7}$ **b** $\frac{3}{7} + \frac{2}{7}$ **c** $\frac{11}{12} - \frac{7}{12}$ **d** $\frac{9}{14} + \frac{11}{14}$

**e** $\frac{21}{25} - \frac{6}{25}$ **f** $\frac{5}{11} - \frac{2}{11}$ **g** $\frac{3}{4} + \frac{5}{12}$ **h** $\frac{4}{5} - \frac{3}{8}$

**i** $\frac{4}{5} + \frac{1}{20}$ **j** $\frac{7}{10} + \frac{11}{30}$ **k** $\frac{7}{9} - \frac{1}{2}$ **l** $\frac{3}{8} + \frac{2}{3}$

**m** $\frac{3}{8} + \frac{2}{7}$ **n** $4 - \frac{2}{13}$ **o** $2\frac{2}{3} + 1\frac{3}{4}$ **p** $3\frac{1}{2} - 2\frac{2}{5}$

**2** Newspaper companies get some revenue from the selling price of the newspaper, but they make much more money from advertising. As of a Thursday, sales reps at a major newspaper had sold the following advertising space for the Sunday paper: 3 full pages, 3 half-pages, 5 quarter-pages, and 7 eighth-pages. If 7 full pages have been reserved for advertisements in the Sunday paper, how much space is still available to be sold?

▶ Continued on next page

**3** Omar eats $\frac{1}{5}$ of a pizza. His friends Tatiana, Maya and Sophie are going to share what is left. If they share it evenly, what fraction of the whole pizza will each girl get?

**4** In a study of university students and their use of technology, students were observed as they walked to the cafeteria.

The observations showed:

- $\frac{7}{16}$ of the students were alone, but either texting or using their phone (but not talking)
- $\frac{1}{8}$ were alone and involved in a conversation on their phone
- $\frac{3}{32}$ were with other people and using no technology whatsoever
- $\frac{3}{16}$ were alone and using no technology.

The rest of the students were all observed using technology with other people present.

**a** Determine the fraction of the observed students who were using technology with other people present.

**b** Determine the fraction of those observed who were using no technology whatsoever.

**c** Determine the fraction of those observed who were using some form of technology while interacting with others (either in person or on the phone).

**d** Do you think these results demonstrate that technology helps face-to-face communication? Explain.

**5** Following a snowstorm, the three Abendroth brothers decide to do a good deed and shovel all the driveways on their street. Paul shovels $\frac{2}{7}$ of the driveways, Max shovels $\frac{1}{5}$ of the driveways and Arthur shovels the rest. What fraction of the driveways does Arthur shovel?

**6** If a third of a cup of water fills one quarter of a container, what fraction of a container will one cup fill? Draw a diagram to represent this problem. Can you determine two different ways to solve this problem using two different operations?

▶ Continued on next page

**7** Look again at the ancestry report on page 154.

**a** What is the sum of all of the amounts? Explain why this makes sense.

**b** What fraction of the total amount comes from African countries?

**c** What fraction of the total amount comes from European countries?

**8** Three-quarters of teenagers use Facebook to communicate with their friends. One half of all teens use Snapchat, while two-fifths use Instagram.

**a** Explain why the sum of these fractions is more than 1.

**b** How has technology helped teens create and maintain relationships? How did people create and maintain relationships *before* cell phones and the internet?

**9** Diwali (or Dipawali) is the Hindu festival of lights (though it is celebrated by people of many different faiths), where families decorate their house with lights and prepare a feast to share with their loved ones. It's a four-day celebration that involves cleaning the house, buying new clothes, giving gifts to family members, and other family activities.

When celebrating the Diwali festival, Devika used $\frac{1}{3}$ of the clay lamps to make a path to the front door, and she made a star-like figure with $\frac{2}{5}$ of them. What fraction were left to decorate a table at the entrance to her house?

## Multiple operations

To perform operations with several fractions you can use the same algorithms that you developed for just two fractions:

- addition and subtraction require a common denominator
- multiplication involves multiplying across (numerators with numerators, denominators with denominators)
- division requires multiplying by the reciprocal of the divisor.

## Activity 8 – Music and math

Music has long been an activity that brings humans together. Celebrations of love, victory, culture, and beliefs all seem to involve music. There is even research indicating that music increases the production of a chemical in the brain associated with increased bonding and trust among people! Human connections seem to be enhanced by the very presence of music.

**1** Music consists of many patterns and is based on set mathematical principles. Notes in music are divided into fractions and organized into a *measure* that consists of 4 beats. If a whole note lasts the duration of a measure, which is 4 beats, determine the number of beats in each of the following notes. Record your answers in a copy of the table.

| Note | Fraction | Number of beats |
|---|---|---|
| 𝅝 | Whole | 4 |
| 𝅗𝅥 | Half | |
| ♩ | Quarter | |
| ♪ | Eighth | |
| 𝅘𝅥𝅯 | Sixteenth | |

**2** If we place a dot after a musical note (e.g. ♩.) this lengthens the note's duration by one half of its usual length. Based on this idea, copy the table below and enter the symbol and the number of beats for each note duration.

| Note | Fraction | Number of beats |
|---|---|---|
| | Three-quarters | |
| | Three-eighths | |
| | Three-sixteenths | |

**3** Create a song or rhythm 6 measures long that consists of different combinations of the notes above. Use each note at least once and try to use more than two notes in each measure.

▶ Continued on next page

4 Write down the score of music, clearly using fractions to show how each measure adds up to 1 whole.

5 Try playing the rhythm of your song by drumming on the table or clapping. Your teacher may ask for demonstrations if any of you would like to share your music with the class!

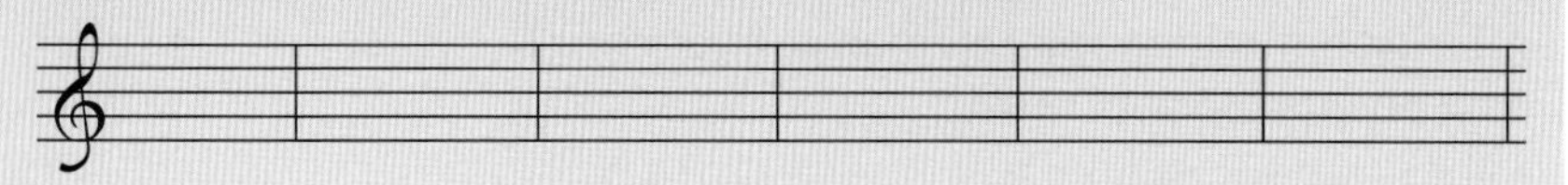

Go to philtulga.com/pie.html and try the Playing Fraction Pies activity, which demonstrates the connection between fractions and musical notes. Choose combinations of fractions that add up to 1 and see them converted to music. You can then click to hear your rhythms of

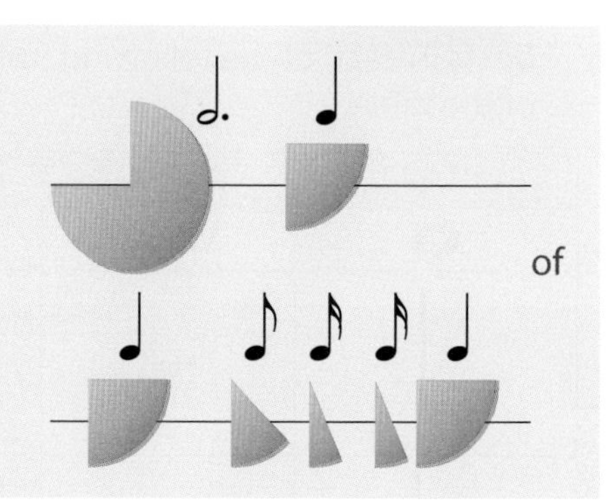

## Example 6

**Q** Perform the following operations.

**a** $\frac{2}{3} + \frac{1}{4} + \frac{1}{8}$

**b** $\frac{6}{7} \div \frac{3}{8} \div \frac{24}{35}$

**A** **a** $\frac{2}{3} + \frac{1}{4} + \frac{1}{8}$

$= \frac{16}{24} + \frac{6}{24} + \frac{3}{24}$

$= \frac{25}{24}$ or $1\frac{1}{24}$

The LCM of 3, 4 and 8 is 24. Write equivalent fractions using this LCM this as the common denominator.

**b** $\frac{6}{7} \div \frac{3}{8} \div \frac{24}{35}$

$= \frac{6}{7} \times \frac{8}{3} \times \frac{35}{24}$

$= \frac{\cancel{6}^{2}}{{}_{1}\cancel{7}} \times \frac{\cancel{8}^{1}}{{}_{1}\cancel{3}} \times \frac{\cancel{35}^{5}}{{}_{3}\cancel{24}}$

$= \frac{2}{1} \times \frac{1}{1} \times \frac{5}{3}$

$= \frac{10}{3} = 3\frac{1}{3}$

Remember to multiply by the reciprocal of each divisor.

Simplify first. It's easier!

## Activity 9 – Domino challenge

In groups of three, you will have two players and a judge. Start with a set of dominoes on the table, facing down. The judge flips over three dominoes and tells the players which number is the numerator and which is the denominator on each domino, and what operation (addition, subtraction, multiplication or division) the players will be performing on the three fractions. Before starting the round, the judge will determine the correct answer either by hand or on a calculator. The first player with the correct answer wins the round.

Play the best of five rounds (using each operation at least once) and then you will rotate jobs so that each student plays two games and is judge once.

ATL2

When you are the judge, look at one player's answer and give constructive feedback by first asking questions to clarify anything that isn't quite clear. Then, give at least one positive comment followed by a suggestion for improvement, if appropriate. Repeat for the other player.

When different operations are involved, remember to follow the agreed order of operations (BIDMAS):

- Brackets
- Indices or Exponents
- Division/Multiplication
- Addition/Subtraction.

For both Division/Multiplication, and Addition/Subtraction, you do whichever operation comes first, when reading from left to right.

## Example 7

**Q** Calculate: $\frac{3}{4} + \frac{2}{3} \div \frac{5}{6}$

**A**

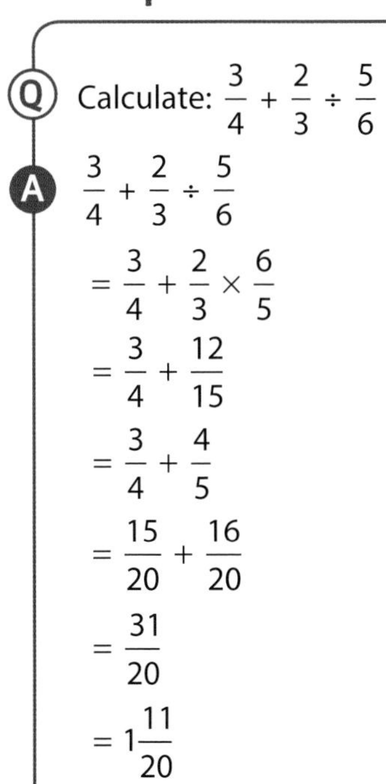

$$\frac{3}{4} + \frac{2}{3} \div \frac{5}{6}$$

Do the division first. Multiply $\frac{2}{3}$ by the reciprocal of $\frac{5}{6}$.

$$= \frac{3}{4} + \frac{2}{3} \times \frac{6}{5}$$

$$= \frac{3}{4} + \frac{12}{15}$$

Simplify the fraction.

$$= \frac{3}{4} + \frac{4}{5}$$

Add, using a common denominator.

$$= \frac{15}{20} + \frac{16}{20}$$

$$= \frac{31}{20}$$

Change to a mixed number.

$$= 1\frac{11}{20}$$

## Practice 5

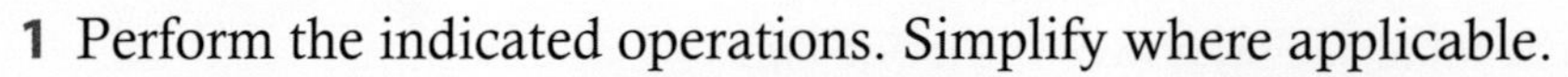

**1** Perform the indicated operations. Simplify where applicable.

**a** $\frac{3}{10} \times \frac{2}{9} \times \frac{5}{7}$ **b** $\frac{5}{9} \div \frac{8}{27} \div \frac{10}{12}$ **c** $\frac{2}{5} \times \frac{15}{28} \times \frac{21}{10}$

**d** $\frac{1}{2} + \frac{5}{6} + \frac{7}{12}$ **e** $\frac{4}{5} - \frac{2}{15} - \frac{1}{6}$ **f** $\frac{5}{9} \div \frac{8}{27} \div \frac{10}{12}$

**g** $\frac{6}{7} + \frac{4}{9} \times \frac{3}{16}$ **h** $2\frac{1}{5} - 1\frac{3}{10} - \frac{2}{15}$ **i** $\frac{3}{4} + 9\frac{4}{5} + \frac{14}{25}$

**j** $\frac{6}{11} \div \frac{9}{22} \times \frac{3}{4}$ **k** $\frac{3}{5} + \frac{8}{15} - \frac{1}{6}$ **l** $\frac{4}{9} \div \frac{8}{21} - \frac{2}{3}$

> In some parts of the world, the memory aid for the order of operations is called BEDMAS, with the E standing for 'exponents' instead of 'indices'. In other places, it is called PEDMAS (P for 'parentheses') or even GEDMAS (G for 'grouping symbols'.)

**2** The following are measures from a typical song sung during Diwali. Use fractions (refer back to Activity 8) to show that the following measure sums to one whole note (four beats).

**a**

In each of the following measures is a new symbol which represents a rest. A rest is a pause in the sound and for every musical note symbol there is a matching rest symbol of the same duration. If each measure sums to one whole note (four beats), find the duration of each of the rest symbols.

**b**

**c**

**3** In Bruck's family tree, two-thirds of his family are from Europe. Three-fourths of those are from England. Of the people from England, two-thirds were born in Leeds. What fraction of Bruck's family tree represents family members born in Leeds?

**4** Which of the following are equivalent to $\frac{2}{3} \div \frac{4}{5} \times \frac{1}{5}$? Explain your reasoning.

| $\frac{2}{3} \times \frac{5}{4} \times \frac{5}{1}$ | $\frac{2}{3} \times \frac{4}{5} \times \frac{1}{5}$ | $\frac{2}{3} \div \frac{4}{5} \div \frac{5}{1}$ | $\frac{2}{3} \div \frac{4}{5} \div \frac{1}{5}$ | $\frac{2}{3} \times \frac{5}{4} \times \frac{1}{5}$ |
|---|---|---|---|---|

## Unit summary

- A fraction is a part of a whole.

$$\frac{4}{7}$$

The numerator is the top number of a fraction. It represents the number of parts of the whole that are included.

The denominator is the bottom number in a fraction. It shows how many equal parts the whole is divided into.

- A mixed fraction is made up of a whole number and a fraction, such as $3\frac{1}{4}$ or $7\frac{2}{5}$.
- In an improper fraction, the numerator is bigger than the denominator, such as $\frac{13}{4}$ or $\frac{37}{5}$.
- In a proper fraction, the numerator is smaller than the denominator, such as $\frac{2}{3}$ or $\frac{9}{16}$.
- When conducting operations with fractions, it is best to convert all mixed numbers to improper fractions first.
- Fractions can be added and subtracted only if they have a common denominator. If the denominators are the same, add or subtract the numerators while keeping the denominator the same. Examples: $\frac{1}{6} + \frac{4}{6} = \frac{5}{6}$ $\quad \frac{9}{13} - \frac{4}{13} = \frac{5}{13}$
- There are two methods for multiplying a fraction:
  (1) Multiply the numerators together, multiply the denominators together, then reduce/simplify.
  (2) Do the reducing/simplifying before multiplying, so you are not working with such large numbers.
- Dividing a fraction is the same as multiplying the fraction by its reciprocal. To divide fractions, change the sign from division to multiplication, invert the second fraction (the divisor) then multiply. For example, $\frac{1}{3} \div \frac{2}{5}$ is equivalent to $\frac{1}{3} \times \frac{5}{2}$.
- The rules of the order of operations (BEDMAS) apply to fraction questions involving multiple operations.

# Unit review

Launch additional digital resources for this chapter

Key to Unit review question levels:

Level 1–2 | Level 3–4 | Level 5–6 | Level 7–8

**1** Compare each pair of fractions using $>$ or $<$.
**Explain** your reasoning.

**a** $\frac{3}{7}, \frac{2}{5}$ **b** $\frac{7}{10}, \frac{5}{8}$ **c** $\frac{4}{11}, \frac{2}{7}$

**d** $\frac{13}{20}, \frac{5}{9}$ **e** $\frac{5}{12}, \frac{9}{16}$ **f** $\frac{17}{21}, \frac{5}{6}$

**2** Arrange each set of fractions from least to greatest.
**Show** all of your working.

**a** $\frac{1}{3}, \frac{5}{12}, \frac{5}{6}, \frac{17}{24}, \frac{7}{8}, \frac{1}{2}$ **b** $\frac{3}{5}, \frac{7}{10}, \frac{27}{50}, \frac{11}{25}, \frac{1}{2}, \frac{3}{4}$

**3** In a recent survey of adolescents, participants read the following statement: 'It bothers me when my friends and family use technology when spending time with me.'

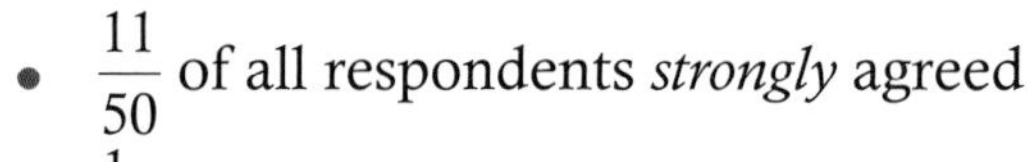

- $\frac{11}{50}$ of all respondents *strongly* agreed
- $\frac{1}{2}$ agreed
- $\frac{1}{5}$ neither agreed nor disagreed
- $\frac{2}{25}$ disagreed.

**a** Rank the responses in increasing order of the number of people who chose each one. **Show** all of your working.

**b** What fraction of respondents agreed or strongly agreed?

**c** Does your answer to **b** surprise you? **Explain**.

**4** Perform the following operations.

**a** $\frac{2}{5}+\frac{4}{5}$ **b** $\frac{12}{17}-\frac{9}{17}$ **c** $\frac{3}{5}\times\frac{1}{7}$ **d** $\frac{6}{11}+\frac{4}{11}$

**5** Perform the following operations.

**a** $\frac{12}{27}\times\frac{18}{15}$ **b** $\frac{6}{21}\div\frac{72}{7}$ **c** $\frac{6}{7}+\frac{2}{21}$ **d** $\frac{3}{5}-\frac{1}{3}$

**e** $\frac{4}{9}+\frac{5}{12}$ **f** $\frac{9}{10}-\frac{3}{8}$ **g** $7\frac{2}{3}\times 2\frac{1}{10}$ **h** $6\frac{10}{11}\div 2\frac{3}{8}$

6 Using only a $\frac{1}{4}$ measuring cup and a $\frac{1}{3}$ measuring cup, how would you make the following amounts for a recipe? **Show** your working.

**a** $\frac{1}{2}$ cup  **b** 1 cup  **c** $\frac{7}{12}$ cup  **d** $1\frac{5}{12}$ cup

7 The festival of Holi, celebrated mainly in India and Nepal, is also called the Festival of Colors because participants chase and color each other with bright, vivid wet and dry colors. It is considered to bridge the social gap as everyone who participates can be chased and colored, regardless of age, race, gender or social status. It is said that even enemies become friends during Holi. It is a time to give blessings and spend time with the community.

Different natural ingredients are mixed with flour to make dry colors. The same ingredients can be mixed with water to make wet colors or pastes.

**a** To make a dry yellow color, you can mix 4 teaspoons of turmeric powder with 7 teaspoons of flour.
To make a blue color, you can mix 3 teaspoons of dried jacaranda flowers with 5 teaspoons of flour.
Which mixture has a higher concentration (fraction) of color? **Show** your working.

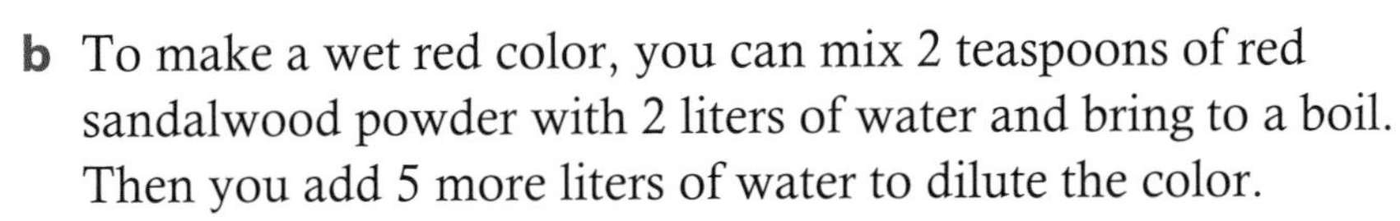

**b** To make a wet red color, you can mix 2 teaspoons of red sandalwood powder with 2 liters of water and bring to a boil. Then you add 5 more liters of water to dilute the color.

To make a wet orange color, you can mix 4 teaspoons of saffron with 3 liters of water, bring it to a boil, and then add 10 more liters.
Which mixture has a higher concentration of color?
**Show** your working.

**c** If $\frac{4}{5}$ of the people of India and $\frac{7}{9}$ of the people of Nepal celebrate Holi, which country has a larger relative participation in the event, and by how much? **Show** your working.

**8** Here are two questions with answers from students.

Evaluate: $\frac{5}{14} \div 3\frac{4}{7}$

Eric's answer:

$$\frac{5}{14} \div 3\frac{4}{7} = \frac{5}{14} \div \frac{19}{7}$$
$$= \frac{5}{14} \times \frac{7}{19}$$
$$= \frac{5}{2} \times \frac{7}{19}$$
$$= \frac{35}{38}$$

Evaluate: $\frac{5}{9} + \frac{6}{15}$

Amanda's answer:

$$\frac{5}{9} + \frac{6}{15} = \frac{{}^{1}\cancel{5}}{{}_{3}\cancel{9}} + \frac{{}^{2}\cancel{6}}{{}_{3}\cancel{15}}$$
$$= \frac{1}{3} + \frac{2}{3}$$
$$= \frac{3}{6}$$
$$= \frac{1}{2}$$

For each student's response:

**a** **explain** any errors in the workings

**b** **explain** what needs to be done to fix the solution

**c** **solve** the question properly, listing the correct problem-solving process steps.

**9** Perform the following operations.

**a** $\frac{1}{6} + \frac{1}{3} - \frac{2}{9}$ **b** $\frac{1}{6} \times \frac{3}{5} \div \frac{3}{10}$ **c** $\left(\frac{3}{4} + \frac{5}{8}\right) \div \frac{24}{33}$ **d** $\left(\frac{1}{5}\right)^2 \div \frac{3}{10}$

**10** Barbecue (or BBQ) is a style of cooking that is very popular in the southern United States. In the 19th century, barbecue was a featured dish at all kinds of activities, from social gatherings to political rallies to family reunions. To this day, barbecue is a centerpiece of activities that promote human connections and relationships in the southern USA. There are many kinds of barbecue sauce; a recipe for South Carolina style sauce is given below.

The recipe makes $2\frac{1}{2}$ cups and you need to make 4 cups.

**a** By how much do you need to multiply each of the ingredients to make 4 cups' worth?

**b** Rewrite the recipe to make 4 cups.

**c** Are there any ingredients that would be difficult to measure for the new recipe? How would you be able to use the correct amount?

**11** Thanksgiving is a holiday celebrated primarily in North America. It was originally a day to give thanks for the harvest and the year.
It is a time when more Americans travel to be with family than any other holiday, including Christmas.

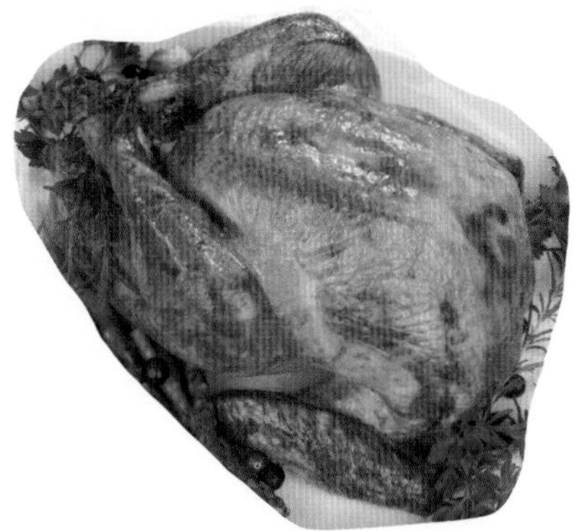

Eating turkey is a tradition on Thanksgiving. In the United States, 50 million turkeys are eaten on the holiday. The total number of turkeys eaten each year in the US is 250 million. The population of the United States is approximately 320 million people, which means roughly $\frac{3}{4}$ of a turkey per person each year!

**a** What fraction of the year's turkey consumption is on Thanksgiving?

**b** Approximately 50 million Americans travel over the Thanksgiving holiday. What fraction of the US population does this represent?

**c** Of those travelling, $\frac{9}{10}$ will be driving to their destination and $\frac{2}{25}$ will be flying. What fraction of those travelling will not be driving or flying? What modes of transportation do you think they will take?

**d** What fraction of the total US population will be flying to their destination for Thanksgiving?

## Summative assessment

criteria C, D

*Kahk*, or Egyptian Eid cookies, are lightly spiced cookies that are very popular in Middle Eastern countries and served during special festive occasions.

Kahk are commonly filled with honey, dates, walnuts, pistachios or Turkish delight, but they are also popular simply dusted with powdered sugar. Making Kahk is a tradition in many houses, when families share recipes that have been passed down from one generation to the next.

The origins of the cookie date back to Ancient Egypt. There are drawings in the Pharaonic temples of ancient Thebes and Memphis illustrating the making of Kahk. In Islamic history, during the Tulunid dynasty, Kahk were made into packs called *kul wishkur*, which means 'eat and say thank you'.

### Part A – Using fractions in cooking

Pairs

**1** Fractions are very important in baking and cooking and are seen in many recipes. Below are the ingredients to make one batch of Kahk, which is roughly 20 cookies. Work in pairs to figure out how much of each ingredient you will need to make 100 cookies, and then 30 cookies. Fill in the appropriate columns on a copy of the table as you go. Explain by drawing or writing how you figured out how to complete the quantities needed for each scenario.

| Ingredient | Measurement for one batch | Ingredients for 100 cookies | Ingredients for 30 cookies | Measuring utensils you will use – set up as a fraction addition |
|---|---|---|---|---|
| Flour | 3 cups | | | |
| Ghee (clarified butter) | 1 cup | | | |
| Milk | $\frac{1}{3}$ cup | | | |
| Salt | $\frac{1}{4}$ teaspoon | | | |
| Granulated (white) sugar | 1 tablespoon | | | |
| Kahk essence – ground mahlab (black cherry stone), anise, fennel | $\frac{3}{4}$ teaspoon | | | |
| Dry yeast | $\frac{1}{2}$ teaspoon | | | |
| Sesame seeds – optional | 2 tablespoons | | | |
| Powdered sugar | $\frac{1}{2}$ cup | | | |

**2** You have looked everywhere in your kitchen and have found measuring utensils in the following sizes.

1 cup $\frac{1}{2}$ cup 1 tablespoon 1 teaspoon $\frac{1}{4}$ teaspoon $\frac{1}{8}$ teaspoon

In the last column of your table, list the measuring utensils you would use to measure your ingredients in the most efficient way possible (with the least number of scoops). Research the conversions between teaspoons, tablespoons and cups to help you.

## Reflect and discuss 8

- Do you think you would go to the trouble of baking exactly 30 cookies? In reality, what could you do that is easier?
- In which countries is the cup a common unit of measurement for volume? Why do you think this form of measurement is so common in recipes? Is this form of measurement common in other areas? Why do you think that is?

## Part B – A special recipe

Choose a recipe that has significance to you and reflects your culture, your heritage or a family tradition. Make sure that the recipe contains some fractions.

**1** Copy down the list of ingredients.

**2** Imagine you would like to make this recipe for your class. Determine how you will alter the recipe to make enough for the whole class. You can round up the recipe to the nearest batch if you like.

**3** Calculate the quantities of each ingredient you will need and what measuring utensil(s) you will use to measure it. These conversions of the ingredients in the recipe are to be submitted on a separate sheet to your teacher, complete with all calculations.

**4** Present your work on **one** Google slide that includes:

- your name
- a picture of your chosen food
- the original recipe of the food
- the background/origins of the food and how it reflects your culture and/or heritage
- why this recipe is significant to you
- images of activities that you think promote human connections and interactions.

Your teacher will use Google Slides to save all of the class's slides together. They will use this to produce a PDF that can then be printed or electronically distributed to everyone.

# 6 Data management

The study of data and data management can help uncover trends in communities, as you will explore in this unit. It can also uncover and represent trends in many other contexts, such as future development or the habits of birds.

## Fairness and development

### Feeling hopeful for the future

How do we know things are improving? How can we tell if there is hope in a difficult situation? While statistics are often used to validate negative trends in both human and natural environments, they can also be used as a tool for optimism.

In 1987, the California condor was declared extinct in the wild. The only living birds were kept in captivity. Since then, the species has been reintroduced and, whilst numbers are still low, there are now a growing number of birds living in the wild. The California condor, along with other endangered species, such as the black rhinoceros, has seen significant improvement in numbers since conservation measures were implemented. Ecologists use statistics to evaluate the factors that led to the decline in their numbers as well as to keep track of growing populations. These success stories give us hope that we can reverse some of the negative effects we've had on our environment.

Data can also be collected and analyzed on a range of human and social issues. When literacy rates improve, when AIDS infection rates decline and when more people have access to clean water, a more hopeful future can be imagined for all humans.

## Orientation in space and time

### Migration and displacement

Every year, millions of animals migrate from one place to another. Many species are constantly moving to find sources of food or water while others return to their breeding grounds once a year.

Arctic terns fly over 71 000km every year as they travel from the Arctic to the Antarctic, accumulating more hours of sunshine than any other animal on the planet and experiencing two summers in the process.

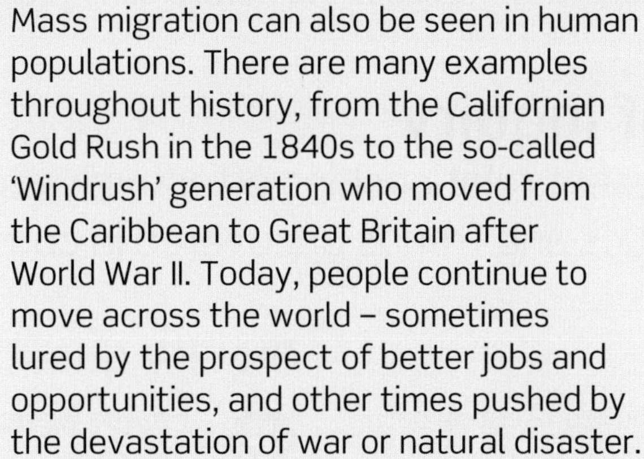

Mass migration can also be seen in human populations. There are many examples throughout history, from the Californian Gold Rush in the 1840s to the so-called 'Windrush' generation who moved from the Caribbean to Great Britain after World War II. Today, people continue to move across the world – sometimes lured by the prospect of better jobs and opportunities, and other times pushed by the devastation of war or natural disaster.

Alongside this, is a constant stream of temporary migration. One of the largest is at Chinese New Year, when millions of people return to their families to celebrate together.

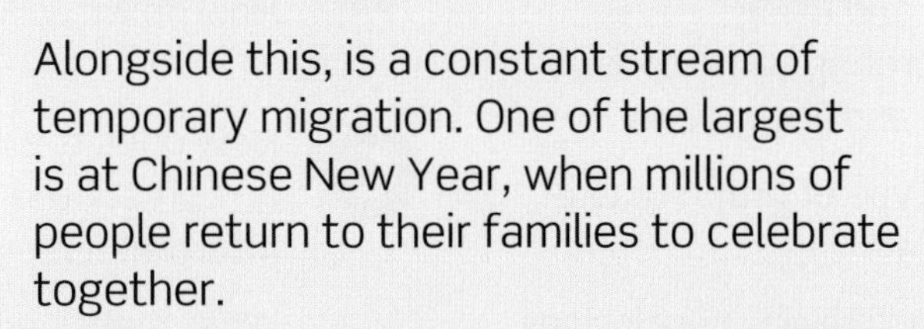

The ability to describe and predict these events requires collecting, representing and then interpreting data.

# 6 Data management

## Trends in communities

## KEY CONCEPT: RELATIONSHIPS

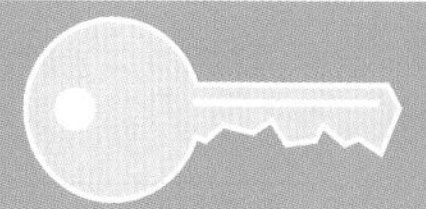

## Related concepts: Representation and Justification

## Global context

In this unit you will learn how to collect, classify and represent information that is gathered from a variety of sources as you explore the global context of **identities and relationships**. Understanding different types of data and the most effective ways to represent them may help you uncover trends and concerns in local, national or even global communities. Equipped with this information, you may discover just how much power you have to make an impact on the world around you.

## Statement of Inquiry

Being able to represent relationships effectively can help justify characteristics and trends uncovered in communities.

### Objectives

- Collecting, classifying and representing data
- Constructing (by hand and using technology) and interpreting bar graphs, histograms, pie charts and line graphs
- Determining the best type of graph to use to represent given data
- Reading, interpreting and drawing conclusions from primary and secondary sources of data
- Applying mathematical strategies to solve problems involving statistics

### Inquiry questions

**F** How do we represent information?

How do we collect information?

**C** How does the way in which information is represented impact our ability to interpret it?

What makes one representation more effective than another?

**D** Whose responsibility is it to identify and help fix problems within a community?

**ATL1 Media literacy skills**

Communicate information and ideas effectively to multiple audiences using a variety of media and formats

**ATL2 Critical-thinking skills**

Identify trends and forecast possibilities

## You should already know how to:

- measure an angle with a protractor
- draw an angle using a protractor
- solve percentage problems
- multiply a fraction by a whole number

# Introducing data management

When trying to establish characteristics or trends in a community, it is often necessary to ask people in the community questions and then analyze their answers. For example, if school administrators want to collect information about the school climate and how safe students feel, they may ask students the following questions (among others).

1. What MYP year are you in? ____________
2. With which gender do you indentify? ____________
3. How many siblings do you have in your school? ____________
4. On average, how many minutes do you spend at school each day? ____________
5. Students at this school are treated fairly by other students, regardless of nationality or ethnicity.
   strongly agree ☐ somewhat agree ☐ somewhat disagree ☐ strongly disagree ☐
6. Students show respect to other students, regardless of their academic ability.
   strongly agree ☐ somewhat agree ☐ somewhat disagree ☐ strongly disagree ☐
7. All students in my school are treated fairly, regardless of their appearance.
   strongly agree ☐ somewhat agree ☐ somewhat disagree ☐ strongly disagree ☐

## Reflect and discuss 1

- Explain how each of these questions relates to school climate and safety.
- What type of answer does each question require?
- How are these answers similar? How are they different?
- What would you need to do in order to establish a trend or pattern? Explain.

## Data

Information that you collect is often called *data*.
Data can be classified in the following ways.

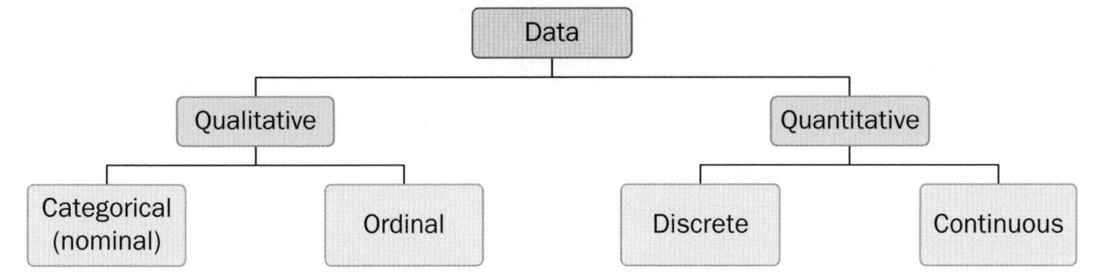


*Quantitative data* contain values or numbers (quantities), whereas *qualitative data* represent words or descriptions (qualities).

### Investigation 1 – Types of data

Here are some examples of each type of data.

**Categorical**
- hometown
- hair color
- music genre
- breed of dog

**Ordinal**
- ranking a movie from 1 to 10
- answering 1, 2, 3, or 4 depending on how much you agree with a statement
- which year of the MYP (1 to 5) you are in

**Continuous**
- hair length of students in the class
- gas mileage of cars
- how far you travel to get to school
- time taken to run 100 m

**Discrete**
- number of students in a class
- how many pets your family has
- points scored in a bossaball game
- shoe size

1. Write down samples of the type(s) of data that could be collected for each example. For example, breed of dog data could be Labrador, Border Collie, Chihuahua, Siberian Husky, etc.
2. Based on the examples, write a definition of each data type: categorical, ordinal, discrete, and continuous.
3. Compare your definitions with those of other people. If necessary, refine your definitions based on what you found.

▶ Continued on next page

4 Classify each of the following types of data as either quantitative or qualitative and state which subcategory it falls under (categorical, ordinal, discrete or continuous).

- Types of game played by children
- Hours of on-screen time spent per week
- Types of boat on a lake
- Ranking your favorite sports
- Eye color of all the students in your class
- Weights of newborn babies in a hospital
- Heights of all bossaball players in a tournament
- Population of a country
- Languages spoken by students at your school
- Number of a person's siblings

5 Give your own example of each type of data.

## Did you know...?

Bossaball is a rather unusual sport, invented in Spain by Filip Eyckmans. The game requires a specially designed inflatable court with a net running down the middle and a circular trampoline on each team's side. Bossaball combines elements of volleyball, gymnastics, football and capoeira, which is a type of martial art. The inflatable surface allows for a whole new dimension in acrobatics, techniques and creative play. Bossaball courts can be set up in less than 45 minutes and can be installed in diverse and unconventional locations.

## Reflect and discuss 2

- Which kind of data do you encounter most often? Explain.
- Which kind of data do you think is the easiest to analyze? Explain.

## Practice 1

**1** Categorize the following data as categorical, ordinal, discrete or continuous. Justify each of your answers.

age | shirt size (S, M, L) | whether you are right-handed or left-handed | color of your favorite hat | the day of the week on which you were born | the results of an election | height | grade on a final exam | your family income | your rating of a movie (1, 2, 3, 4 or 5 stars) | your year of birth | your phone number

**2** If a survey contained the following questions, what kind of data (categorical, ordinal, discrete or continuous) would be collected?

**a** How many siblings do you have?

**b** How much time did you spend studying last night?

**c** How many texts do you send in a typical day?

**d** What is your favorite movie genre?

**e** What social media site do you use most often?

**f** How many times did you go to the cinema last year?

**g** Which class in school is your favorite?

**h** How long do you spend brushing your teeth?

**3** Write down a question that would allow you to collect each of these types of data. Be sure not to repeat questions.

**a** quantitative **b** qualitative **c** ordinal

**d** discrete **e** categorical **f** continuous

**4** Decide on a characteristic about your class, family, school or community that you would like to investigate.

**a** Design a short survey (3 to 5 questions) that you could use to collect the required data.

**b** What type of data would you be collecting for each question?

**c** Describe the steps you would need to take to determine a trend or pattern in the characteristic over time.

## Using graphs to represent data

*Data management* is the process of collecting data, organizing it, and then finally displaying the data so that you can interpret it and hopefully draw conclusions. The different types of data all need to be represented somehow, and each method of representation is more effective for some types of data than others.

The most common way to represent statistical information is to use something called a *graph*. In statistics, a graph is an organized picture that represents data.

### Activity 1 - What makes a good graph?

1 a Summarize in one sentence what this graphic is telling you.

b What trends or characteristics do you notice in the data that is being represented?

c Do you think this is an effective way to display the information? What aspects about it are effective? What aspects are not so effective? Explain your answers.

d Can you suggest another way to represent this data? Would this new way be better? Explain.

How much sleep do you need?

Newborns: 16–18 hours
1-year-olds: 15 hours
2-year-olds: 13–14 hours
4-year-olds: 12 hours
10-year-olds: 10 hours
Teenagers: 8–9 hours
Adults: 7–8 hours

2 Gender parity is where the percentage of girls enrolled in or attending school is equal to the percentage of boys. This has been reached in primary schools in most countries, but gender parity at the secondary level is still an issue. Use the graphic to answer the questions below.

**Gender parity at the secondary level**

0.94 or less (girls less likely than boys to be enrolled)
0.97–1.00 (gender parity)
0.95–0.96 or 1.04–1.06 (near gender parity)
1.06 or more (boys less likely than girls to be enrolled)
Data not available

Continued on next page

**a** Is there gender parity in your country? If not, who is more likely to be enrolled in secondary school, boys or girls?

**b** Give a summary (no more than a few sentences) of what this map is showing you.

**c** What trends or characteristics in this information stood out when you were first analyzing it? Justify your answer.

**d** Could this representation be considered a graph? Give your reasoning.

**e** What aspects of the map are effective in representing the data? What aspects are less effective?

**f** List the five top characteristics of a good graph.

## Practice 2

Students in one city answered the questions at the beginning of this unit (page 204), and their responses are shown in several different representations here.

**1** *What MYP year are you in?*

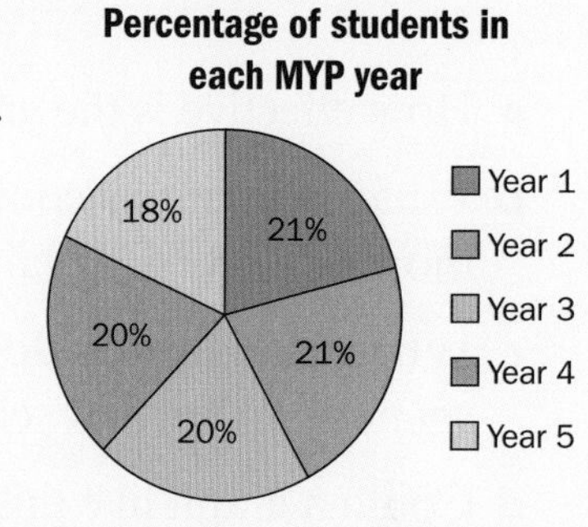


**a** How effectively does the graph represent the data? Explain.

**b** In this graph, the size of the sections represents the percentages. If you were drawing this graph by hand, how would you achieve this?

**c** What trend seems to be happening in this city from Year 1 to Year 5? Explain how the trend is visible in the graph.

**d** Does the graph explain why this trend is happening? Explain.

**e** Suggest reasons for the trend.

**2** *With which gender do you identify?*

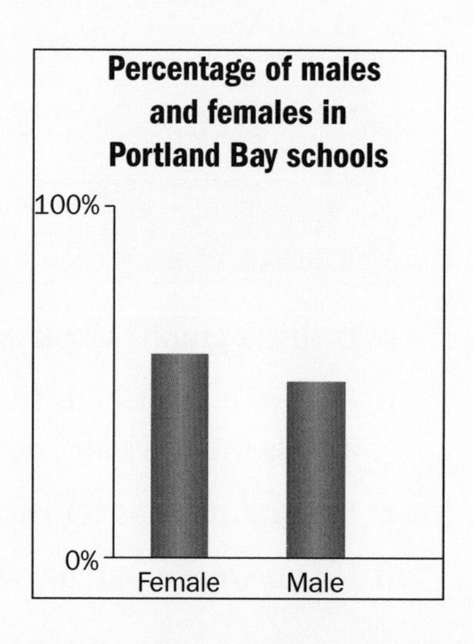


**a** Based on the graph, what can you conclude about the number of males and females in schools in Portland Bay?

**b** Can it be concluded that every school in the city has more female students than male students? Explain.

**c** Can the results of the graph be applied to schools across the country? Explain.

**d** What other information could be provided to make the graph more helpful or clearer?

**e** What might be a more effective way to represent the information in this graph?

Continued on next page

**3** *How many siblings do you have in your school?*

| Number of siblings | Number of responses |
|---|---|
| 0 | 1434 |
| 1 | 1509 |
| 2 | 878 |
| 3 + | 122 |

**a** Is this a graph? Explain.

**b** What conclusions can you draw from this data?

**c** Is the way in which the question is worded likely to cause any confusion? How could the question be rephrased?

**d** Why do you think this question was included in a survey on school climate and safety? Explain.

**e** What suggestions would you make to improve the way the data is represented? Explain your reasoning.

**4** *On average, how many minutes do you spend at school each day?*

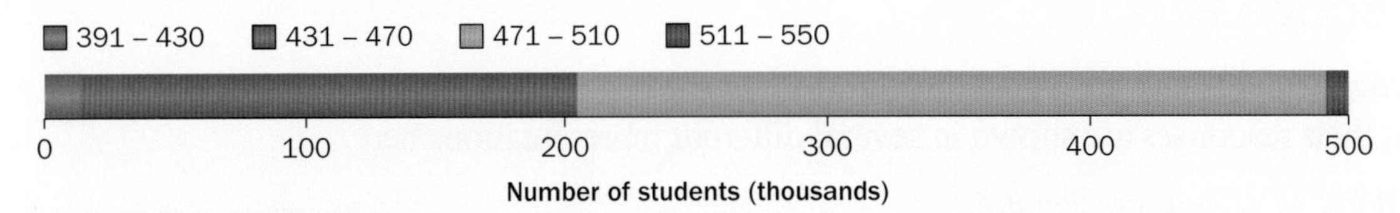


**a** How effective is the graph in representing the data? Explain.

**b** What suggestions could you give to make the representation more effective? Explain.

**c** What relationships do you think could be found between time spent at school and school safety? Explain.

**d** Could a graph like this show trends in the city? Explain.

ATL2

**5** *Students at this school are treated fairly by other students regardless of nationality or ethnicity.*

| | Year 1 | Year 2 | Year 3 | Year 4 | Year 5 |
|---|---|---|---|---|---|
| Strongly agree | 40.1% | 33.3% | 28.3% | 22.8% | 20.1% |
| Somewhat agree | 35.8% | 38.9% | 43.1% | 46.5% | 45.8% |
| Somewhat disagree | 15.3% | 19.1% | 19.0% | 20.1% | 22.7% |
| Strongly disagree | 8.8% | 8.7% | 9.6% | 10.6% | 11.4% |

**a** Is this a graph? Explain.

**b** Describe any trends that you see in the data. Remember to look both across the rows and down the columns.

**c** Suggest reasons for each of the trends that you described.

**d** If you were a parent, would this data trouble you? Justify your answer.

▶ Continued on next page

6 *Students show respect to other students, regardless of their academic ability.*

Suppose the following data was collected from the Year 1 students' responses to this statement.

- 11% strongly agree
- 22% somewhat agree
- 43% somewhat disagree
- 24% strongly disagree

**a** What would you need in order to be able to identify trends with Year 1 students? Explain.

**b** Does this data trouble you? Explain.

7 *All students in my school are treated fairly, regardless of their appearance.*

Suppose the following data was collected from the Year 1 students' responses to this statement.

- strongly agree – 88 students
- somewhat agree – 146 students
- somewhat disagree – 51 students
- strongly disagree – 19 students

**a** Represent this data using a graph.

**b** Describe the advantages and disadvantages of the graph you created.

**c** Are there certain kinds of data that are easier to represent? Explain.

## Types of graph

There are many different types of graph that you can use to organize and represent data. The goal is to organize data in the best possible way so that you can interpret it, draw conclusions from it, and perhaps even make predictions.

### Frequency tables

In medieval Europe, people used a 'tally' or a 'tally stick' to keep track of things, like money. A *frequency table* uses tally marks in one column to keep track of items that are being counted. When the counting is done, the tallies are added up and the totals are entered as numbers in another column.

Year 5 students in a school were surveyed regarding how many hours they slept the previous night. Their answers were collected and are summarized in this table.

| Number of hours of sleep | Tally | Frequency |
|---|---|---|
| $<4$ | | 0 |
| 4 | II | 2 |
| 5 | 𝍸 | 5 |
| 6 | 𝍸 IIII | 9 |
| 7 | 𝍸 III | 8 |
| 8 | III | 3 |
| 9 | II | 2 |
| 10 | I | 1 |
| $>10$ | | 0 |

## Reflect and discuss 3

- What types of data would be useful to represent in a frequency table? Explain.
- When would using a frequency table be difficult? Explain.
- Where would your response to the question of how many hours you slept last night fit in the frequency table?
- Based on the table, it was concluded that most Year 5 students sleep between 5 and 7 hours per night. Do you agree or disagree? Explain.

Go to sciencekids.co.nz/gamesactivities/math/frequencytables.html and scroll down to the activity - you will see 'Try this' at the top of the activity box. Click the OK button to start. You will create frequency tables, and then represent the data in bar charts and pictograms.

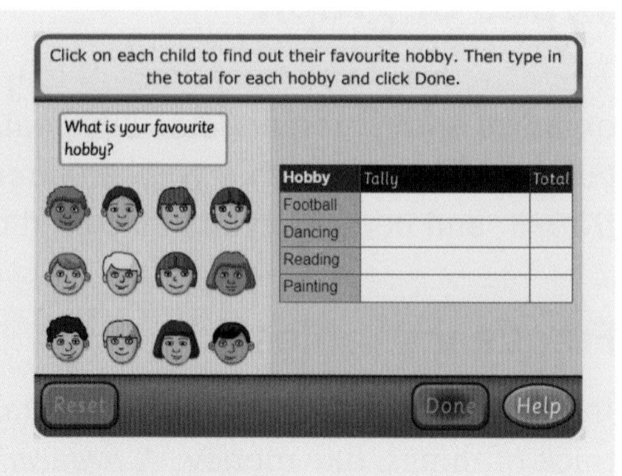


Once you have collected data, you can represent it using a graph like the ones that follow.

## Bar graphs

The data from the frequency table at the top of page 212 is shown here as a *bar graph*.

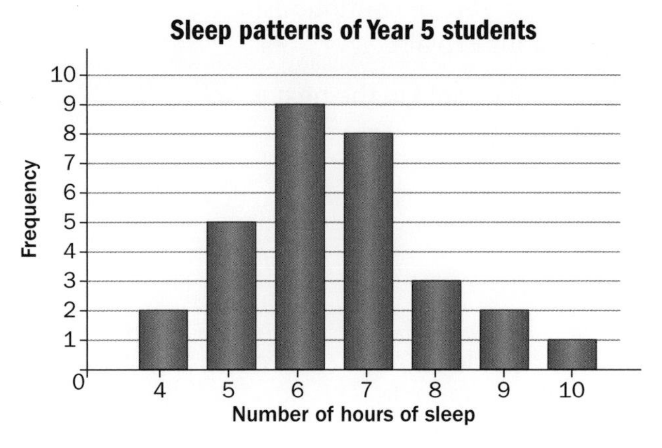


### Reflect and discuss 4

- How do the widths of the columns compare to each other? Is the width of the column important? Explain.
- What do the heights of the bars represent? How do you know how tall to draw them?
- What type of data (categorical, ordinal or discrete or continuous) could be easily represented in a bar graph? Explain.

## Line graphs

*Line graphs* are created by joining data points together with lines. The lines help you see how the data changes, and indicate trends in the data, which you can often use to make predictions about how the data will continue to change.

One common type of line graph is called a *time series* graph. This is when time is on the horizontal axis.

The data in the table shows approximate worldwide smartphone sales in millions. Here is the line graph.

| Year | Smartphone sales (millions) |
|---|---|
| 2007 | 120 |
| 2008 | 140 |
| 2009 | 170 |
| 2010 | 300 |
| 2011 | 470 |
| 2012 | 680 |
| 2013 | 970 |
| 2014 | 1250 |
| 2015 | 1420 |
| 2016 | 1495 |

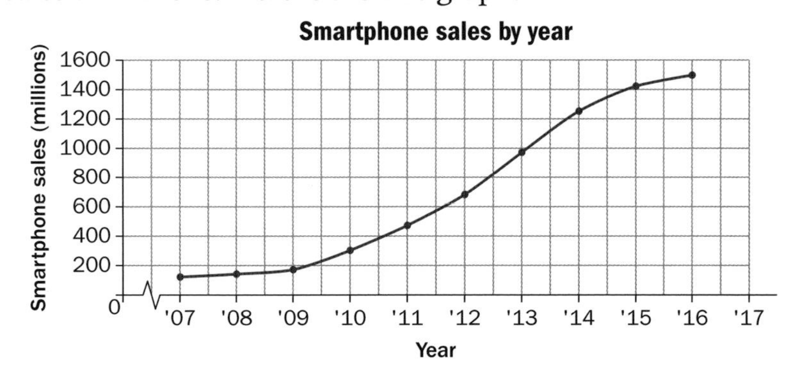


Label each axis and evenly distribute the numbers on the axes. However, the horizontal axis can have a different scale than the vertical axis.

## Reflect and discuss 5

- Could a frequency table be used when collecting the smartphone data? Explain.
- What type of data (categorical, ordinal, discrete or continuous) could be easily represented in a line graph? Explain.
- How many smartphones do you think will be sold in the next year? And in the year after that? Justify your answer.

## Activity 2 - Time series

*Time series* is a common form of line graph in which time is represented on the horizontal axis and some other variable on the vertical axis. Such a graph displays how something changes over time.

**1** Match each photo description with one of the graphs, and state what units are on the $y$-axis.

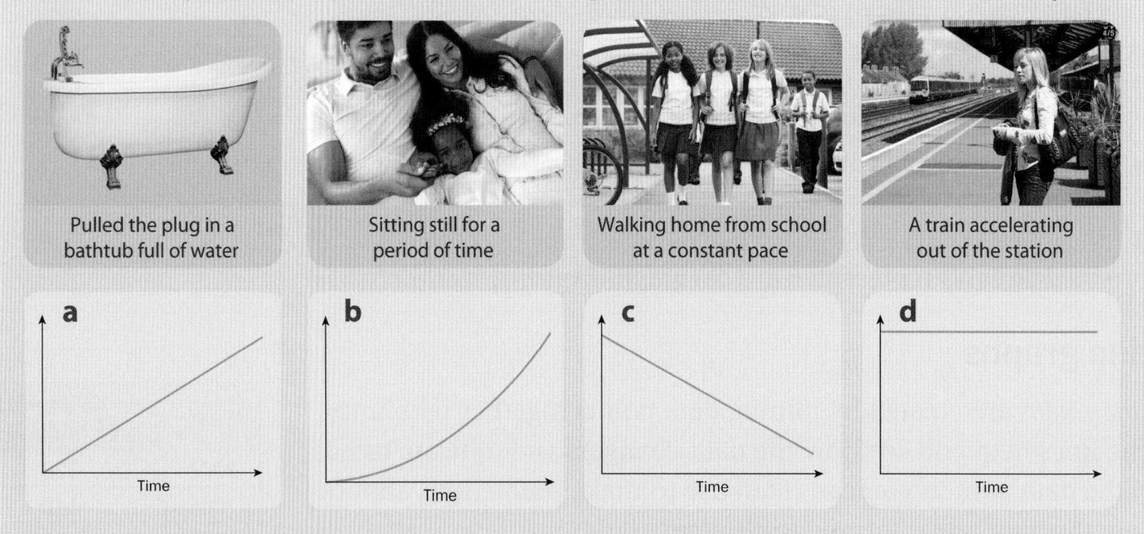


**2** Given the time series graph below, think about what story might explain it. Time is along the horizontal axis but you need to decide the units (seconds, hours, etc.). You also need to decide the measure and scale of the vertical axis.

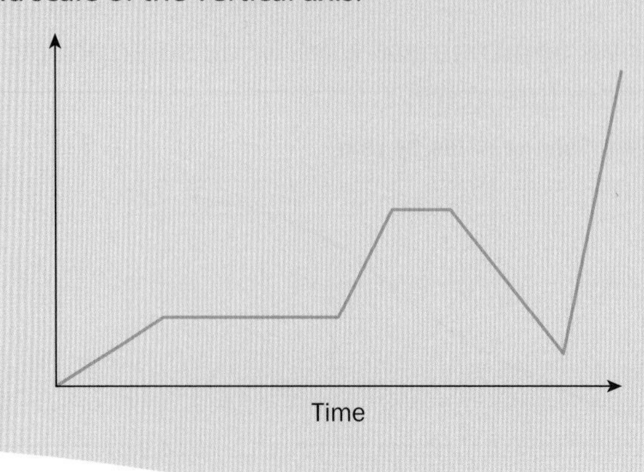

## Reflect and discuss 6

- What advantages does a time series graph have over the other types of graph you have seen so far?
- What type of data (categorical, ordinal, etc.) could be easily represented in a time series graph? Explain.
- Give an example of data that changes over time, but for which a time series graph might be a less effective representation. What kind of graph would be more appropriate? Why?

## Circle graph (pie chart)

A *circle graph* is a useful way of displaying data when you want to show how things are divided up. A full circle represents all of the data, and you divide the circle into *sectors* to show each category. The circle graph here represents how Year 1 students in England use their mobile phone.

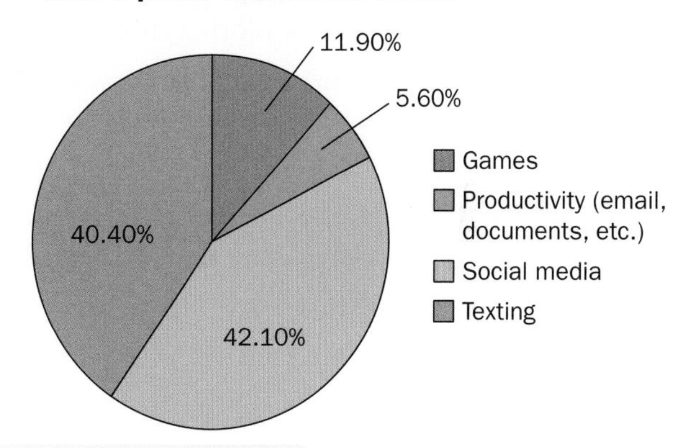


## Reflect and discuss 7

- When constructing a pie chart starting with raw data, how would you know how big to make each sector?
- What type of data (categorical, ordinal, etc.) could be easily represented in a pie chart? Explain.
- Can you find a trend in a pie chart? Explain.

## Activity 3 – Creating a pie chart by hand

When producing a pie chart by hand, you will need a protractor and a compass. The 'pie' is divided into sectors. As there are 360° in a circle, the degree measure of each sector must be calculated so that it is proportional to the amount of data that it will represent.

For example, a survey of 100 students found that the average adolescent consumes 2000 calories per day, and those calories come from the types of food shown in the table on the next page.

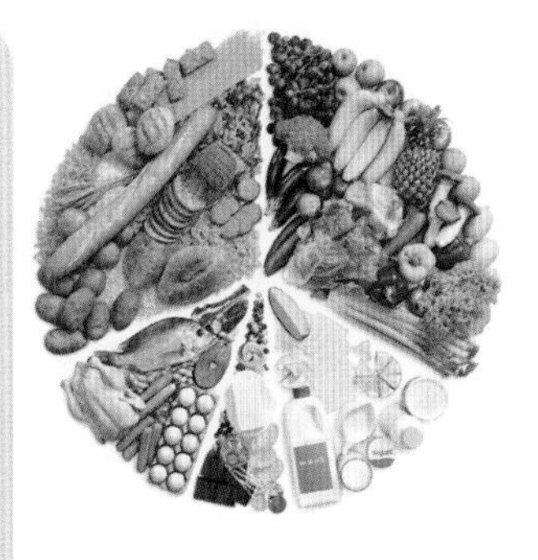

▶ Continued on next page

| Type of food | Number of calories | Fraction of total | Degree of sector |
|---|---|---|---|
| home cooking | 1400 | | |
| fast food | 400 | | |
| junk food | 200 | | |

1 For each category of data (in this case, type of food) calculate the fraction of the total that the data represents. Enter this in the appropriate column.
2 Multiply each fraction by 360 and enter the value in the last column. Make sure that the column total is 360.
3 Draw a circle using a compass. Be sure to clearly mark the center.
4 Draw a line from the center to the top of the circle. This will be your starting point for plotting the results.
5 Starting on this line, draw the home cooking sector so that it spans the number of degrees in your table.
6 Repeat step **5** for fast food and junk food, starting from the last line you drew each time.
7 Add a title and any necessary numeric information that you wish. Create a key and color code the pie chart.

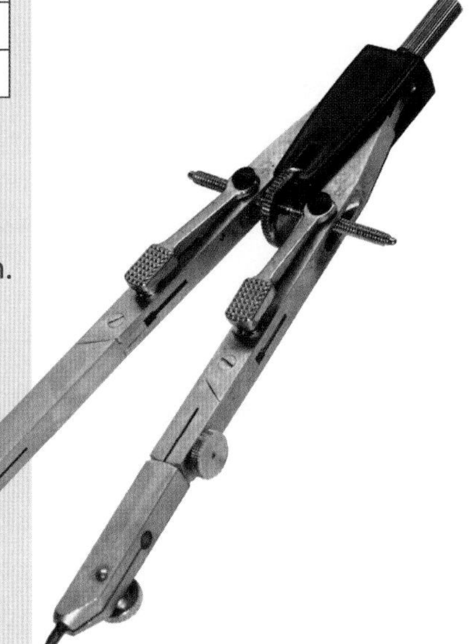

## Practice 3

ATL2 1 The graph below shows the literacy rate for most of the world's population.

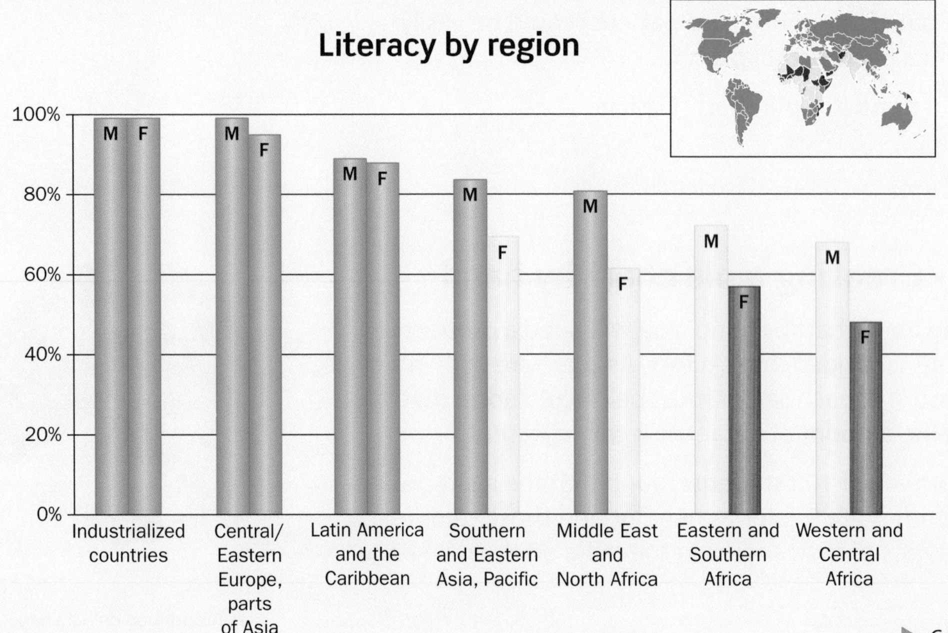


▶ Continued on next page

**a** What type of graph is this?

**b** What characteristics can be seen in this representation?

**c** It has been reported that the number of people who are illiterate has gone down significantly over the past 30 years. Is this represented in the graph? Explain.

**d** If you were to see graphs of the same data, collected 5 and 10 years later, what trend would you hope to see? Explain how that trend would be visible in the graphs.

**2** The graphs below represent how Year 5 students at a particular school spend time on weekends. They were first surveyed in 2005 and then again in 2010.

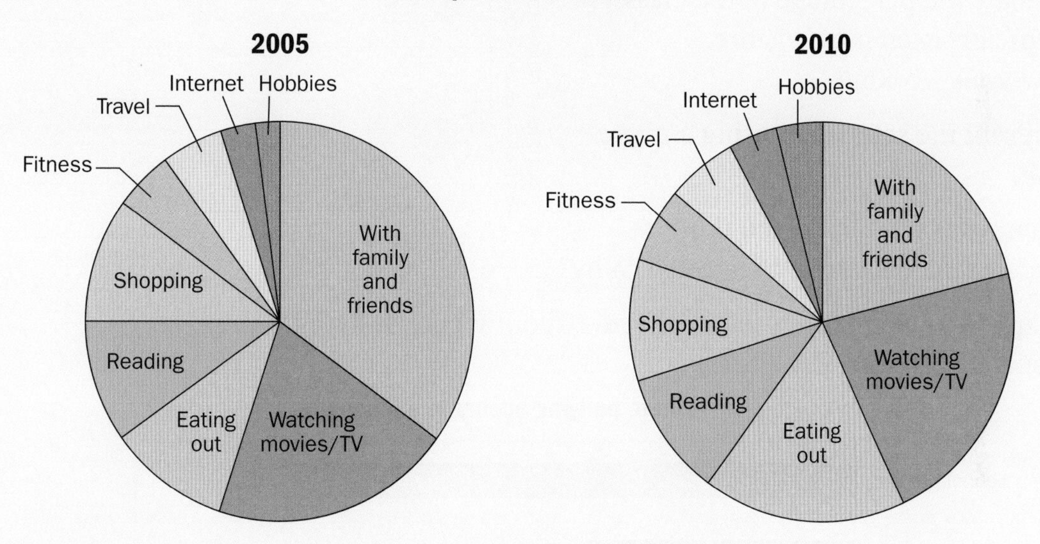


**a** Suggest percentages for each sector of the graphs. Make sure they total 100% for each graph.

**b** What trends do you notice between 2005 and 2010? Suggest reasons for each of these trends.

**c** Write down a question that would collect the data necessary to create these graphs. What kind of data (categorical, ordinal, discrete or continuous) would this question collect?

**d** Pose your question to at least 20 Year 5 students and collect the data. You may add new categories if you need to.

**e** Create a pie chart for the data you collected.

Programs like MS Excel can help you draw graphs like those in this question.

▶ Continued on next page

ATL1 **3 a** As a class, decide what movie genres most people like, such as comedy, sci-fi, action, animation, etc.

**b** Create a frequency table to help you fill in the data on favorite movie genre as you collect it.

**c** Write down a question that you could ask in order to collect data on movie genre preferences.

**d** Conduct a survey of your class and ask students to vote for the *one* movie genre that is their favorite.

**e** Tally the responses and then fill in the frequency column.

In part **e**, add up all the numbers in the frequency column to determine the total, and double check that your total is the same as the number of students in your class.

**f** Now create a bar graph to represent the movie preferences in your class.

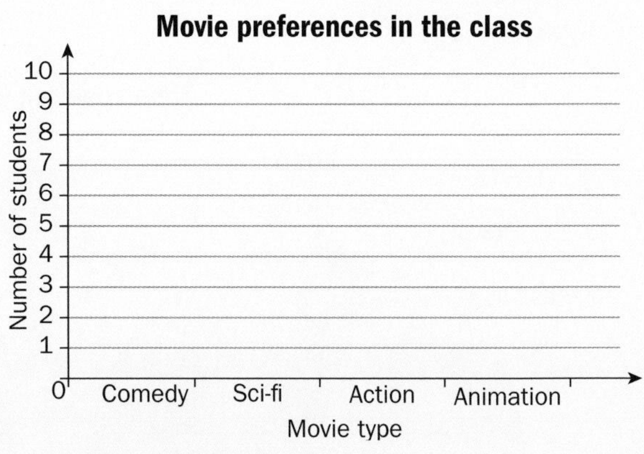


**g** Calculate the percentage of the class that prefers each movie genre. Show your working.

**h** Represent the same data using a circle graph.

**i** Which type of graph do you think represents the data most effectively? Why?

**4** The graph below represents data collected about where bullying occurs in schools in Ontario, Canada.

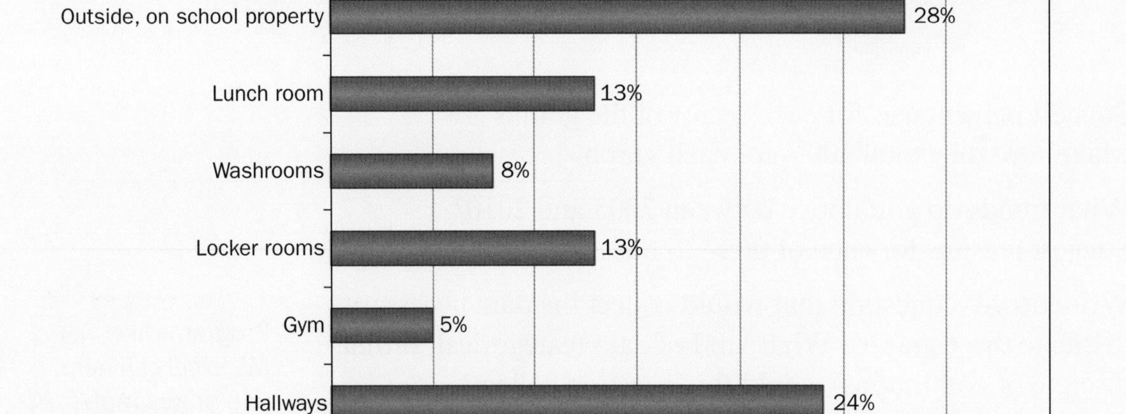


▶ Continued on next page

| Location | Percentage |
|---|---|
| Classrooms | 9 |
| Hallways | 24 |
| Gym | 5 |
| Locker rooms | 13 |
| Washrooms | 8 |
| Lunch room | 13 |
| Outside, on school property | 28 |

**a** Which aspects of this graph are effective? Which ones are less effective? Explain.

**b** Represent the same data using a pie chart.

**c** What are the advantages and disadvantages of a pie chart compared to the original graph? Explain.

**d** How do these representations compare to just representing the data in a table like this one?

**e** Which representation (bar chart, circle graph or table) allows the data to be interpreted most effectively? Explain.

**5** Residents in a beachside community gathered data on the types of trash that were picked up by community members during 42 weekend clean-ups. The data from the table is represented on the line graph.

Total number of pieces of trash by type

| | Cigarettes | Styrofoam | Plastic | Paper | Wood | Metal | Glass | Cloth | Rubber | Other | Total |
|---|---|---|---|---|---|---|---|---|---|---|---|
| Total of 42 trips | 13546 | 6278 | 5324 | 4876 | 1080 | 663 | 603 | 278 | 257 | 359 | 33264 |

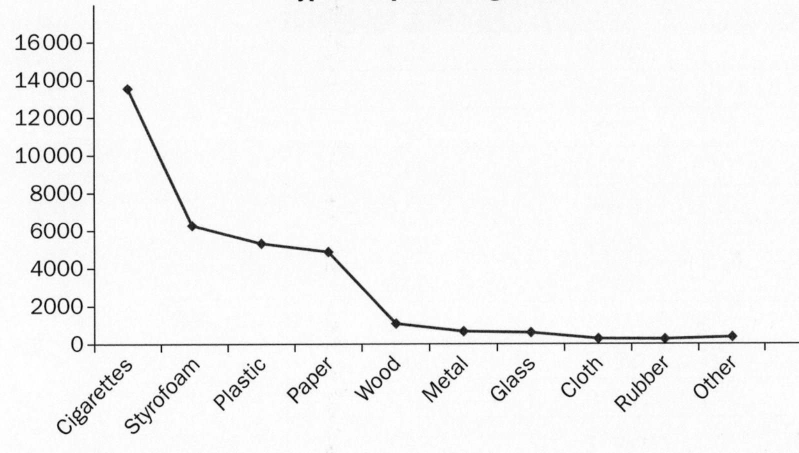

For each question below, be sure to provide an explanation.

**a** Is a line graph an effective way to communicate this data?

**b** What mistakes were made when creating the graph?

**c** Draw a bar graph and a pie chart to represent the percentage of trash collected in each category. Which graph is the most effective way to represent this data?

**d** What characteristics of beach trash does this data suggest?

**e** What characteristics of beachgoers does this data suggest?

## Using the different graphs

Knowing when to use each type of graph can help you to both present and interpret any data that you collect. In the following investigation, you will explore the advantages and disadvantages of the three graphs in this unit.

### Investigation 2 – When to use the different graphs

You will be given a packet of candy that comes in a variety of different colors.

1 Open your packet and sort the candy into colors. Set up a frequency table to organize your data. Once your table is complete, double check your total against the actual number of candies that you have.

2 Create a bar graph to represent the data. Think about what the scale on the vertical axis will be and clearly label both the chart and the axes.

3 Create a pie chart to represent the same data. When calculating how big each sector will be, you may round your answers to the nearest degree. Double check that all of your answers add up to 360. Using a protractor, create the pie chart, complete with titles and a colored key.

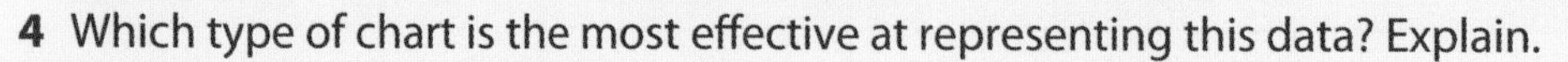

4 Which type of chart is the most effective at representing this data? Explain.

5 Explain why a line graph is not appropriate for this kind of data.

The table below details the cost of a 1 oz unit of a popular candy from 1936 until 2016 (USA currency)

| Year | Cost per oz (cents) | Year | Cost per oz (cents) |
|---|---|---|---|
| 1936 | 5 | 1977 | 16.7 |
| 1941 | 5 | 1978 | 20.8 |
| 1947 | 5 | 1980 | 23.8 |
| 1955 | 5 | 1983 | 24.1 |
| 1960 | 5 | 1986 | 27.6 |
| 1965 | 5 | 1991 | 29 |
| 1966 | 5.7 | 1995 | 32.3 |
| 1968 | 6.7 | 2003 | 51.6 |
| 1970 | 7.3 | 2007 | 54.5 |
| 1973 | 7.9 | 2010 | 61.2 |
| 1974 | 10.7 | 2013 | 63.9 |
| 1976 | 12.5 | 2016 | 70 |

6 Represent this data as a graph and give an explanation as to why your choice of graph type is the most effective for this data set.

7 In no more than a few sentences, give a summary of what this graph is showing you and explain the trend.

8 Based on this graph, predict what the price of this candy will be in 2025.

9 Create a table that lists the different types of graph and presents the main characteristics of each, as well as what type(s) of data each type of graph best represents.

## Reflect and discuss 8

- Which type of graph, if any, is most effective for representing the following types of data? Explain.
  - categorical
  - ordinal
  - discrete
  - continuous
- Which type of graph is most effective at representing a trend? Explain, perhaps by including some examples.

## Activity 4 – Analyzing different graphs

**1** The following data shows the hourly breakdown of activities of a typical Year 1 student.

As a bar graph, the data is represented like this:

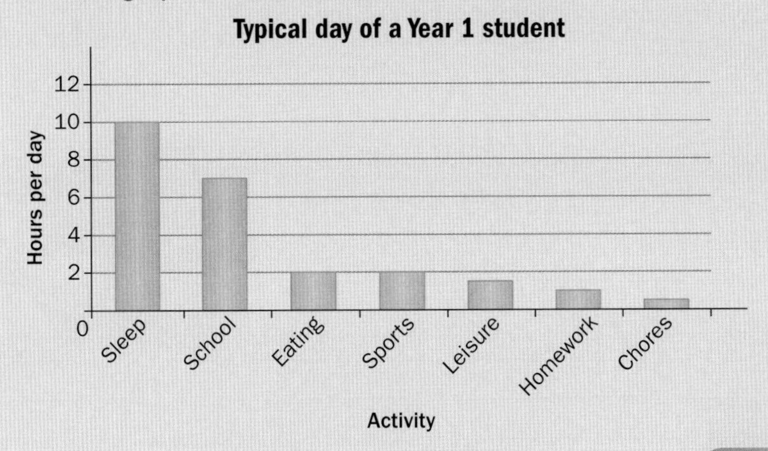


As a pie chart, the data is represented like this:

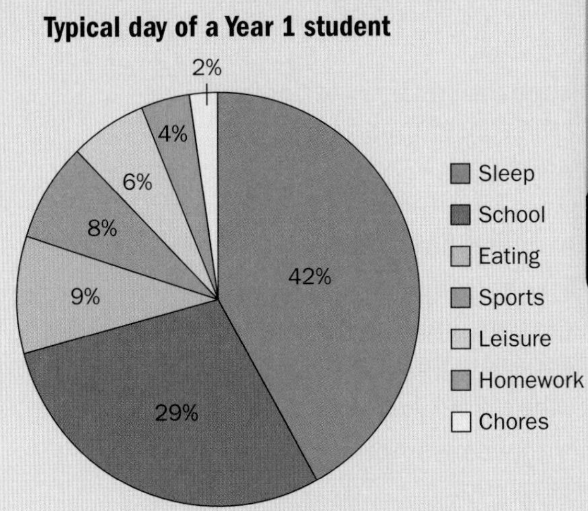

▶ Continued on next page

Using both graphs from page 221, answer these questions.

**a** Which activity takes up the least amount of a typical Year 1 student's time? Which activity takes up the most amount of time?

**b** How much time does homework take in a day?

**c** You can list the actual number of hours on top of each bar in the bar graph. Do you think this would have been more or less effective? Explain.

**d** You could put the number of hours on the pie chart instead of the percentage. Do you think this would have been more or less effective? Explain.

**e** Which graph do you think represents the data more effectively? Explain your choice.

**f** Why was a line graph not used? Justify your answer.

**2** Below is a line graph that shows the sugar consumption of people in the United Kingdom since the 1700s.

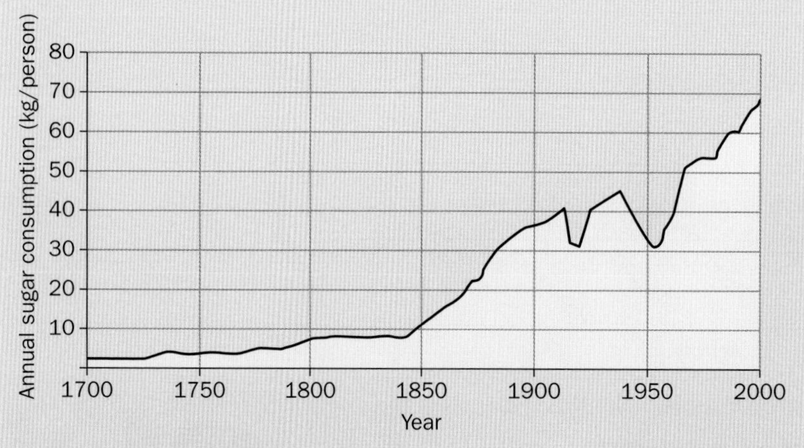

**a** Describe the trend in sugar consumption.

**b** In 1874, William Gladstone abolished the sugar tax in England. How do you think that impacted the consumption of sugar? Where do you see that on the graph?

**c** There are two large dips in the graph after 1900. What world events might have been going on at that time? Why would that have affected sugar consumption?

**d** Give an example of when the sugar consumption doubled over a span of years. (The years will be approximate.)

**e** What general trend is indicated by the graph? Justify your reasoning.

**f** When would you predict annual sugar consumption to be 80 kg per person? Explain your reasoning.

**g** Explain why you could not represent this data in a pie chart.

**h** Could you represent this data in a bar graph? Explain.

## Reflect and discuss 9

- Why is being aware of the different types of data (categorical, ordinal, discrete and continuous) important?
- Is it always possible to represent the same set of data with each of the three graphs? Explain.
- How does the way in which information is represented impact our ability to interpret it?

## Practice 4

**1** The graph shows the drink preferences of a group of high school students who were surveyed at their school cafeteria.

**Drink preferences of high school students**

Number of students: 5, 10, 15, 20, 25

tea, coffee, soft drink, water, juice, smoothie

Preferred drink

**a** What is the most popular drink?

**b** How many students prefer to drink juice?

**c** How many students prefer to drink tea or coffee?

**d** How many students were surveyed to collect this data? How many responded in each category?

**e** Represent this data using a circle graph.

**f** Which is the most effective representation for this data? Explain.

**g** What trends, if any, are visible in this graph? Justify your reasoning.

**2** The pie chart on the next page illustrates the number of times adolescents in developed countries go online each day for personal use.

**a** What is the most common number of times an adolescent goes online for personal use?

**b** Calculate the sector angle that would be needed for each category if you were to draw this pie chart by hand.

▶ Continued on next page

c What characteristics of adolescent internet usage are revealed in the pie chart? Explain.

d Represent this data using a different kind of graph. Which graph is more effective? Explain your reasoning.

e Create a question to collect similar data from your peers and then ask at least 20 people.

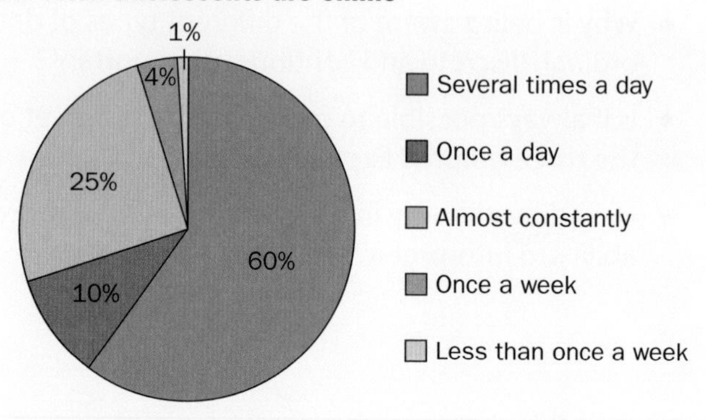


3 In a national poll in the United States conducted by CS Mott Children's Hospital, adults aged 18 and over were asked to categorize a list of concerns as 'a big problem', 'somewhat of a problem' or 'not a problem'. The following list represents the top ten 'big problems' among the adults who responded.

| Concern | Percentage |
|---|---|
| Bullying | 57% |
| Obesity | 56% |
| Drug abuse | 52% |
| Internet safety | 49% |
| Stress | 46% |
| Child abuse and neglect | 45% |
| Sexting | 44% |
| School violence | 42% |
| Suicide | 41% |
| Depression | 40% |

a What do you think is the best way to present this information? Justify your choice and graph it.

b Do you think these are large concerns in your community? Would you rank these concerns differently?

c Do any of these concerns stand out to you? Why?

d Your teacher will poll the class and ask for your top 'big problem'. You may choose a concern of your own, or one from the table. Once all the data is collected, graph the results of the class using the most appropriate form of representation.

e Are the class results similar to or different from the data above? Explain why you think this is the case.

Continued on next page

**4** At least once every 10 minutes a child is bullied and the resulting interventions are listed in this table.

| Adult intervention | 4% |
|---|---|
| Peer intervention | 11% |
| No intervention | 85% |

**a** Which type of graph would be the most effective way to represent this information? Justify your choice and create the graph.

**b** Summarize in one sentence what the statistics are representing.

**c** Do any of these percentages concern you? Why?

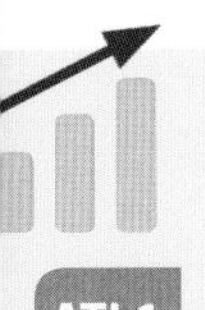

## Formative assessment – My neighborhood profile

ATL1

In this task, you will investigate your own neighborhood and create three different graphs that reveal characteristics of your community. You will include one line graph, one bar graph, and one circle graph. You will choose characteristics that are of interest to you, and choose the most appropriate type of graph to illustrate the data that you gather.

**1** In your inquiry book, answer these questions in complete sentences.

- What data would be appropriate for a bar graph?
- What data would be appropriate for a line graph?
- What data would be appropriate for a circle graph?

**2** Choose three characteristics of your neighborhood and collect the data necessary to create the graphs. You may be able to collect the data by doing research or by walking around your community. If you need to, write questions that pertain to your topic and collect data by asking people in your neighborhood.

**3** Create a bar graph, a line graph, and a circle graph from your data. Your teacher will tell you if you are creating these graphs by hand or using a computer programme such as Excel or Google Sheets. Remember to choose an appropriate scale, and do not forget to label the axes.

**4** Organize your work by characteristic. Display the collected data in a table (along with the question asked, if appropriate). Indicate whether the data is categorical, ordinal, discrete or continuous, and include the graph used to represent the data.

**5** Draw a conclusion related to each characteristic. What does your data reveal about your community?

▶ Continued on next page

6 Search online for 'what toothbrushes tell us about inequality'. Look at the photos produced by Dollar Street to see how different people around the world clean their teeth. What category would you fit into to? Choose one of the other categories and predict what the three graphs about their lives and communities would look like. Sketch the three graphs and justify your reasoning.

**Now reflect.**

7 How do you think your graphs would differ from those of a student in another country?

8 How does the way in which information is represented impact our ability to interpret it?

9 What makes one representation more effective than another?

## Making comparisons and showing trends

For comparisons, you can create a multiple bar graph, or multiple pie charts, using the same scale, or put two or more lines on the same set of axes (multiple line graphs), as shown below. Using several bars, pie charts or lines can help demonstrate trends and show similarities and differences between groups.

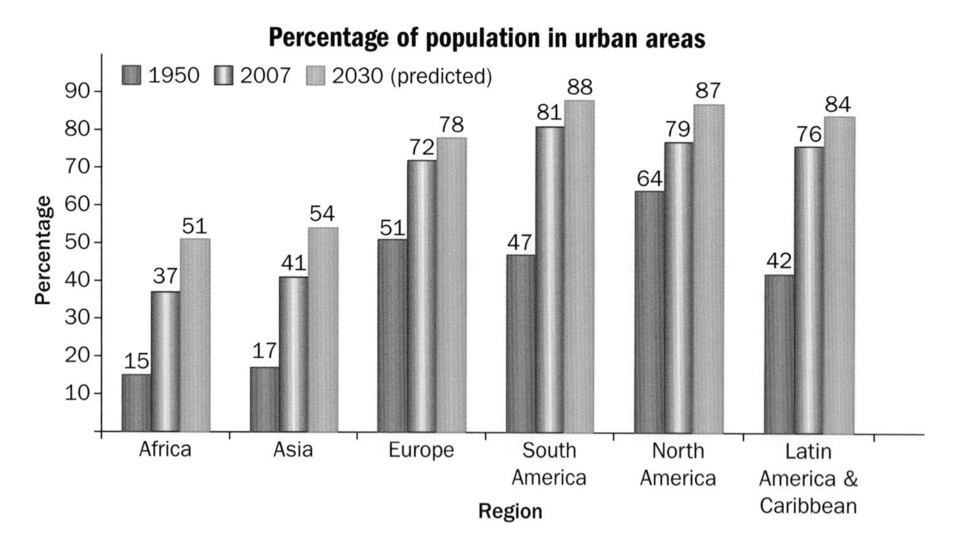


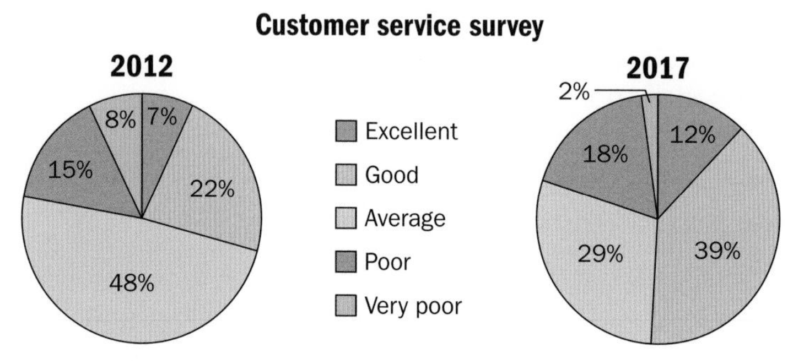

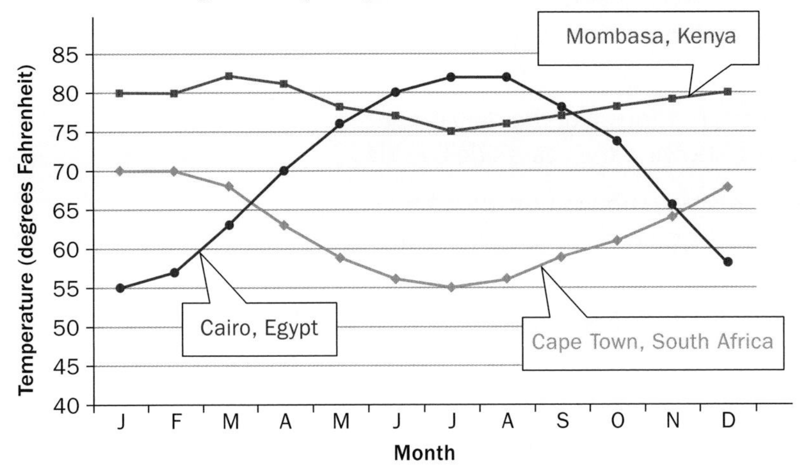


## Reflect and discuss 10

- Which graph(s) show trends in a community? Describe these trends.
- Which graph(s) reveal characteristics about a community? Describe these characteristics.
- Could the data in the multiple bar graph be represented using a multiple pie chart or a multiple line graph? Explain.
- Could the data in the multiple pie chart be represented using a multiple bar graph or a multiple line graph? Explain.

## Activity 5 - Comparison graphs

ATL2

Some Year 1 students were asked, 'What is the greatest source of stress in your life?' Their responses are shown in this table.

| Stressor | Year 1 | Year 3 | Year 5 |
|---|---|---|---|
| School | 36% | | |
| Family | 45% | | |
| Friends | 15% | | |
| Dating/relationships | 4% | | |

1 Fill in the rest of the table based on how you think students in each year would respond to the same question.

2 Compare your table to four of your peers. What trends do you predict will happen as you get older? Justify your answer.

3 Create a pie chart for each year, putting them side by side.

In part **1**, make sure each column totals 100%.

▶ Continued on next page

4 Create a bar graph for the stressors, putting all of the data on one graph. Group the data by year, showing all of the stressors side by side.

5 Create a bar graph for the stressors, putting all of the data on one graph. Group the data by stressor, showing the years side by side.

6 Create a line graph for the four stressors, showing the data for Years 1, 3 and 5. Put the three lines on one graph, but use a different color for each year group.

7 In which graph are the trends you predicted most visible? In which are they the least visible? Explain.

8 Which representation do you think is the most effective? Explain your choice.

## Reflect and discuss 11

- List the advantages and disadvantages of each type of graph.
- If you wanted to show that a trend existed in data you collected, which graph would you choose? Explain.
- If you wanted to hide a trend, which graph would you use? Explain.

## Practice 5

**1** The graph shows the average heights of Japanese adolescents aged 5 through 17.

**a** Estimate:

**i** the average height of a male at age 8

**ii** the average height of a 10 year old.

**b** In your own words, compare the heights of girls and boys over the 12 year period.

**c** How would these statistics differ around the world?

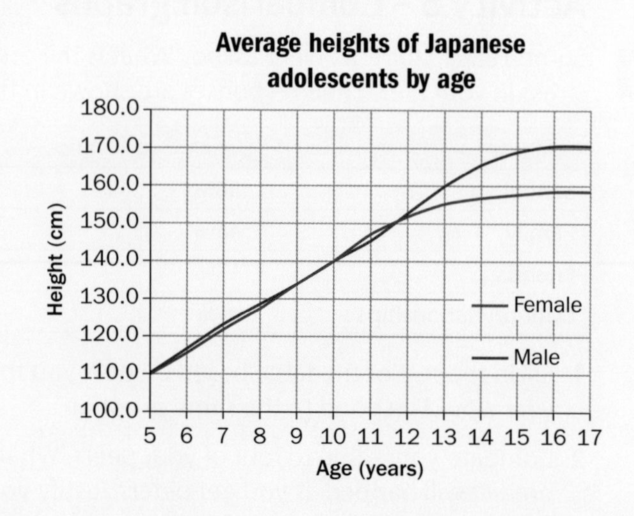


▶ Continued on next page

**d** Is this the most effective way to represent this data? Explain.

**e** What characteristics of Japanese adolescents, if any, are visible in this graph? Justify your answer.

**f** What trends among Japanese adolescents, if any, are visible in this graph? Justify your answer.

**2** The following graph shows the favorite sports of students at an international school.

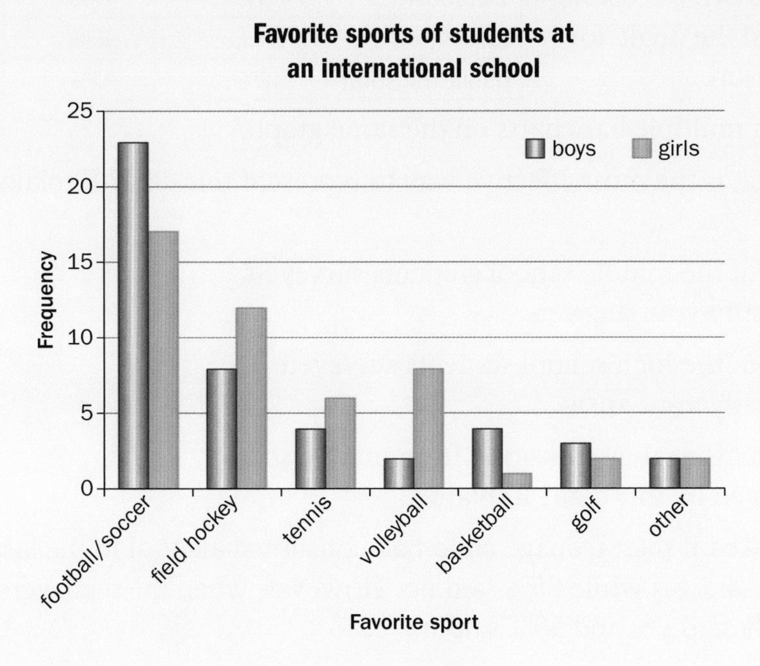


**a** What is the most popular sport?

**b** What sports are more popular with girls than boys?

**c** How many boys said golf was their favorite sport?

**d** How many students said tennis was their favorite sport?

**e** How many students were surveyed to collect this data? Explain.

**f** How many students had a favorite sport that was not listed as an option in this survey?

**g** Do you think it is acceptable to have a column for ‘other’? Why or why not?

**h** Is this the most effective way to represent this data? Explain.

▶ Continued on next page

**3** The table presents data about the modes of transportation used by a group of middle school students and a group of high school students to and from school. 60 middle school students said they take public transport.

| Mode | Middle school | High school |
|---|---|---|
| bike | 11% | 8% |
| walk | 12% | 12% |
| skateboard | 5% | 12% |
| car | 24% | 36% |
| public transport | 48% | 32% |

**a** How many middle school students were surveyed? How many middle school students walk to school?

**b** Represent this data using two pie charts. Show how you calculated the angle for each mode of transportation.

**c** Represent this data using multiple bar charts on the same graph.

**d** Which graph do you think is the most effective way to represent this data? Explain your reasoning.

**e** What characteristics about the middle school students surveyed does this data reveal? Justify your answer.

**f** What characteristics about the high school students surveyed does this data reveal? Justify your answer.

**g** What trends among students as they transition from middle school to high school are suggested by this data? Explain.

**4** In a recent survey, when asked if their teenage child had consumed alcohol in the last 12 months, 15% of parents said yes while 85% said no. However, when the teenagers themselves were asked, 50% said yes and 50% said no.

**a** Design a graph to represent this data.

**b** If you were to write an article that discussed the results of this survey, what headline would you use to summarize these statistics?

**c** Based on this information, would you be able to identify a trend in family communication between parents and teenagers? Explain. Would you need other information?

**d** Does this data trouble you? Explain.

## Misleading graphs

Graphs are meant to represent data in a fair and meaningful way. However, it is possible to create graphs that are misleading or that make trends less visible. Part of being media literate is being able to determine when a graph is misleading or is hiding information and trends.

## Activity 6 – Interpreting scale

Trends in social media are showing that teenagers enjoy a large variety of choices, from texting and photo apps to micro-blogging and even self-destructing apps.

**1** Below is a graph showing the percentage of teenagers (who have access to and use social media) who use Facebook.

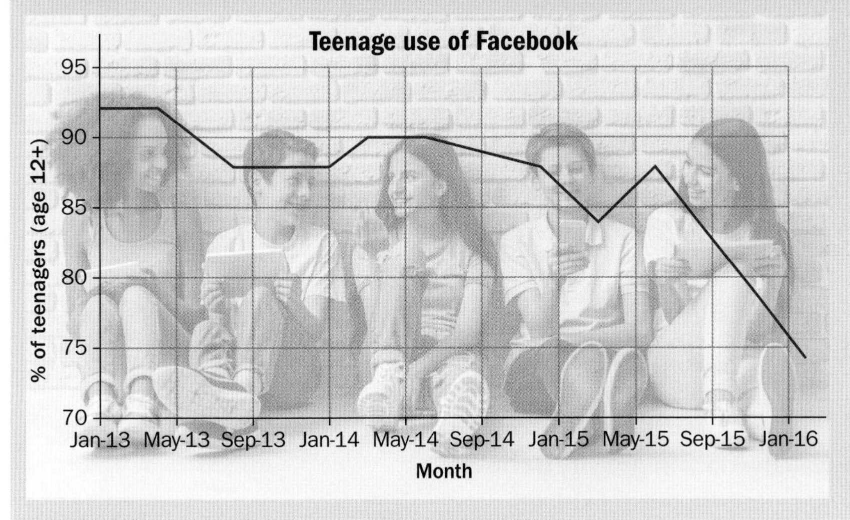


The title of the article where this graph appeared was: 'Facebook no longer popular with teenagers'.

- **a** Do you agree with the title of the article?
- **b** What vertical scale was used in this graph?
- **c** Is the scale appropriate? Explain.
- **d** Why do you think this scale was chosen?

**2** Here is the exact same data displayed differently.

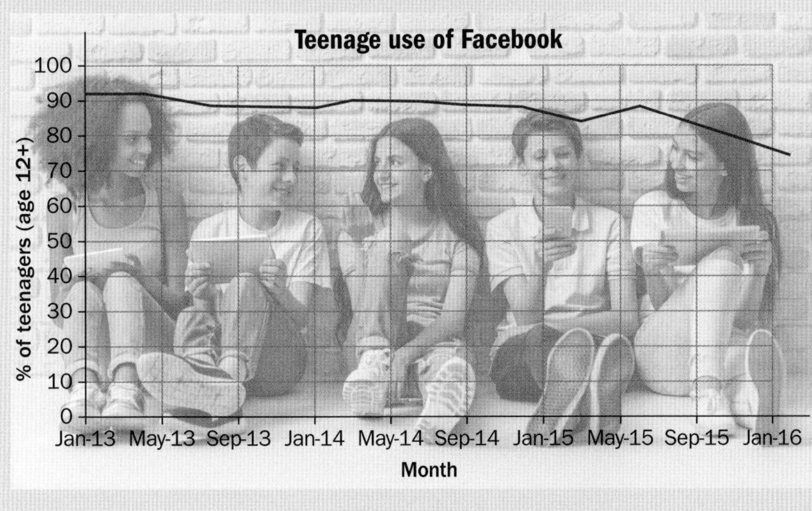


▶ Continued on next page

a How has the vertical scale changed?

b Explain why the line does not look as steep.

c Which scale do you think is more effective, and why?

d Looking at the graph now, what conclusion can you draw about the percentage of teenagers (who have access to and use social media) who use Facebook?

ATL2

e Based on the trends above, what do you predict will happen to teenage usage of Facebook over the next few years?

f How did the different representations impact how you interpreted the data?

## Reflect and discuss 12

- Explain how data can be misrepresented.
- Can you think of another way the data in Activity 6 could be displayed that would also be inappropriate? Explain.

## Practice 6

1 These graphs show the homicides per 100 000 people in each of three countries.

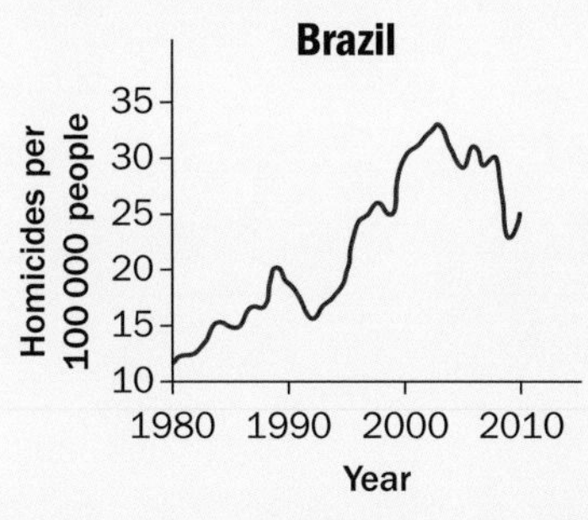


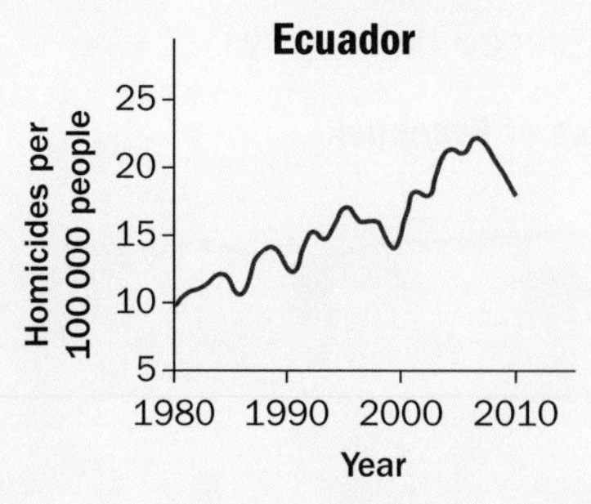


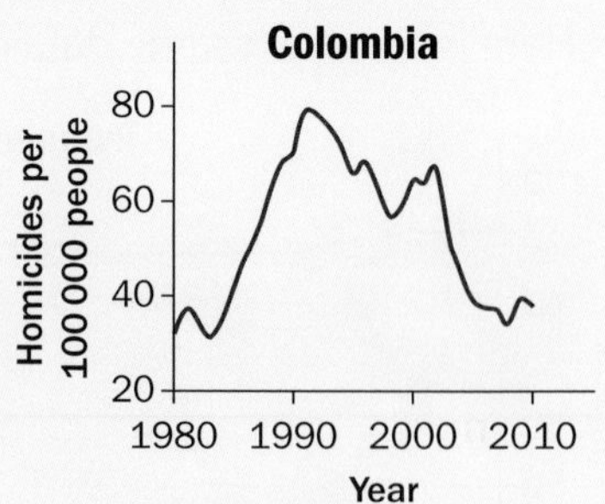


a Summarize the homicide rate of each country over the 30-year period.

b Do you think it's fair to use these graphs as a way to compare crime rates of the three countries over the 30-year period? Justify your response using features of the graphs.

▶ Continued on next page

**2** Obesity prevalence in youth is represented in this line graph.

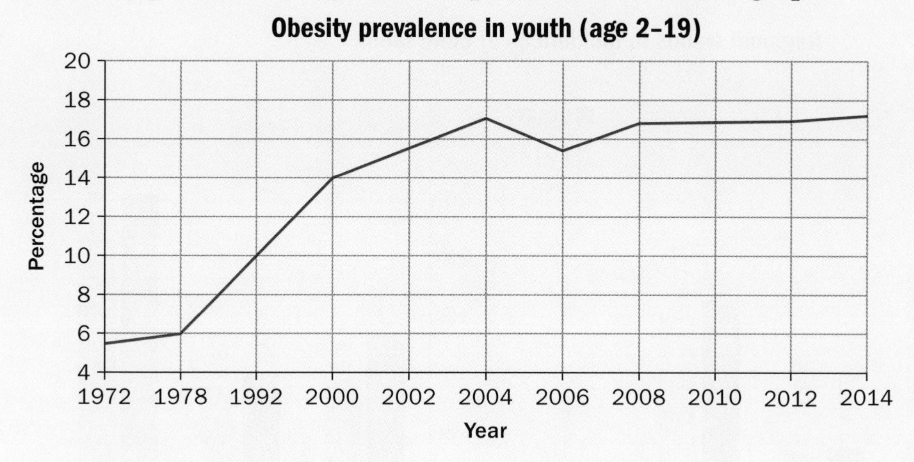


**a** Summarize what the statistics are representing. Is there any information missing?

**b** What is the general trend in this data?

**c** Based on this trend, what do you think the obesity rate among youth would be today? In 10 years? Justify your predictions.

**d** List how you would improve the representation of this data.

**3** The number of violent crimes per 100 000 residents for six large American cities is represented in this multiple line graph.

**a** What is the general trend in violent crime for these cities?

**b** Do you think it is fair to compare these cities in this way? Explain.

**c** The city of San Diego used the graph to advertise, 'We have always been one of the safest cities in America.' Is that a fair interpretation of the data? Justify your answer.

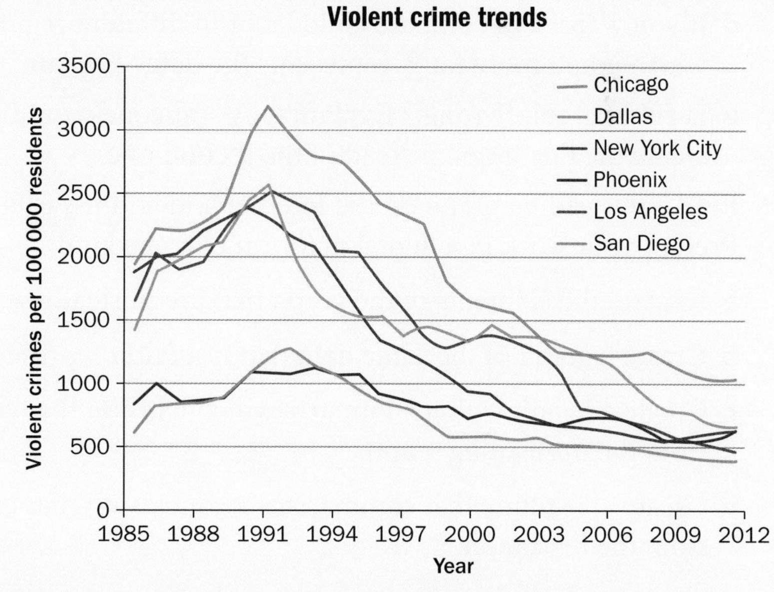


**d** Suppose the mayor of Chicago used the graph to explain, 'Our violent crime rate has fallen so dramatically that it is now essentially equal to that of other large cities'. Would that be a fair interpretation of the data? Explain.

▶ Continued on next page

**4** The two graphs shown here represent the same data regarding child labor.

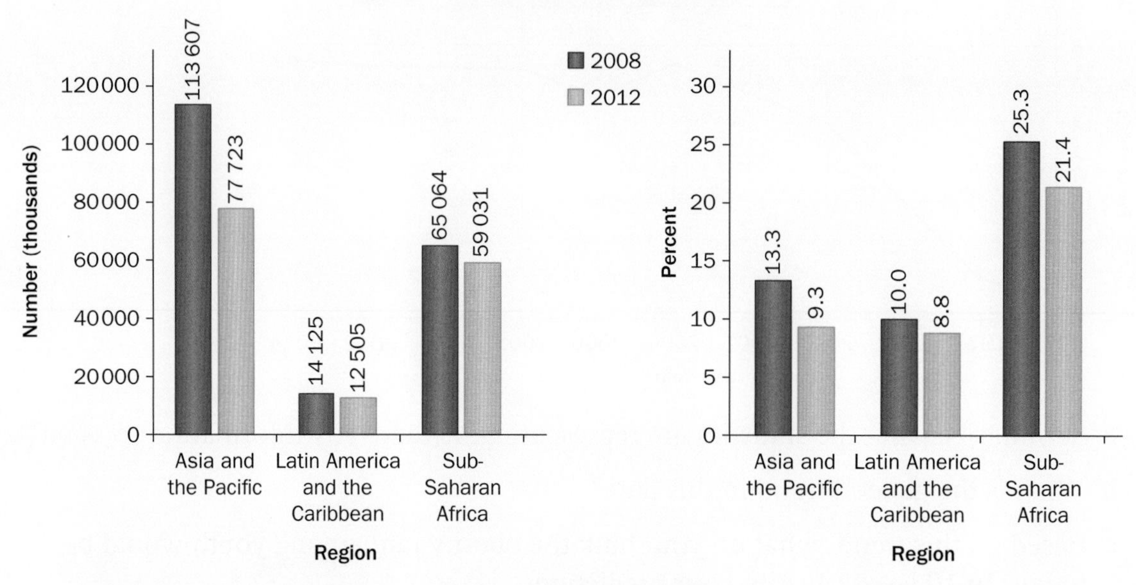


**a** Explain any differences between the graphs.

**b** Write a conclusion that would be true for the first graph but not the second one.

**c** Write a conclusion that would be true for the second graph but not the first one.

**d** If you wanted to compare child labor in different regions of the world, which graph more fairly represents the data? Explain.

**e** Is either graph 'wrong'? Explain why someone would prefer to create the first graph instead of the second one.

**5** Find a misleading graph on the internet and address each point below. Present your work in a digital platform of some kind.

**a** Explain the elements of the graph that are misleading.

**b** Create a graph of the same data that is a fairer representation.

**c** Create a headline that summarizes a conclusion that can be drawn from the misleading graph.

**d** Create a headline that summarizes a conclusion that can be drawn from the fair graph.

▶ Continued on next page

**6** This chart was used in an article that claimed, 'High school graduation rates have increased dramatically under new government!'

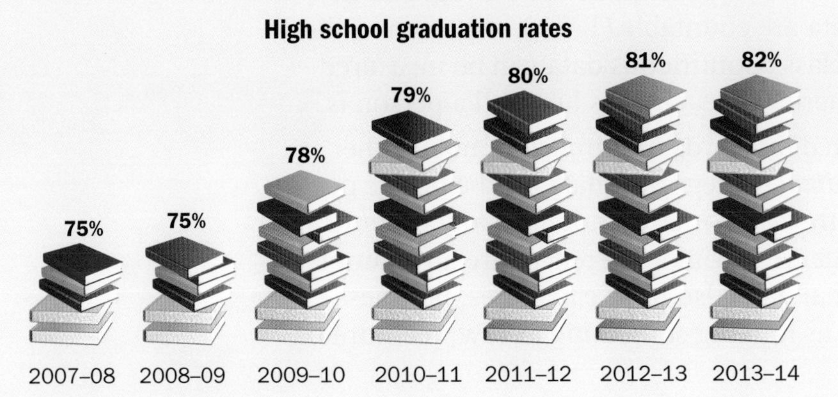


**a** Explain whether or not you think the claim is justified from the graph.

**b** List any factors that make the graph misleading.

**c** Draw a graph that would represent the data more fairly and effectively.

**d** Write down a conclusion that could be drawn from your new graph.

**e** Explain how your conclusion is different than the original.

Suppose data on graduation rates were gathered from several decades and represented in this line graph.

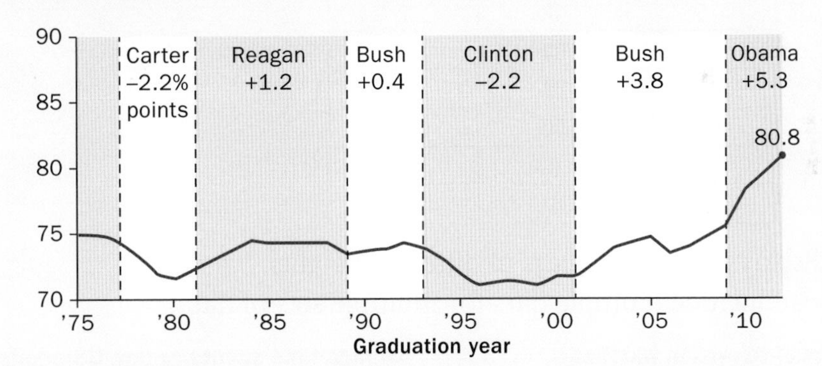


**f** What in this graph, if anything, is misleading?

**g** How could this graph be used to justify the original conclusion about there being a dramatic increase in graduation rates?

# Unit summary

*Quantitative data*, represented by quantities, can be either *discrete* or *continuous*. Discrete data are countable (1, 2, 3, ...), such as the number of students in a class. Continuous data can be measured and can take on any numeric value, such as how tall a person is.

*Qualitative data*, represented by words or numbers, can be either *categorical* (*nominal*) or *ordinal*. Categorical data can be of one or more categories, though the categories do not have any order to them. Hair color, music genres, drink preferences are all examples of categorical data. Ordinal data also have categories, but these now have an order, such as ranking something as low, medium or high.

The most common way to represent statistical information is to use a *graph*. In statistics, a graph is an organized picture that represents data.

A *frequency table* uses tally marks in one column to keep track of items that are being counted. When the counting is done, the tallies are added up and the total is entered in another column.

| Mark | Tally | Frequency |
|---|---|---|
| 4 | II | 2 |
| 5 | II | 2 |
| 6 | IIII | 4 |
| 7 | ~~IIII~~ | 5 |
| 8 | IIII | 4 |
| 9 | II | 2 |
| 10 | I | 1 |

A *bar graph* displays data using bars of different heights.
A bar graph can include multiple bars for multiple sets of data.

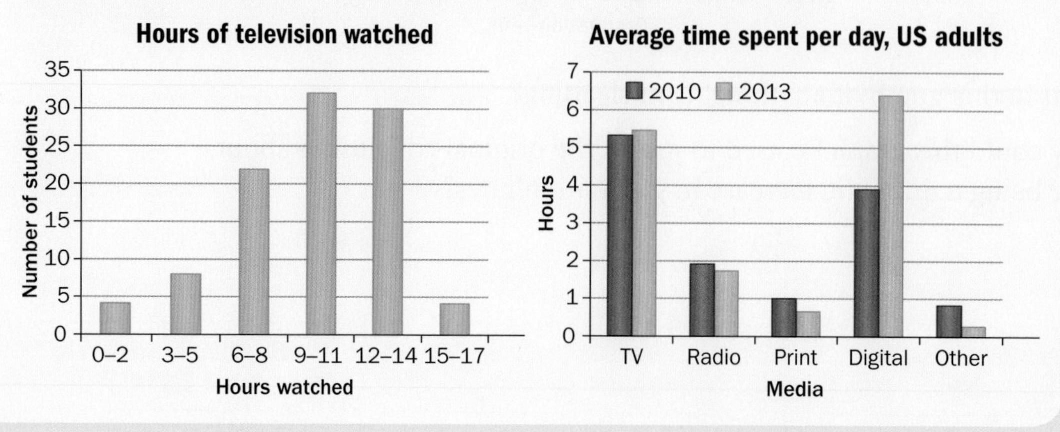

A *line graph* is created by joining data points with lines. A line graph can include one or more lines.

*Time series* is a common form of line graph in which time is represented on the horizontal axis.

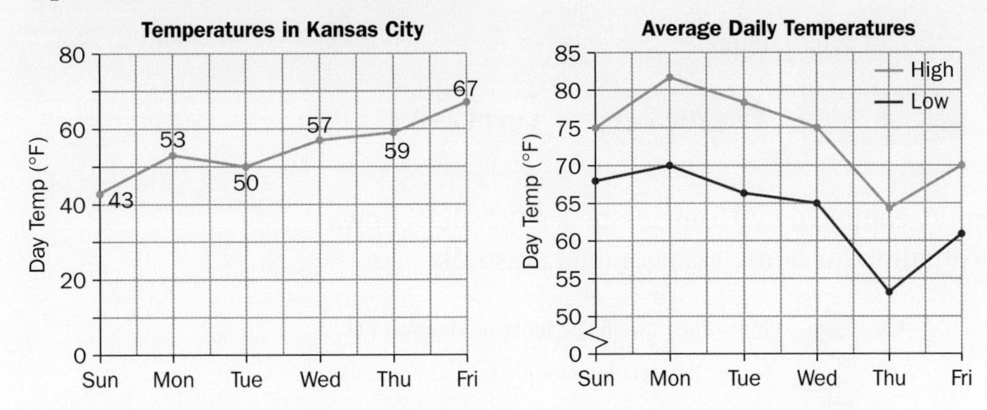


A *circle graph* (pie chart) uses a circle divided into sectors to show the percentage of each category represented. Several pie charts together may be used to compare data.

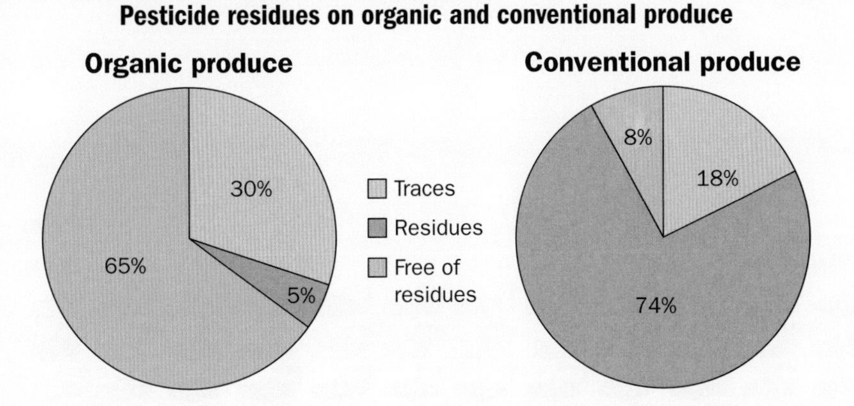


When creating a pie chart by hand, you divide the pie into sectors. Since there are 360° in a circle, the degree measure of each sector must be calculated so that it is proportional to the amount of data it represents. For example, 10% would be represented by a sector spanning 36°.

Graphs can be misleading if the scale is not properly chosen or if an ineffective graph is selected to represent data.

# Unit review

Launch additional digital resources for this chapter

Key to Unit review question levels:

Level 1–2 | Level 3–4 | Level 5–6 | Level 7–8

**1** This infographic, compiled by Homelessness Australia, shows the dwelling situation for homeless people in Australia.

| | |
|---|---|
| 6% | Improvised dwellings, tents or sleeping out |
| 20% | Supported accommodation for the homeless |
| 17% | Staying temporarily with other households |
| 17% | Boarding houses |
| 1% | Other temporary lodging |
| 39% | Severely overcrowded dwellings |

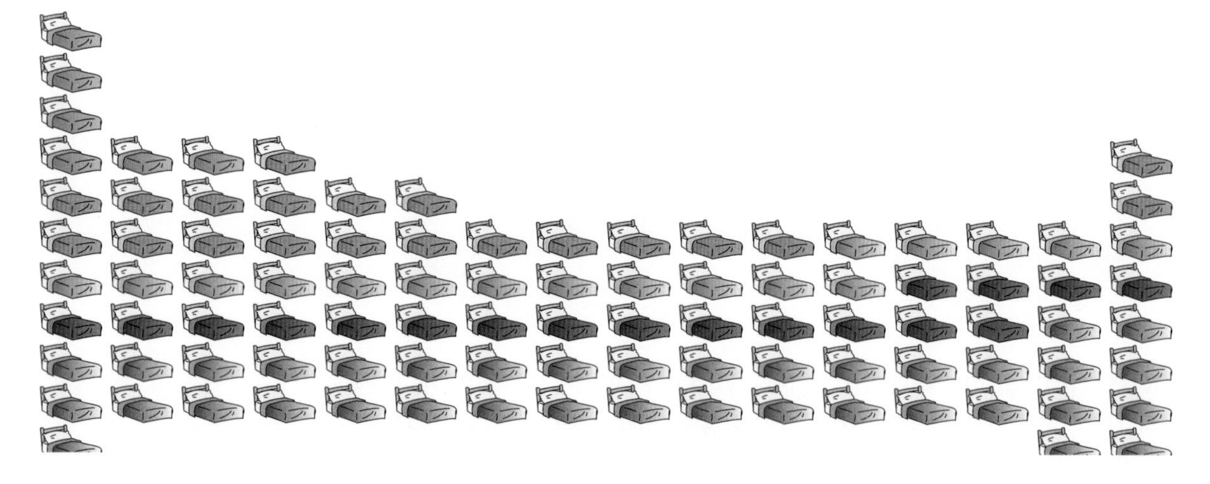

**a** What kind of data (categorical, ordinal, discrete or continuous) is being collected? **Explain** your reasoning.

**b** **Describe** what this graph is displaying.

**c** What aspects of this graph make it effective? What aspects make it less effective? Give reasons for your answers.

**d** What type of graph do you think would be more effective at representing this data? **Explain** your choice. **Draw** the graph.

**2** In a survey carried out in the US on youth risk behavior, it was found that of the 68% of high school students who rode bikes in the last year, 81% of them rarely or never wore helmets.

**a** What kind of data (categorical, ordinal, discrete or continuous) was collected? **Explain** your reasoning.

**b** What aspects of this graph make it effective? What aspects make it less effective? Give reasons for your answers.

**c** What does this tell you about youth risk behavior?

**3** Data was gathered on cyberbullying in a school.

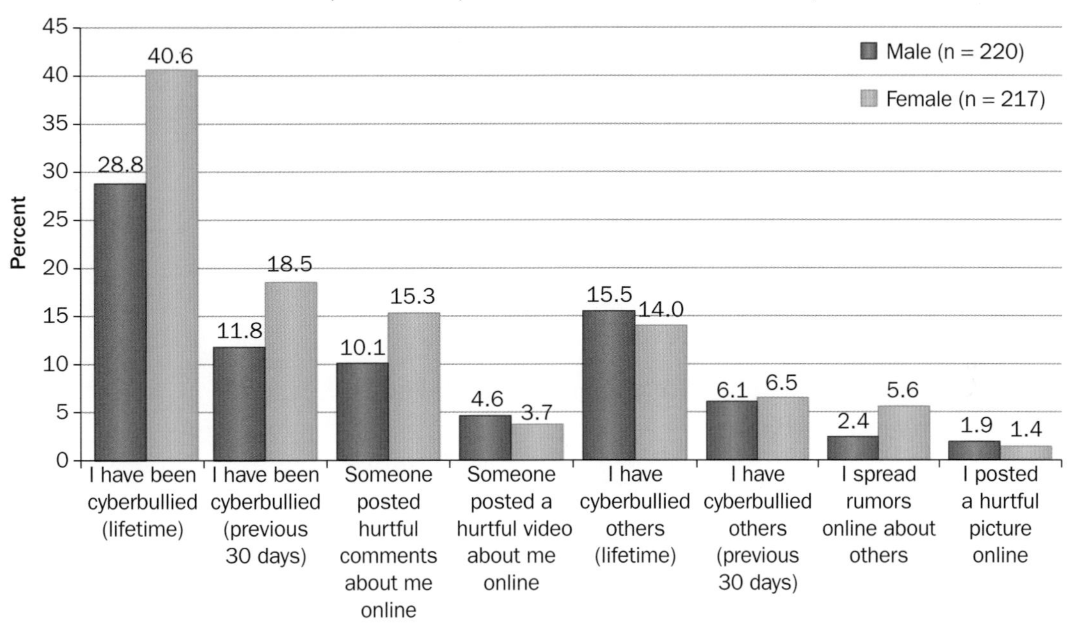


**a** What characteristics do you see in this school? **Explain**.

**b** Why would a pie chart not be an effective way to represent this data?

**c** Based on the graph, does cyberbullying seem to be a problem in this school? **Justify** your answer.

**4** The following was published in a newspaper about the increase in Cooper's hawks in a local community.

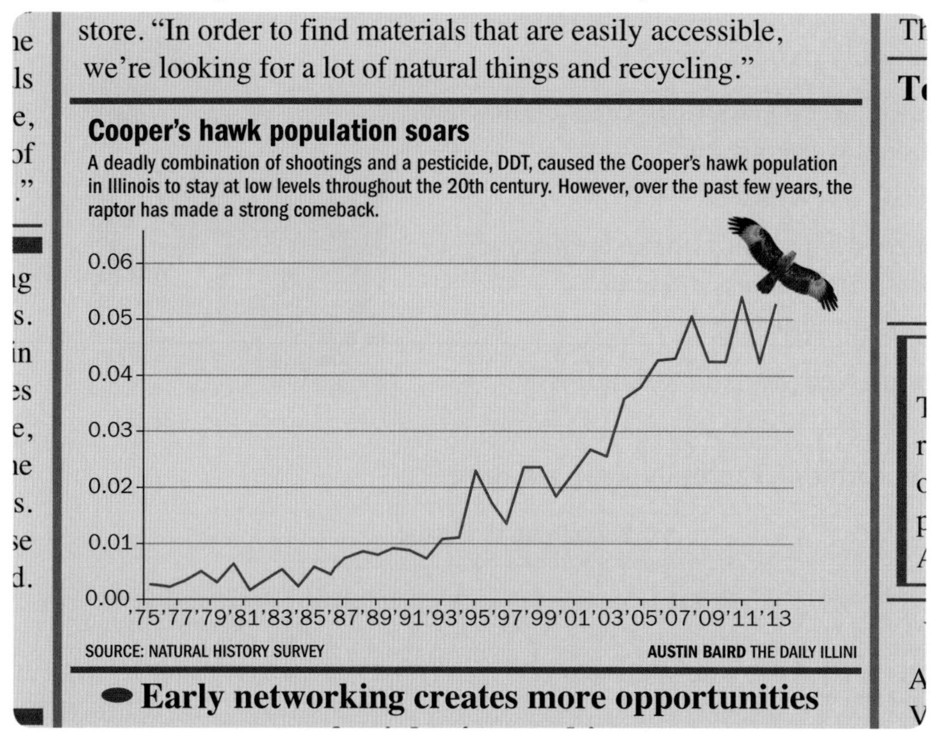
store. "In order to find materials that are easily accessible, we're looking for a lot of natural things and recycling."

**Cooper's hawk population soars**

A deadly combination of shootings and a pesticide, DDT, caused the Cooper's hawk population in Illinois to stay at low levels throughout the 20th century. However, over the past few years, the raptor has made a strong comeback.



SOURCE: NATURAL HISTORY SURVEY

AUSTIN BAIRD THE DAILY ILLINI

**Early networking creates more opportunities**

**a** **Explain** how the graph matches the headline.

**b** What aspects of the graph are misleading or unclear? What would you need to know in order to create a graph that more clearly represented the data?

**c** **Write down** a new headline for the graph that is more representative of the data.

**5** As a class, fill in a copy of this tally chart.

**a** What type of data (categorical, ordinal, discrete or continuous) was collected?

**b** **Draw** a bar graph to represent the data.

**c** **Draw** a pie chart to represent the data.

**d** Which representation is the most effective? **Explain**.

**e** What characteristics about your class does this reveal?

| Hours of homework completed last night | Tally | Frequency |
|---|---|---|
| 1 | | |
| 2 | | |
| 3 | | |
| 4 | | |
| 5 | | |
| 6 | | |

**6** A study was done to collect data on the amount of homework required from students at a school in Canada.

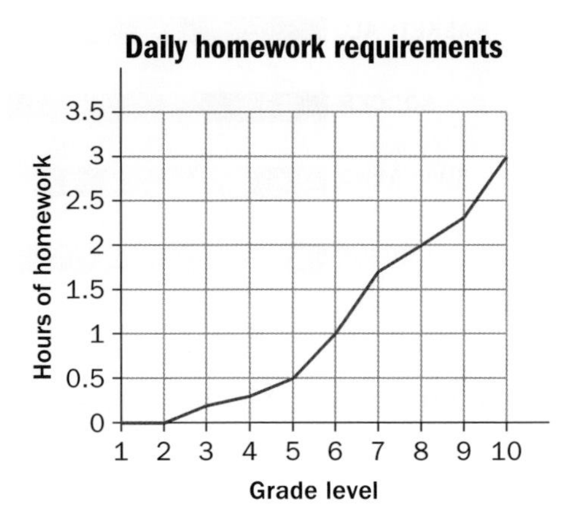


**a** What type of data is 'Grade level'?
What type of data is 'Hours of homework'?

**b** What is the average amount of homework required daily from a Grade 6 student?

**c** How much does that increase in Grade 7?

**d** In what grades are there no homework requirements?

**e** In what grade are the homework requirements double those in Grade 6?

**f** Between which two grades is there the greatest increase in homework requirements?

**g** What is your prediction for the number of hours of homework that a Grade 11 student will have?

**h** **Explain** why you could not represent this data in a pie chart.

**i** Should this data be represented in a line graph? **Explain**.

**j** What would be the most effective way to represent this data? **Draw** the graph using the given data and explain why it is the most effective representation.

**7** Year 3 students in a school were asked for their grade in a science class, and about the use of their cell phone in the class.

| Frequency of cell phone use per class | Average grade in class (1–7) |
|---|---|
| 0 | 6.5 |
| 1 | 6.3 |
| 2 | 6.2 |
| 3 | 5.8 |
| 4 | 5.5 |
| 5 | 5.2 |
| 6 | 4.7 |

**a** **Draw** two different types of graph to represent the data. **Discuss** the effectiveness of each graph.

**b** What trends do you see in the data? **Justify** your answer.

**c** Propose an explanation for the trends you see in the data.

**d** **Explain** how this data would impact your use of a cell phone in class.

**8** The data below represent the participation rate in various sports based on the family's income level.

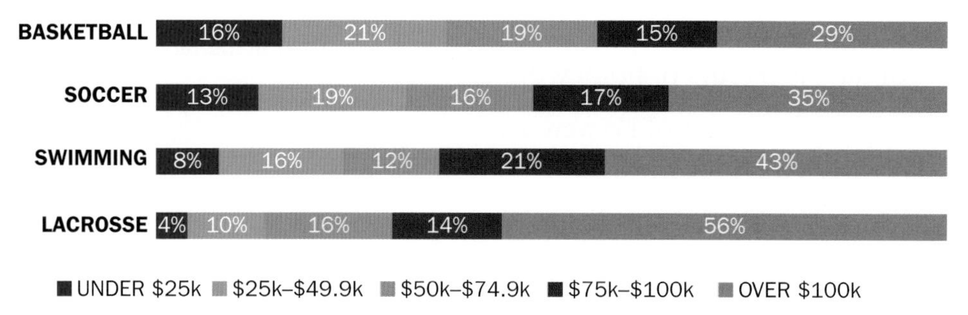


**a** **Draw** multiple pie charts to represent the data (one pie chart for each sport).

**b** Which representation do you think is more effective, the original bar graph or the pie charts? **Explain**.

**c** What characteristic about families do you see in this data? **Justify** your answer.

**d** Based on the graph, which sport do you think is the most expensive to play? **Explain**.

**e** How can a community increase participation in sports by families that don't necessarily have the means to afford it?

**9** An earnings ratio shows how much someone earns in comparison to a standard. The standard is usually represented by 1.0, so an earnings ratio above 1.0 means earning more than the standard. An earnings ratio less than 1.0 means earning less than the standard. The earnings ratios for a range of education levels are represented in this bar chart.

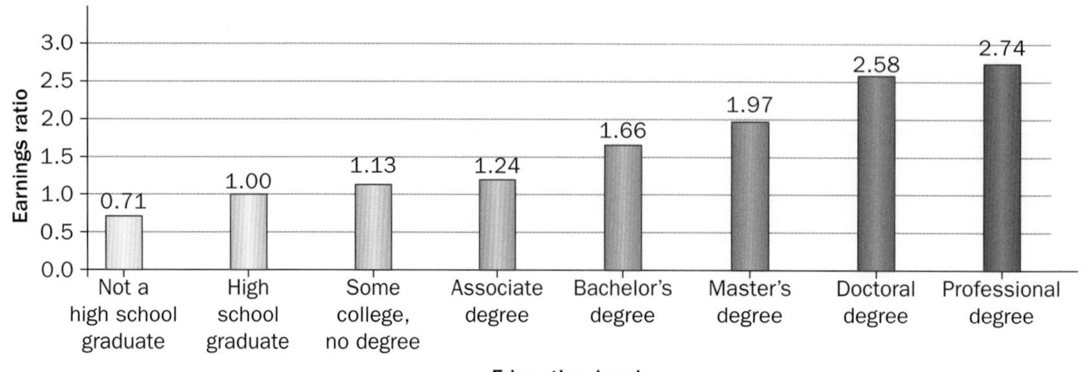

**a** Which education level is the standard against which others are compared?

**b** What information does the graph convey? **Explain** your reasoning.

**c** How much education would you need in order to earn at least twice as much as a high school graduate?

**d** Is this the most effective way to represent this data? **Explain**.

ATL 1&2

**10** A study of the number of texts sent per day produced the following data.

| Number of texts sent per day | Percentage of teen responses | Percentage of adult responses |
|---|---|---|
| 0 | 2% | 9% |
| 1 – 10 | 22% | 51% |
| 11 – 20 | 11% | 13% |
| 21 – 50 | 18% | 12% |
| 51 – 100 | 18% | 7% |
| 101 or more | 29% | 8% |

**a** **Use** technology to represent each column of responses with a circle graph.

**b** **Use** technology to represent this data using multiple bar graphs.

**c** Which representation is more effective? **Explain** your answer.

**d** What characteristics about each age group does this reveal regarding texting? **Explain**.

**e** **Draw** a graph for how you think the same age groups would have responded ten years ago. **Explain** your graph.

**f** **Draw** a graph for how you think the same age groups will respond ten years from now. **Explain** your graph.

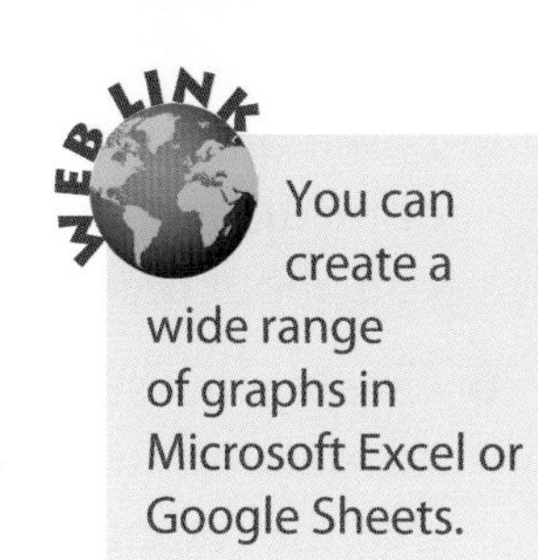


You can create a wide range of graphs in Microsoft Excel or Google Sheets.

**11** A study collected data on media usage and other elements of life for a group of of 8- to 18-year-olds. The data is represented in the table below.

**Media, grades and personal contentment**

| | **I am a heavy user of media** | **I am a moderate user of media** | **I am a light user of media** |
|---|---|---|---|
| I get good grades | 51% | 65% | 66% |
| I get fair/poor grades | 47% | 31% | 23% |
| I have a lot of friends | 93% | 91% | 91% |
| I get into trouble a lot | 33% | 21% | 16% |

**a** Based on the table, describe characteristics of heavy users of media compared to light users. Does anything surprise you about the data?

**b** Create a question that could have been asked to collect the data on students' grades. **Explain** how it is possible that the responses for a media user's grades do not add up to 100 percent.

**c** Represent this data in a more effective form. **Explain** why that form is more effective.

**d** Collect the same data from your class. Represent the class data in the most effective form. **Describe** differences between your class data and the data given above.

**e** Would you change your media usage based on this information? **Explain** your reasoning.

**12** The following data is for the five largest African countries by population.

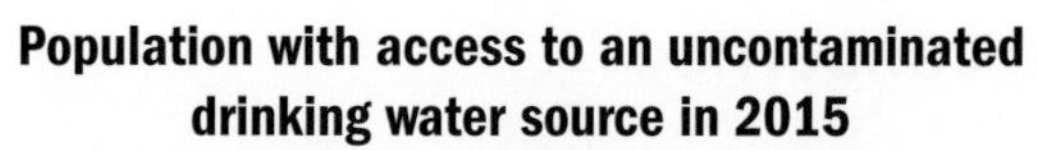


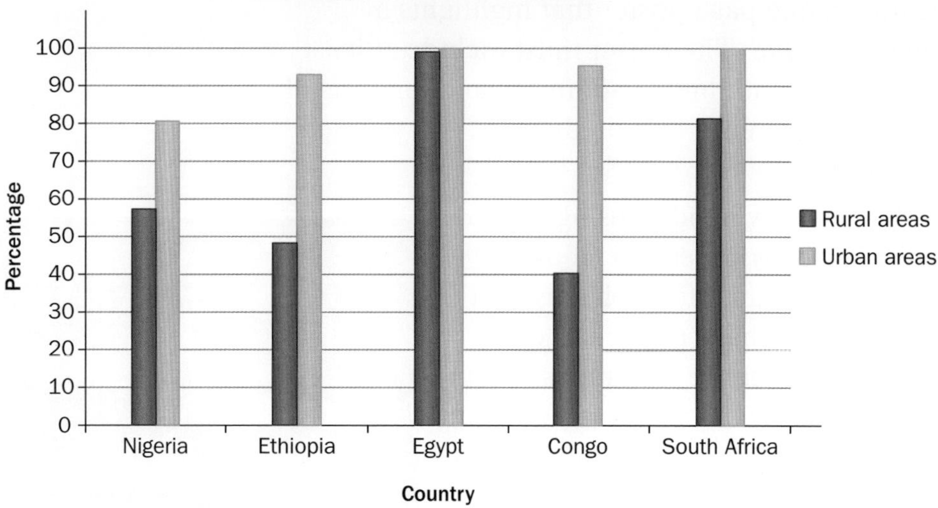


**a** What is the difference between a rural and an urban area?

**b** Which country has the lowest percentage of access to safe drinking water in rural areas?

**c** Which country has the lowest percentage of access to safe drinking water in urban areas?

**d** Overall, which country has the most access to safe drinking water?

**e** Represent this data using multiple pie charts. Which representation is more effective? **Explain**.

**f** Search UNICEF data (data.unicef.org) and, under 'data by topic', select 'water' and download 'drinking water, sanitation & hygiene' data. Search for your country and compare these results to those of the African countries. Are the percentages higher or lower in your country? Why do you think that is?

# Summative assessment

ATL 1&2

## Should your community be concerned?

You will be creating a one-page poster that highlights a particular community issue. The poster must include at least three graphs (bar, circle, and line), a summary and analysis of the graphs, key statistics, trends and predictions that you see, and possible next steps.

You will create the poster by hand or using technology, as directed by your teacher. If using technology, you can create an infographic using a free program such as Piktochart (piktochart.com) or Canva (canva.com) and create the graphs within the program. Alternatively, you can import all your graphs from graphing software such as Excel or Google sheets.

### Part 1–Collecting the data

You will be researching a variety of community issues, such as school attrition (drop out) rates, homelessness, traffic safety, environmental problems, poverty, transportation, drug use or crime that occur in your community, city or town. Once you have done some research, you may choose a particular area to focus on, or your teacher may give you a particular issue to look at. You may collect the data yourself using a survey or you may collect data from online sources. Make sure you collect data to display in:

- a bar graph or multiple bar graphs
- a circle graph (pie chart) or multiple circle graphs
- a line graph or multiple line graphs.

At least one graph needs to represent multiple data sets (for example, a multiple bar chart, multiple line graph or multiple pie charts).

## Part 2-Displaying the data

Create at least one of each type of graph, making sure each has a relevant title, all axes are labelled, and legends are clear and color-coded where appropriate. The graphs must be of an appropriate size to ensure all the information is clear.

## Part 3-Analyzing the data

Use your graphs to analyze the data and summarize your findings in paragraph form. Present your data and findings in a digital format that includes:

- a headline for the data/conclusion
- a general statement (summary) of your data
- any patterns or trends that you see
- changes that have taken place over the years
- predictions for the future based on your data, and justifications for these predictions.

## Part 4-Reflect and discuss

Also, in your presentation, include a reflection that answers the following questions, making clear reference to your data and findings.

- Should the community be concerned? If so, what could you do to help address the community concern you identified?
- Whose responsibility is it to identify and fix problems within your community? How about other communities around the world?

# 7 Perimeter, area and volume

In this unit, you will see that being able to define and calculate the amount of space taken up by an object is an important step in figuring out how to minimize its impact on the environment. However, shapes and space also play an important role in understanding a wide range of other global contexts.

## Fairness and development

### Questions of power and privilege

How much space does a human being need in order to live? Why do some people have so much space while others have so much less? Perimeter, area and volume can be used to analyze housing settlements around the world, and to illustrate the stark differences between those who have power and privilege and those who do not.

At the same time, it has been said that less than 5% of the world's population uses at least 25% of its natural resources. What problems might this pose?

Calculating the volume of resources that are available and the extent of the damage caused to the planet in obtaining them can help to further demonstrate the inequality.

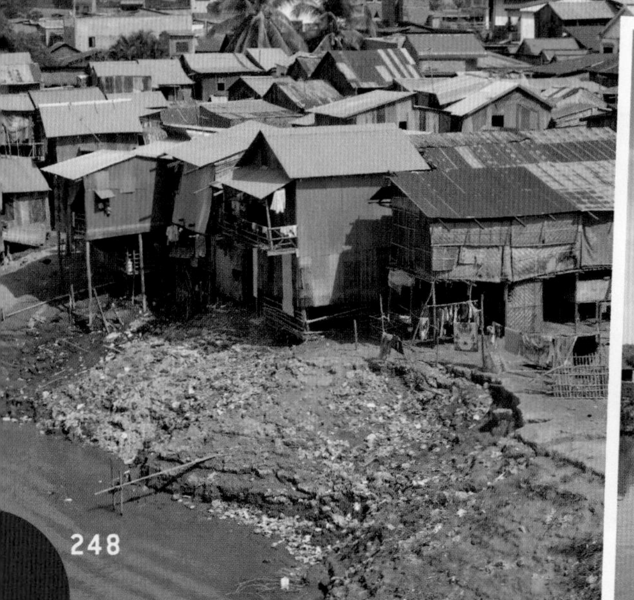

## Orientation in space and time

### Urban planning in ancient civilizations

We can study the great ancient civilizations through what remains of their architecture and the products that they created. The Ancient Egyptians and Chinese produced impressive structures that can be analyzed by looking at the space they occupy and shapes chosen. Whether it's an obelisk covered in ancient writing or the Great Wall of China, looking at perimeter, area and volume can help us understand the magnitude of a project and give important insight into the civilization that created it.

The Rama empire existed in what is now Pakistan and India. The empire was so advanced that its cities, including Mohenjo-daro, were carefully planned before their construction. The use of rectangular structures for the city grid and the housing predates evidence of the basic principles of geometry. More impressive still is the fact that the sewage system they developed was more sophisticated than those found in many cities today!

# 7 Perimeter, area and volume

## Environmental impacts

## KEY CONCEPT: RELATIONSHIPS

**Related concepts:** Generalization and Measurement

## Global context

In this unit you will explore the global context of **globalization and sustainability** as you discover how being able to calculate perimeter, area and volume can help improve decisions that impact the environment. From the tiny house movement to alternative energy sources, knowing how to analyse shapes and the space they occupy can help you do your part to make the planet a liveable place for future generations.

## Statement of Inquiry

Generalizing the relationship between measurements can influence decisions that impact the environment.

### Objectives

- Generalizing the relationship between the area and perimeter of a rectangle, and between the areas of a triangle and a rectangle
- Finding the perimeter and area of compound 2D shapes
- Generalizing the relationship between the area of 2D shapes and the volume of their corresponding prisms
- Finding the surface area and volume of regular 3D prisms
- Applying mathematical strategies to solve problems involving 2D and 3D geometric shapes

### Inquiry questions

**F** What can be measured?

What is the relationship between area and volume?

**C** How are relationships between measurements generalized?

**D** What is the cost of being environmentally friendly?

What is really influencing decision making?

**ATL1 Creative thinking skills**

Design improvements to existing machines, media and technologies

**ATL2 Reflection skils**

Consider ethical, cultural and environmental implications

## You should already know how to:

- calculate the perimeter of simple shapes
- find the area of simple shapes
- solve problems involving percentages
- round numbers correctly

# Introducing perimeter, area and volume

Why is there so much empty space in a bag of potato chips? Why isn't a box of rice filled all the way up to the top? What does it mean when we see statements such as, 'This package is sold by weight, not by volume'?

Some products fit fairly tightly into their packaging, but there is often an excess of packaging anyway. How are packaging decisions made? What impacts do they have? The answers to these questions will be the focus of this unit.

ATL2

## Reflect and discuss 1

- Why do you think bags of snack foods and boxes of rice aren't filled all the way up to the top?
- Why might a company put a small product in a large package?
- Explain what impact these packaging practices have on the environment.

The issues related to packaging and the environment can be analysed using mathematics. After generalizing relationships between measurements of various shapes, you can explore how to optimize packages so that there is as little waste as possible.

## Two-dimensional shapes

### Venn diagrams

One way of representing the relationship between several items or ideas is by using a *Venn diagram*. For example, the relationship between characteristics of plants and people can be represented with this Venn diagram.

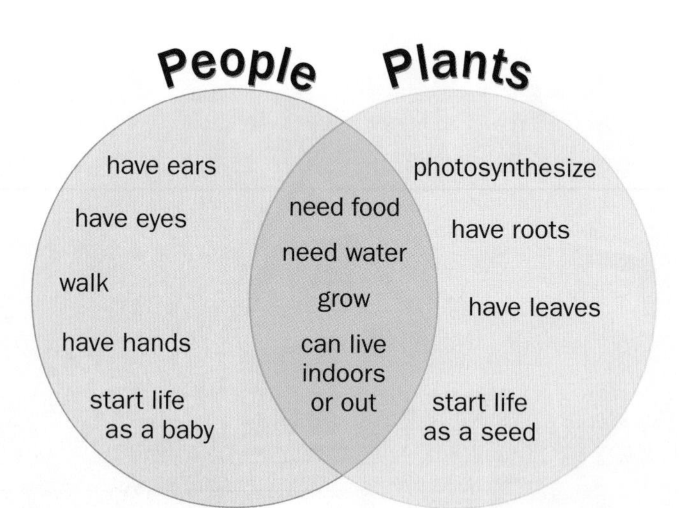

Two circles that overlap show that both groups share some things in common, but not everything. The characteristics that the groups have in common are placed in the area where the circles overlap. The remainder of each circle is filled with characteristics that are unique to that group.

A circle that is completely inside another circle indicates that everything in the smaller circle is also inside the larger one. For example, in the Venn diagram below, all of the animals included in the 'have fins' circle can also swim. Therefore, that circle is fully inside the 'can swim' circle.

However, there are animals in the 'can swim' category that do not have fins, so they are placed in a way so that they are in only the 'can swim' circle.

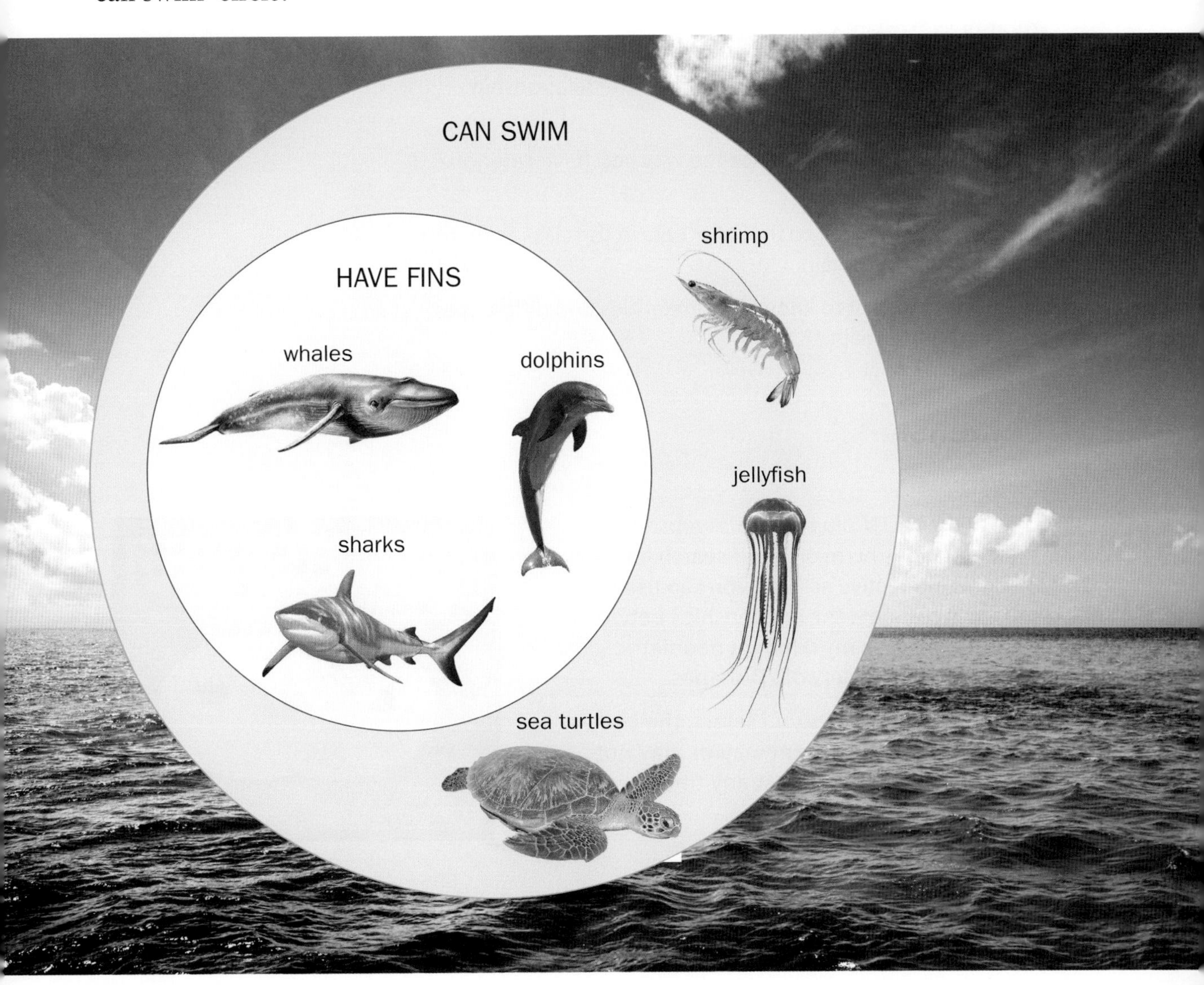

## Activity 1 – Shapes in mathematics

**1** Draw a Venn diagram to represent the relationship between the following groups of items. Be sure to write at least two characteristics in each section. Explain your reasoning in one or two sentences.

**a** triangles circles

**b** squares rectangles

**2** Research the definitions and properties of the following five shapes and write them in your notebook.

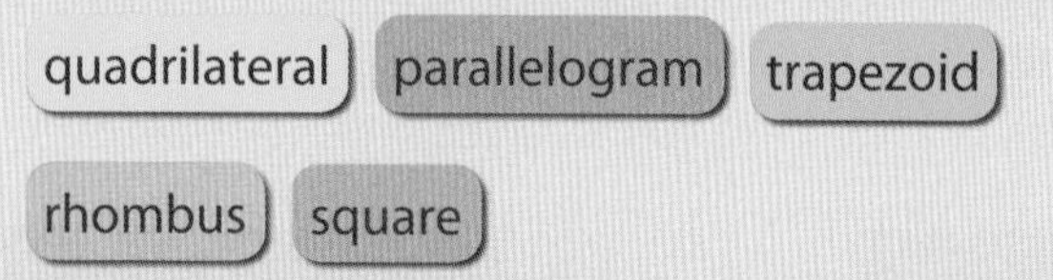


**a** Create a Venn diagram to demonstrate the relationship between these different shapes.

**b** Write down an explanation for the way you have categorized the shapes.

**c** Compare your answer to that of others in the class. Is there a correct answer? Explain.

**d** Why is it important to know the similarities and differences between these shapes?

WEB LINK

Go to the NCTM Illuminations website (illuminations.nctm.org) and search for 'Shape sorter'. In this interactive activity, you can use Venn diagrams to investigate the relationships between shapes, and explore many different geometric properties of a wide variety of 2D shapes.

Depending on which rules you select, the circles in the resulting Venn diagram may or may not overlap. Click the tick button at any time to check your work.

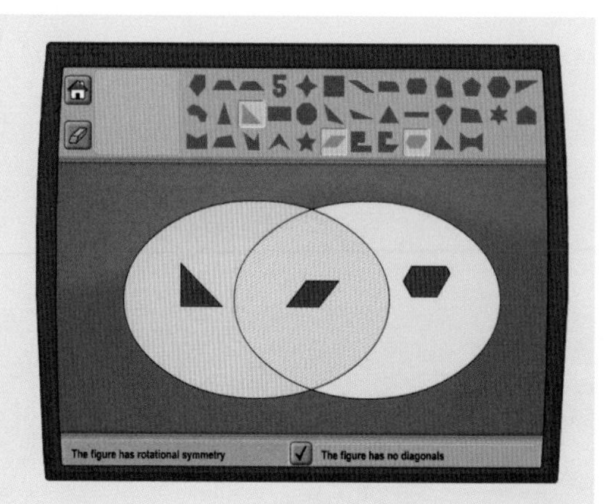

## Area of 2D shapes

Area is an important concept in mathematics. It is the foundation upon which many other mathematical concepts are built. Area is also an important idea in the environment. The amount of space that something occupies can be a vital factor in whether or not it is good for the planet. Area can also be used to calculate energy requirements, as you will see in Activity 2.

### Activity 2 - Solar power

Solar panels are environmentally friendly devices used for creating and storing electricity. Owners of residential houses and buildings are seeing the benefits of having solar panels installed on their rooftops. Some people have reduced their electricity bill to zero, while others actually *make* money by selling their unused power back to the utility company!

**Part A: Your school's energy requirements**

1 Brainstorm what factors you think would affect the efficiency and effectiveness of the use of solar panels.

The energy output of solar panels can be calculated using the following formula:

$E = 0.15AH$, where:

$E$ = Energy (kWh)

$A$ = Total solar panel area ($m^2$)

$H$ = Annual average solar radiation on the panel

$H$ is a value between 200 $kWh/m^2$ (Norway) and 2600 $kWh/m^2$ (Saudi Arabia).

2 Why do solar radiation values vary so much around the globe?

3 What is the solar radiation value where your school is? You can find this value by looking at 'Solar potential maps' in the 'Solar irradiance' section on Wikipedia.

4 Your teacher will give you the annual energy requirements of your school in kWh. What area of solar panelling would you need to install on the school roof or in the school grounds to meet these needs?

5 If a typical solar panel is a rectangle measuring 1.65 m by 1 m, how many panels would your school need?

▶ Continued on next page

**Part B: Is there enough space to install solar panels at your school?**

Pairs

1. As a class, conduct a survey of the physical space of the school, looking for places where solar panels could be installed. In pairs, create a design of where the panels could be located, keeping in mind the dimensions of a solar panel. Draw scale diagrams of your designs to share with the class.
2. Calculate the total perimeter and area of the solar panel designs. What is the maximum area that can be achieved?
3. What percentage of your school's electricity requirements could be met by solar power if the panels were placed in the spaces you indicated?
4. Running your school on renewable energy, such as solar power, has many advantages. Can you think of any restrictions or limitations that would hinder the installation of solar panels at your school?

**Part C: Could you run your home on solar energy?**

1. Look at your electricity bill from home and determine how much energy you use each year (in kWh).
2. How many typical solar panels would you need to install on or near your home to provide for your home's energy needs?
3. What limitations might hinder the installation of solar panels at your home? Do the benefits of using solar panels outweigh the problems of installing them? Justify your response.

Your electricity bill will be for a certain number of days, so determine how much energy is used on average each day and multiply it by 365.

The formula for the area of a rectangle is fairly simple:

$$\text{Area} = \text{length} \times \text{width (or base} \times \text{height)}$$

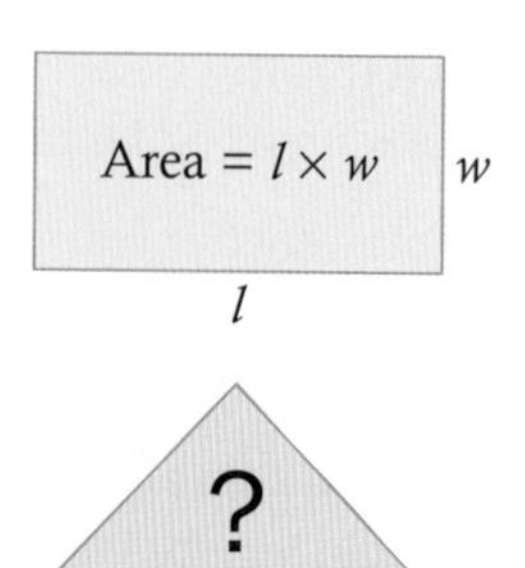


What is the formula for the area of a triangle?

## Investigation 1 – Area of a triangle

criterion B

**1 Dimensions and area of a triangle**

- **a** You will work in small groups for this investigation. Each person in your group will draw a rectangle with a height of 4 units and a base of 6 units on a piece of graph paper.
- **b** Pick any point on the base of the rectangle to be the vertex of a triangle.
- **c** Draw diagonal lines from this vertex to the top two corners of the rectangle to create a triangle.
- **d** Count the total number of grid boxes (including whole and partial grid boxes), and hence find the area of the triangle.
- **e** Compare the area of each group member's triangle. What do you notice?
- **f** Why did that happen?
- **g** Complete this process for another sized rectangle with dimensions decided by your group. Do you get the same result?

**2 Formula for the area of a triangle**

In your group, you will now develop a formula for the area of a triangle.

- **a** Calculate the area of the $4 \times 6$ rectangle you drew in step **1**.
- **b** Compare its area to that of the triangles you created inside the rectangle.
- **c** Do the same calculations for the second rectangle and triangle you drew.
- **d** What is the relationship between the area of a triangle and the area of a rectangle with the same base and same height?
- **e** Given that the formula for a rectangle is $A = bh$, write down the formula for the area of a triangle.

WEB LINK

For greater accuracy of calculations, Investigation 1 can be done using dynamic geometry software (e.g. GeoGebra), or on an interactive geoboard like the one on the Math Playground site (mathplayground.com/geoboard.html). When using the geoboard on this site, you will need to click the yellow grid twice so you can see the scale of the diagram and determine the measures of each side.

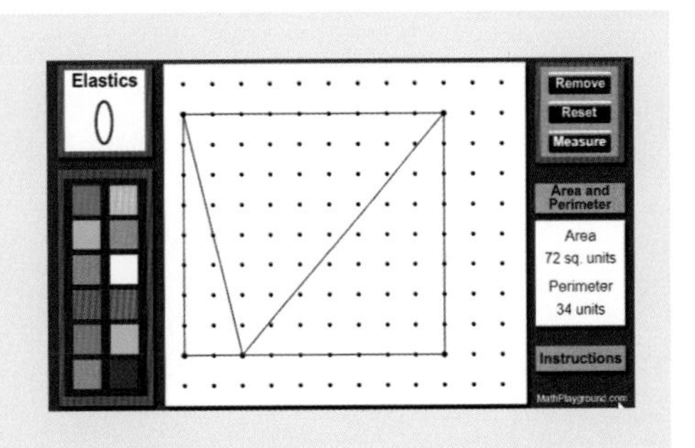

## Reflect and discuss 2

- Martijn says that the base of a triangle is always the side on which the triangle is resting. Anika says that any side can be the base. Who is correct? Explain.
- In the formula for the area of a triangle, how are the base and height of the triangle related to each other? Can *any* two sides be used as the base and the height? Explain.
- What is the relationship between the areas of the two triangles shown here?
- What difference can be said about measuring the height of the triangle on the left versus the one on the right?

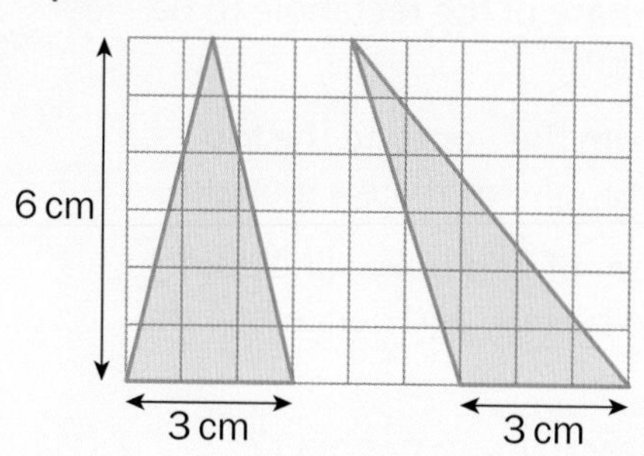


## Example 1

**Q** Find the area of these triangles.

**a**

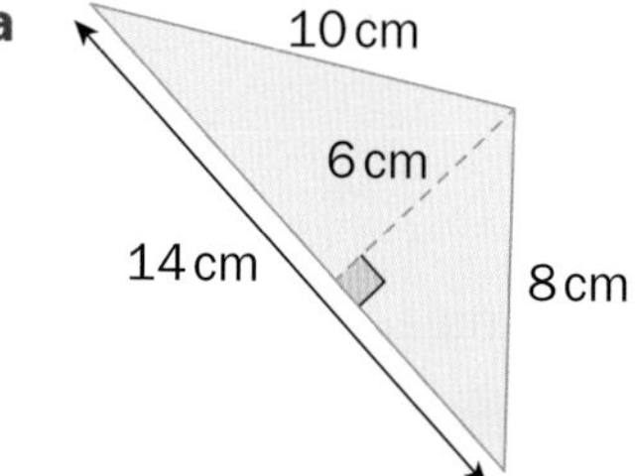


**b**

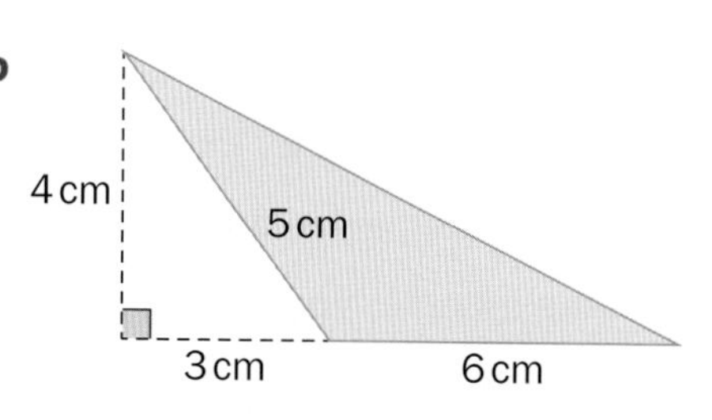


**A** **a** Area of a triangle $= \frac{1}{2}bh$

$= \frac{1}{2}(14)(6)$

$= 42\,\text{cm}^2$

You can decide which is the base and the height, as long as they are perpendicular to each other.

**b** Area of a triangle $= \frac{1}{2}bh$

$= \frac{1}{2}(6)(4)$

$= 12\,\text{cm}^2$

The height is 4 cm. The base is 6 cm. It is the side that is perpendicular to the height.

Note: 3 cm is *not* the measure of any side of triangle **b**.

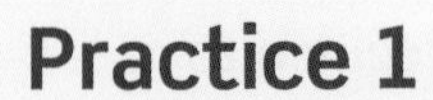

## Practice 1

**1** For each triangle below, which measurement(s) do you **not** need to calculate the triangle's area?

**a**

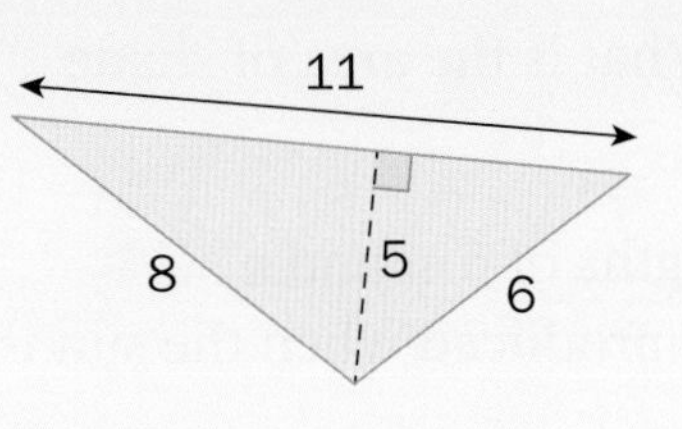


**b**

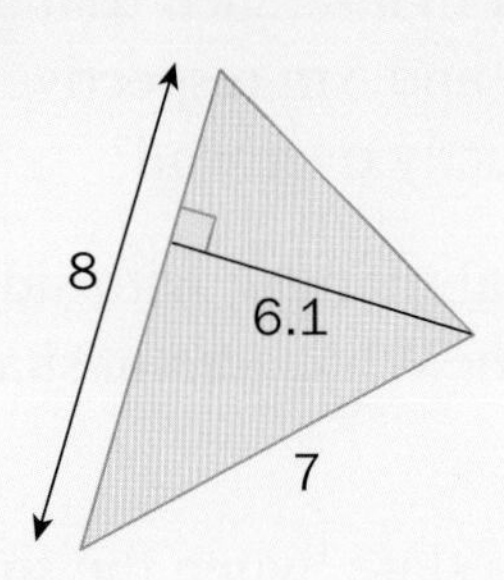


**c**

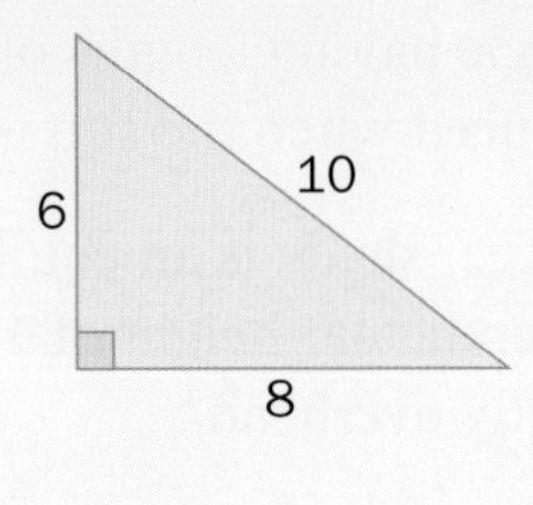


**2** Find the area of each triangle. Round your answer to the nearest tenth if appropriate.

**a**

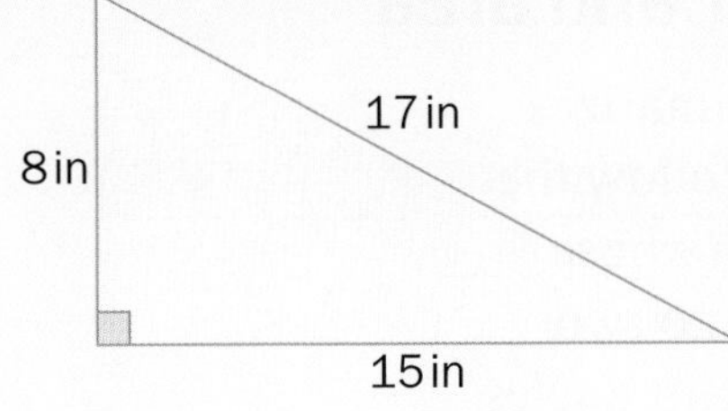


**b**

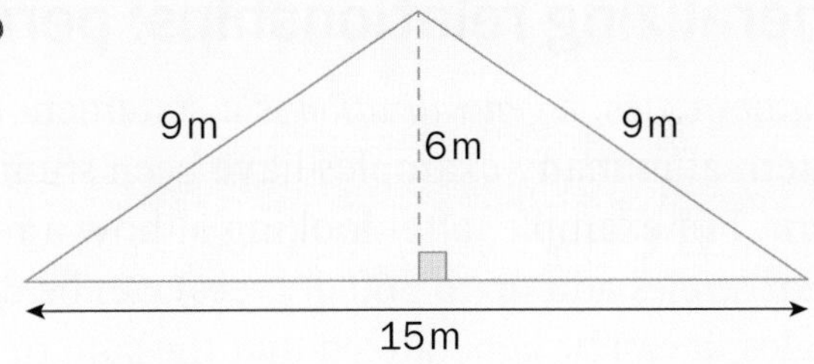


**c**

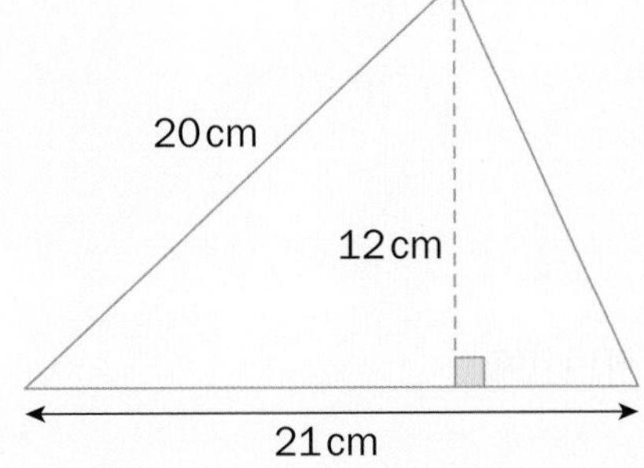


**d**

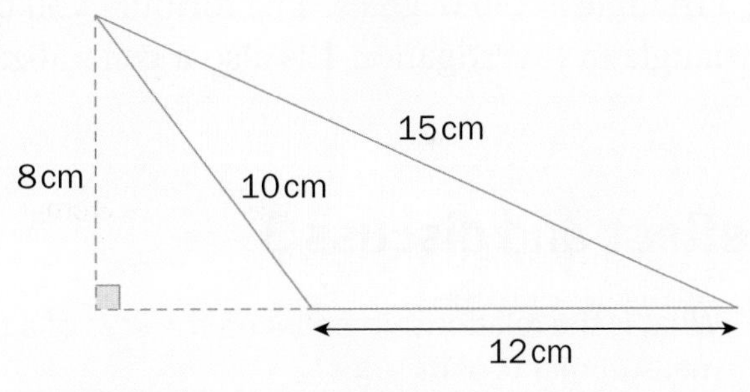


**3** Fill in a copy of this table. Round numbers to the nearest hundredth where appropriate.

| Triangle's base (cm) | Triangle's height (cm) | Triangle's area ($cm^2$) |
|---|---|---|
| 10 | 12.6 | |
| | 14.5 | 132 |
| 8.4 | | 210 |
| | 16.25 | 84 |
| 10.5 | | 48 |

▶ Continued on next page

**4** In an attempt to reduce their air conditioning usage, Lavinia's family decided to put up a triangular shade outside, with seating underneath. These shades come in a variety of sizes and shapes, which cover different areas.

**a** One shade is in the shape of a right-angled triangle, with the two legs of the triangle having lengths of 4 m and 5 m respectively. What is the area of shade produced when the sun is directly overhead?

**b** Another shade is an equilateral triangle, with side lengths of 5 m and a perpendicular height of 4.33 m. What area of shade is produced when the sun is directly overhead?

**c** Explain why is it important to state 'when the sun is directly overhead'.

## Generalizing relationships: perimeter and area

In mathematics, a *generalization* is a statement that summarizes a pattern after many examples have been studied that follow the pattern. For example, after looking at how a rectangle (whose interior angles add up to 360 degrees) can be divided into two triangles, it can be generalized that the sum of the interior angles of a triangle is 180 degrees. The formula you created for the area of a triangle in Investigation 1 is also a generalization.

### Reflect and discuss 3

- What is the relationship between the area of a rectangle and the measurements of its sides?
- What is the relationship between the area of a right-angled triangle and the measurements of its sides?
- How would you help a younger student to make these same generalizations? What activities could you do with them?

### Perimeter and area of a rectangle

In Investigation 2 you will generalize a relationship between the perimeter and area of a rectangle.

## Investigation 2 – Perimeter and area of a rectangle

criterion B

While planning for new residential developments, city planners want to designate a rectangular portion of a large forest as a protected area for native plants and animals. The plan is to use 20 km of fencing to enclose the area, and city planners are investigating how to make the enclosed area as large as possible. Fill in a copy of this table and generalize what you find. Add more rows as necessary and be sure to use both whole numbers and decimal values.

| Perimeter (km) | Length of rectangle (km) | Width of rectangle (km) | Area of protected region ($km^2$) |
|---|---|---|---|
| 20 | 1 | | |
| 20 | 2 | | |
| 20 | | | |
| 20 | | | |
| 20 | 5 | | |
| 20 | | | |
| 20 | | | |
| 20 | 9.8 | | |

1 Perform the same calculations with other values for the perimeter.

2 Which shape seems to maximize the area of the protected region?

3 Write down a generalization about the relationship between the area of a rectangle and its perimeter, when the perimeter remains constant.

4 Perform similar calculations, but this time, instead of a constant perimeter, use a constant *area* of 36 $km^2$. Set up a table similar to the one above and calculate the perimeter of each configuration.

5 Describe the rectangle that has the largest perimeter. Describe the rectangle that has the smallest perimeter.

6 Write down a generalization about the relationship between the area of a rectangle and its perimeter, when the area remains constant.

7 Give a real-life example of when you would use these generalizations, like the scenario given at the beginning of this investigation.

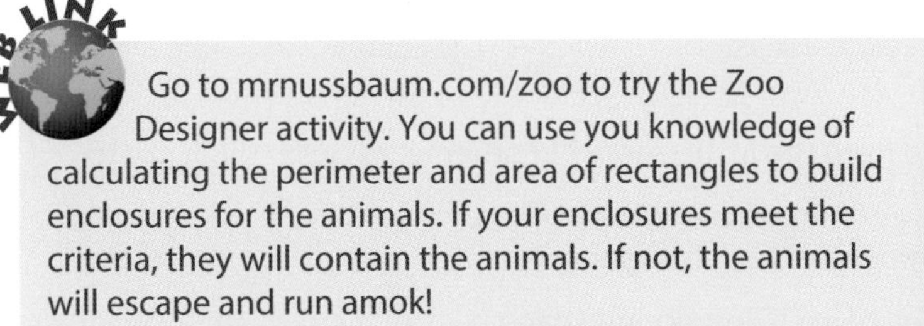

WEB LINK

Go to mrnussbaum.com/zoo to try the Zoo Designer activity. You can use you knowledge of calculating the perimeter and area of rectangles to build enclosures for the animals. If your enclosures meet the criteria, they will contain the animals. If not, the animals will escape and run amok!

## Reflect and discuss 4

- Do you think that the generalization you wrote in step **3** of Investigation 2 will be true for *any* rectangular perimeter? Explain.
- Do you think your generalization in step **6** will be true for *any* rectangular area? Explain.
- Given what you discovered in Investigation 2, what type of triangle do you think would maximize area, when given a fixed perimeter?
- How do you think these patterns could be used to help people have a positive impact on the environment?

## Perimeter and area of compound shapes

While there are many different kinds of shapes, finding the area of a more complicated shape (called a *compound shape*) often involves breaking it down into components that are themselves simple shapes.

### Activity 3 - Compound shapes

Look at these compound shapes and decide how to divide each one into simpler shapes. Copy the diagrams in your notebook and use dotted lines to divide each compound shape. Be sure to use only rectangles and triangles when dividing up each compound shape.

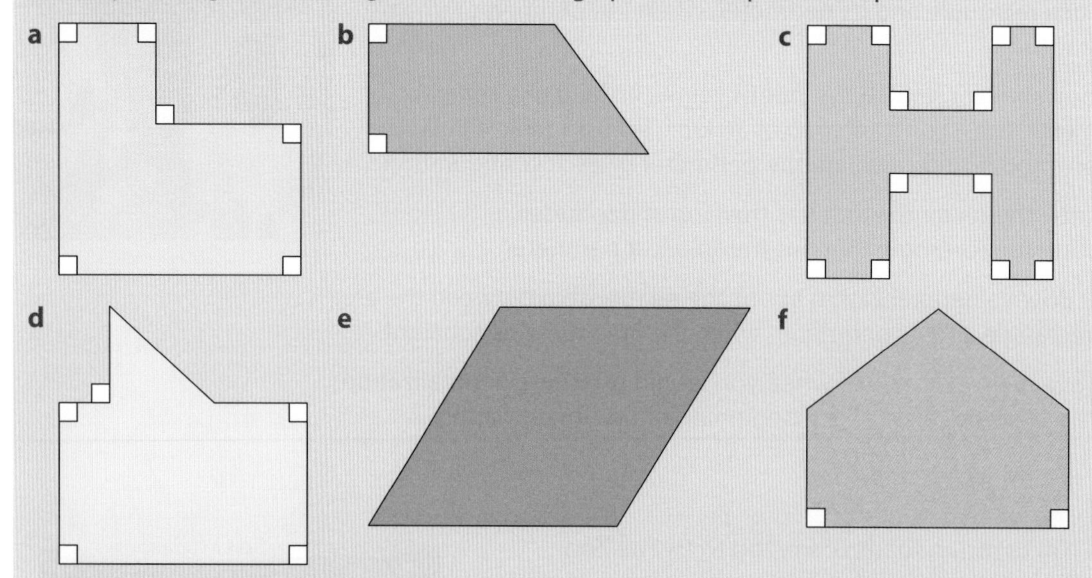


1 Why was the word 'square' omitted from the instructions? In other words, why did it say to 'use only rectangles and triangles'?

2 Explain how you would find the perimeter of each compound shape.

3 Explain how you would find the area of each compound shape.

## Example 2

**Q** Find the perimeter and area of this shape.
All measurements are in meters.

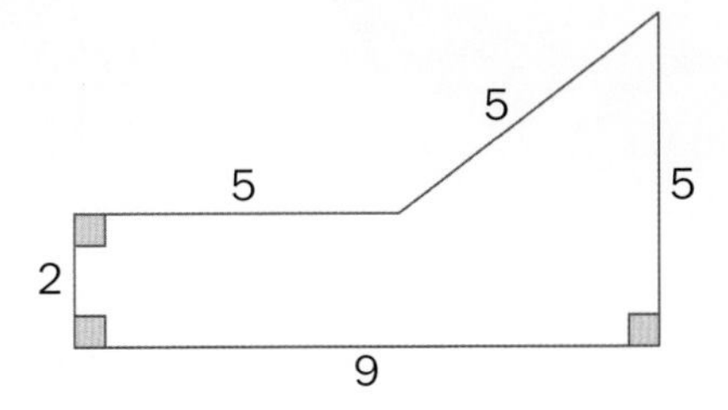


**A** Perimeter $= 9 + 2 + 5 + 5 + 5 = 26$ m

To find the perimeter, add the measurements of the sides.

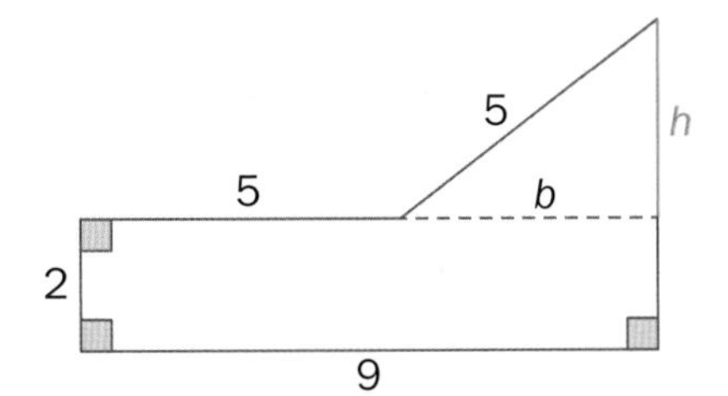


To find the area, divide the shape into two simpler shapes, and calculate the measurements of the missing sides.

Area = area of rectangle + area of triangle

$= l \times w + \frac{1}{2}(b \times h)$

$= 9 \times 2 + \frac{1}{2}(4 \times 3)$

$= 18 + 6$

$= 24\ \text{m}^2$

The height of the triangle is $h$.
$h = 5 - 2 = 3$

The base of the triangle is $b$.
$b = 9 - 5 = 4$

Remember to include the correct units.

## Practice 2

**1** Find the perimeter and area of each of these compound shapes.
All measurements are in centimeters.

**a**

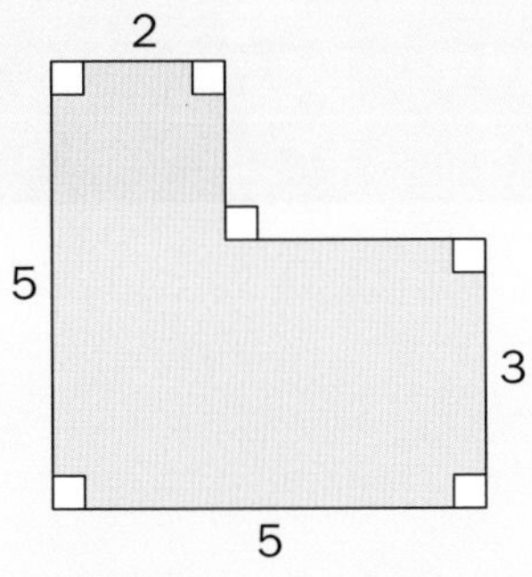


**b**

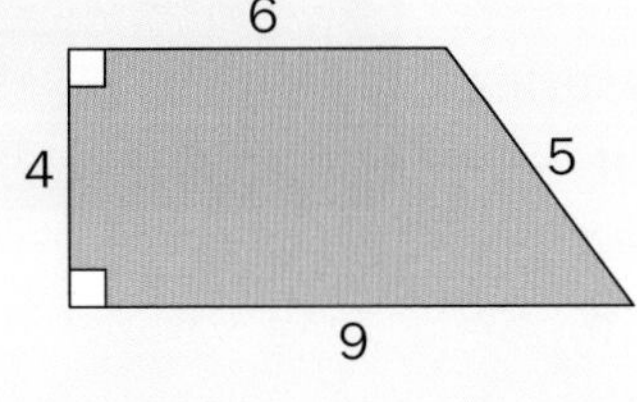


**c**

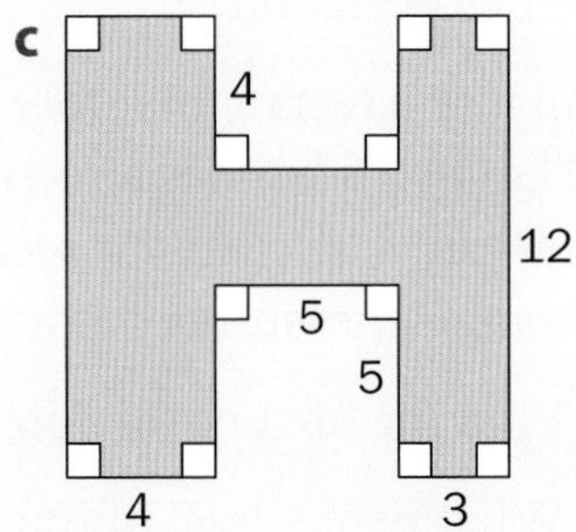


▶ Continued on next page

**d**

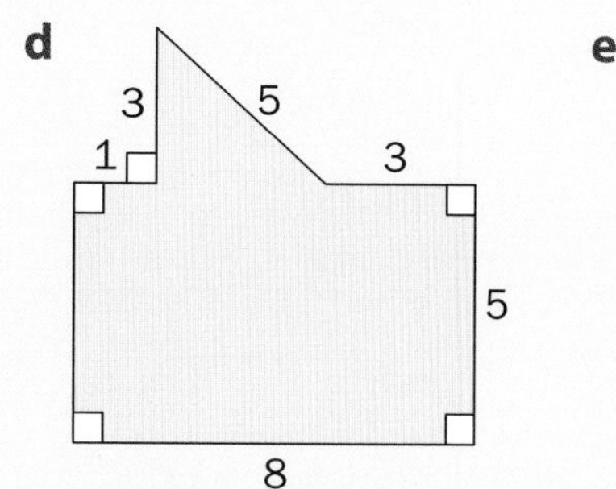


**e**

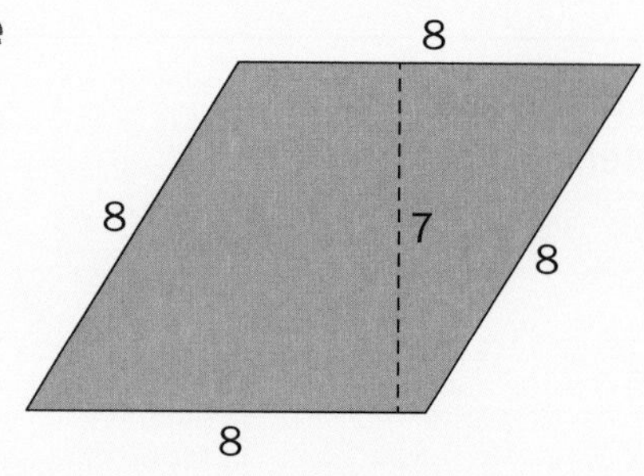


**f**

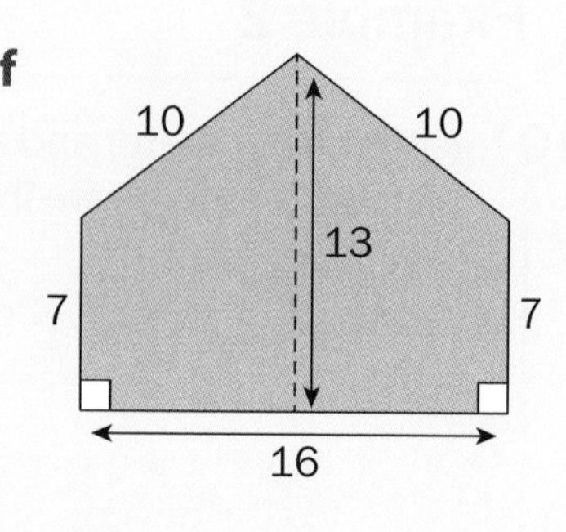


**2** Calculate the area of each orange region. All measurements are in millimeters.

**a**

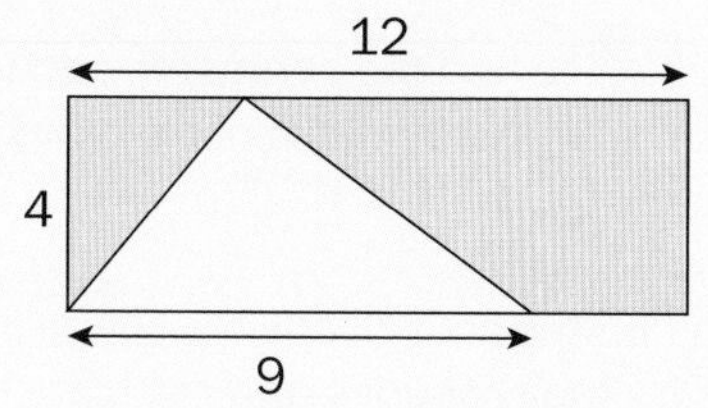


**b**

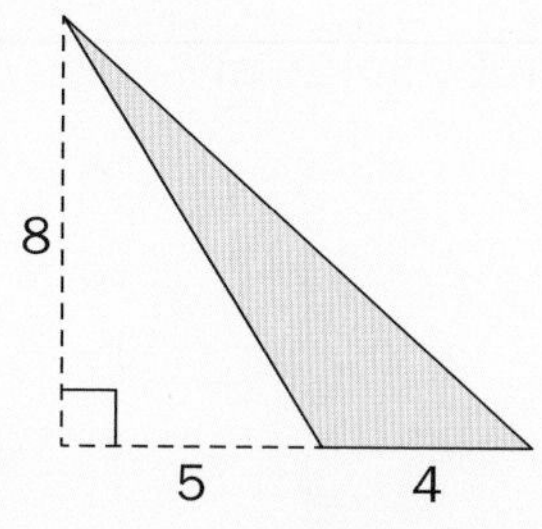


**c**

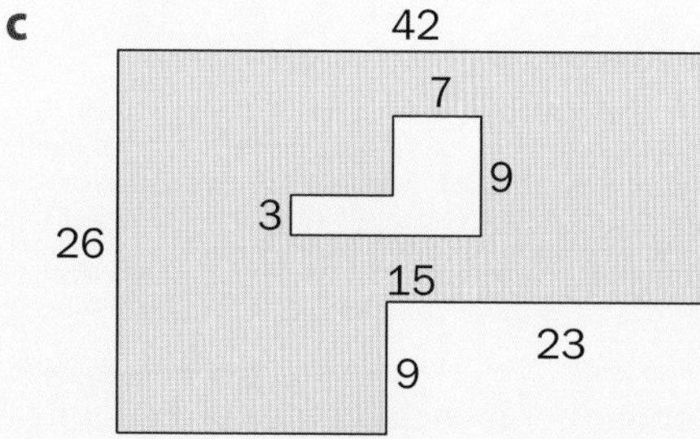


**d**

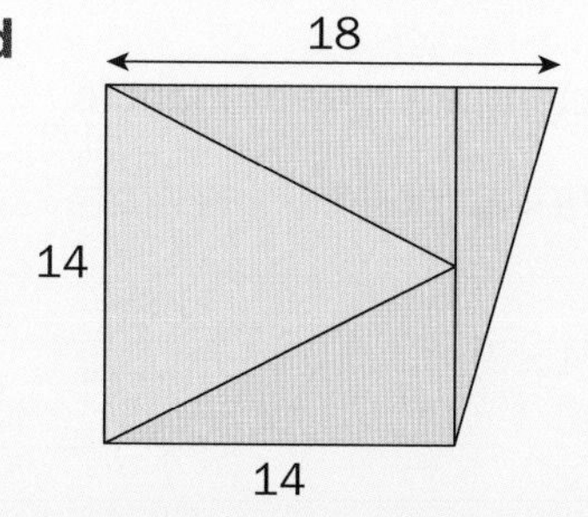


ATL2 **3** Clear-cutting is a practice used in the logging and forestry industries where all of the trees in a specific area are cut down. The damage can be vast, and often an aerial photo, like this one, can emphasize the scale of the deforestation.

Suppose a rectangular forest measuring 10 km by 15 km has a section of it clear-cut. The clear-cut section is a rectangle measuring 5 km by 6 km.

**a** Find the area of the forest that is still intact.

**b** What percentage of the original forest was cut?

**c** Research the environmental impacts (both positive and negative) of clear-cutting.

WEB LINK

Go to learnalberta.ca and search for the 'Exploring Composite Figures' activity. You will explore how to find the area of composite figures by breaking them into smaller figures and adding their areas together. Some shapes have sections cut out, so you will need to subtract the appropriate areas.

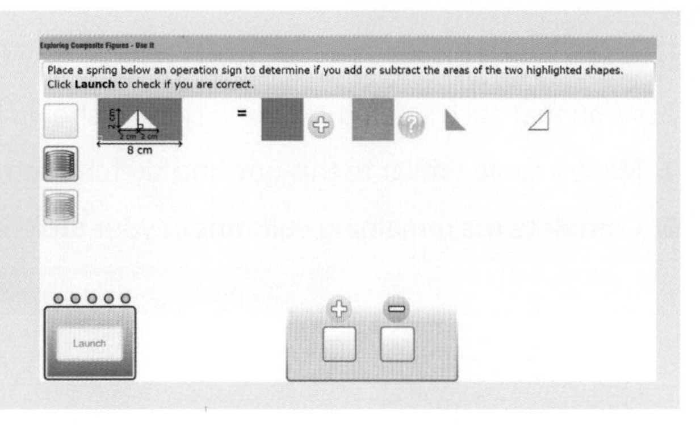


# Moving from 2 dimensions to 3 dimensions

## Defining volume

A two-dimensional object has only two dimensions or measurements, but no thickness. A three-dimensional object has three dimensions or measurements, such as a length, width, and height.

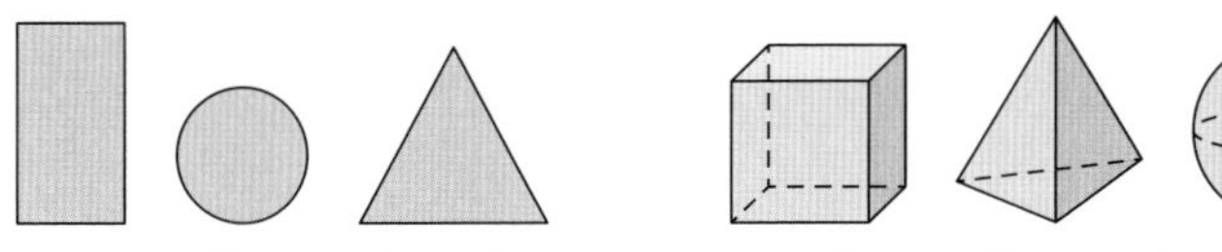

two-dimensional three-dimensional

An extra measurement means that objects can now be filled, and this is referred to as *volume*. The volume of a 3D shape is defined as the amount of space it occupies. While area is a measure of the outside (or surface) of a 3D shape, volume is a measure of the amount that can fit inside it.

The most basic unit of volume is the cube. If each side of a cube measures 1 unit, then the volume of the cube is said to be 1 cubic unit. The volume of larger shapes can be found by counting cubes. For example, the volume of the rectangular prism below is 12 cubic units, since it is made up of 12 cubes.

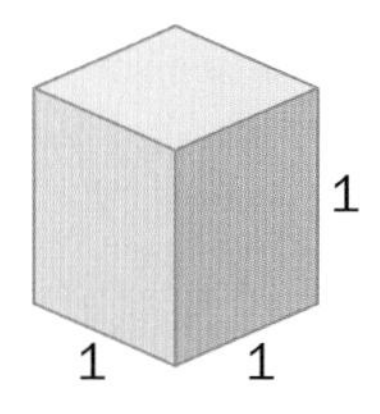


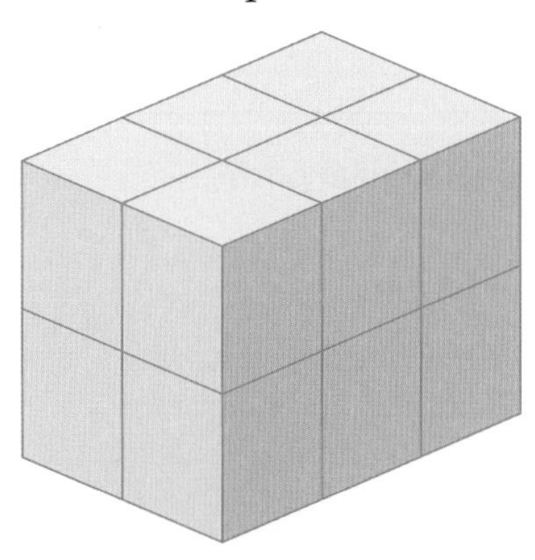

# Investigation 3 - Volume of a rectangular prism

criterion B

Assume that each small cube in every solid shown here is one unit on each side.

1 Make a table similar to this one and sketch each 3D shape in the first column.

2 Complete the remaining columns in your table by counting the cubes.

| | Length | Width | Height | Volume |
|---|---|---|---|---|
| | | | | |
| | | | | |
| | | | | |
| | | | | |
| | | | | |
| | | | | |

3 How could you calculate the volume of each rectangular prism without counting individual cubes? Explain.

4 Verify your result for other sizes of rectangular prisms.

5 Write down a formula to calculate the volume $V$ of a rectangular prism if you know the measure of its length $l$, width $w$, and height $h$.

## Reflect and discuss 5

- Why do you think it is important to have a general rule or formula for finding the volume of a prism? Explain.
- If the length, width, and height are all measured in centimeters, what would the units of volume be? Explain.
- Explain why the volume of a cube measuring 1 cm on each side is 1 $cm^3$.

The volume of a shape (the amount of space inside a 3D shape) can be determined by:

$$\text{Volume} = (\text{area of its base shape}) \times \text{height}$$

## Activity 4 – Three-dimensional prisms

In each prism here the base shape is highlighted.

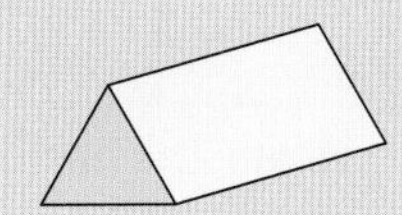
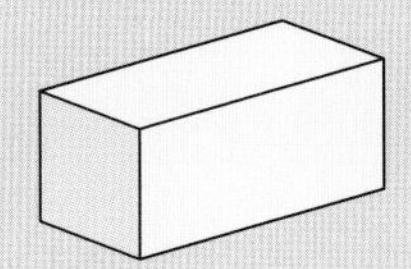
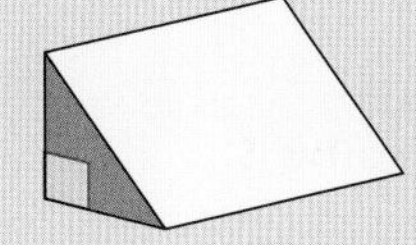

1 How would you define the 'base' of a prism?

2 Marcos says, 'The base is the side that the shape sits on.' Is Marcos correct? Explain.

3 Make a generalization in words about the relationship between the area of a base shape and the volume of its corresponding prism.

4 You can determine the specific volume formula for any kind of prism if you know the formula for the area of its base shape. Simply multiply the base shape's area by the third dimension, which we have called **height** below.

5 Focusing on triangular and rectangular prisms, create a table like this one and determine the specific formula to calculate the volume of rectangular and triangular prisms.

| 3D shape | Name of base shape | Area formula of base shape | General volume formula | Specific volume formula of the prism |
|---|---|---|---|---|
| Rectangular prism | | | Volume = Area of base × **height** | |
| Triangular prism | | | Volume = Area of base × **height** | |

6 Is a cube a rectangular prism? Can you apply the same process to determine the formula for calculating the volume of a cube? What would that formula be?

## Example 3

**Q** Find the volume of this prism.

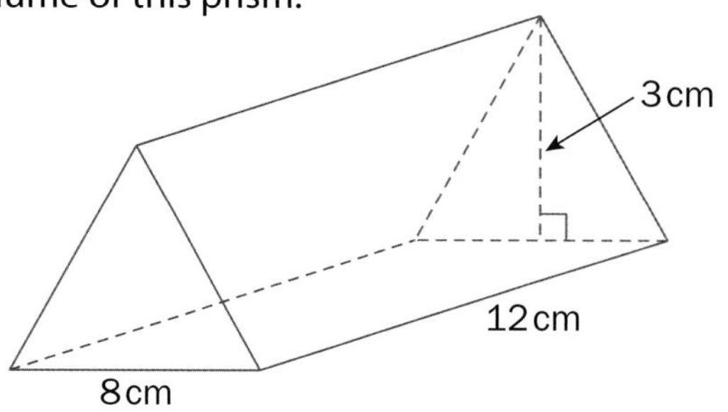


**A** The base shape is a triangle whose area is:

Volume = area of base × height

$$A = \frac{1}{2}bh$$

$$= \frac{1}{2} \times 8 \times 3$$

$$= 12 \text{ cm}^2$$

Therefore, the volume is:

Note: the base shape is not always at the bottom of the prism.

$$V = 12 \text{ cm}^2 \times 12 \text{ cm}$$

$$= 144 \text{ cm}^3$$

## Practice 3

**1** Find the volume of each prism.

**a**

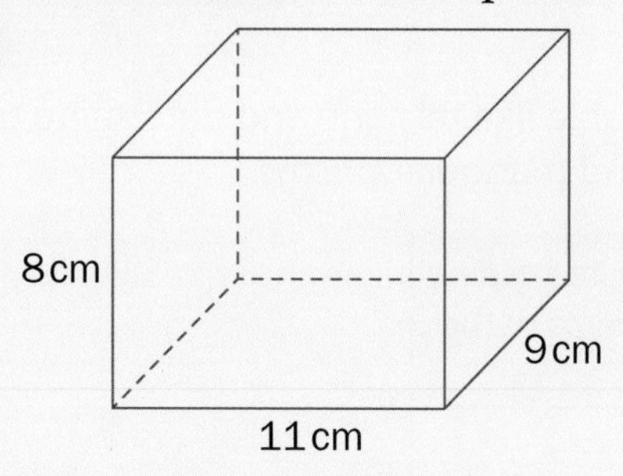


**b**

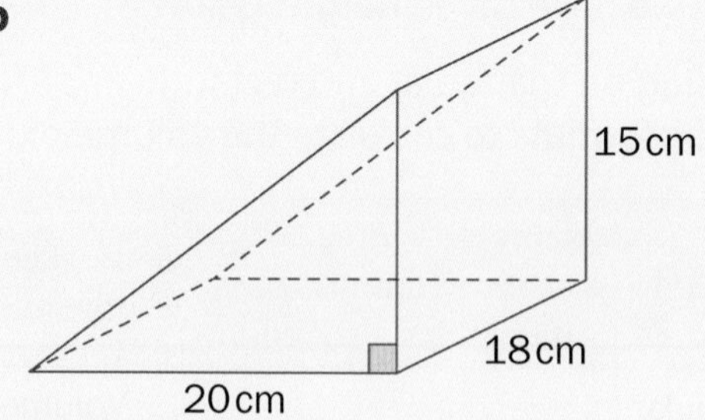


**c**

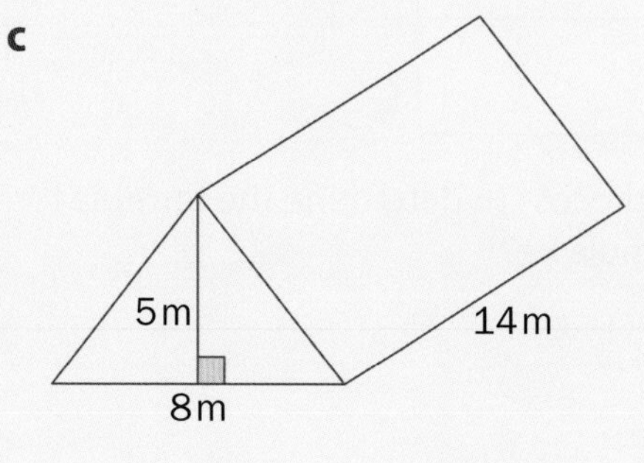


**d**

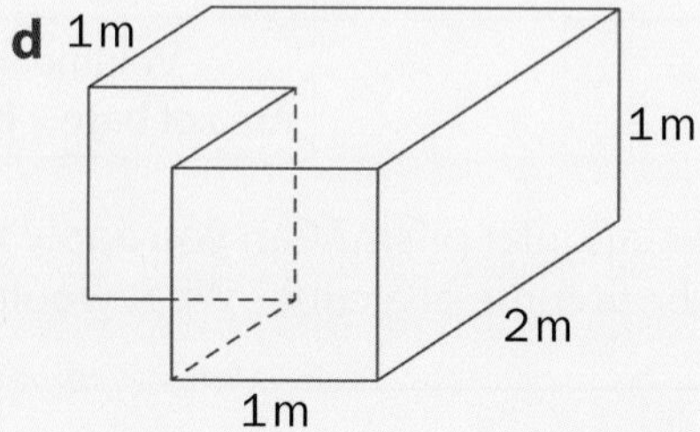


▶ Continued on next page

**2** The volume of each prism is shown below it. Find the missing measurement in each prism, as indicated by the pink edge.

**a**

6 cm, 4 cm

$V = 168\text{ cm}^3$

**b**

6, 6, 5.2, 6

$V = 156\text{ cm}^3$

**c**

10 cm, 8 cm, 15 cm

$V = 720\text{ cm}^3$

**3** Complete your own copy of this table, where all shapes are rectangular prisms.

| | Length (cm) | Width (cm) | Height (cm) | Volume (cm$^3$) |
|---|---|---|---|---|
| Shape 1 | 10 | 12 | | 600 |
| Shape 2 | | 4 | 9 | 90 |
| Shape 3 | 3 | | 25 | 300 |
| Shape 4 | 7 | 4 | 11 | |
| Shape 5 | 8 | 20 | | 560 |

ATL1 **4** A box of rice in the shape of a rectangular prism has a length of 12 cm, a height of 18 cm and a width of 2.5 cm.

**a** Find the volume of the box.

**b** The box says that it contains 7 ounces of rice. If 1 ounce has an approximate volume of $30\text{ cm}^3$, approximately how many cubic centimeters of rice are in the box?

**c** Find the volume of empty space in a box filled with 7 oz of rice.

**d** Design a rectangular prism that would be a better fit for the given volume of rice.

ATL1 **5** New designs for products are constantly being created in an effort to find more ecologically friendly packaging and reduce waste of resources. A clever design for packaging toothpaste is a 'tetra pack' which is shaped like a triangular prism. This design not only cuts out the need for an exterior box, but also allows for packs to fit together compactly for more efficient shipping.

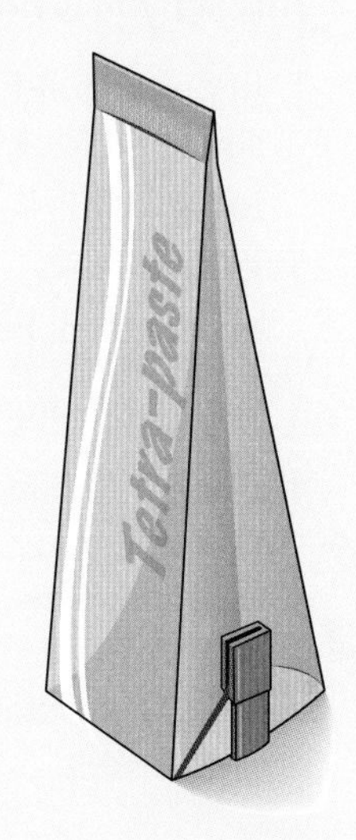


**a** A traditional toothpaste box (rectangular prism) measures 3.8 cm by 4.2 cm by 19 cm. Draw a diagram of this, complete with dimensions, and calculate the volume.

**b** The new tetra pack design has a base width of 3.5 cm, height of 16.4 cm and depth of 3.5 cm. Draw a diagram of this, complete with dimensions, and calculate the volume.

**c** Both of these prisms contain the same amount of toothpaste. Compare the volume of space the packaging of each takes up.

**d** If you were to package 10 tetra packs together, how could you configure them to minimize space? Draw a diagram of the configuration, including dimensions.

## Surface area

### Activity 5 - Nets

Every 3D shape can be flattened to two dimensions. As long as all of the faces are connected somehow, these two-dimensional versions are called *nets*. The net for a cube is shown below left. This is a correct net because it can be folded into a cube, as shown below right.

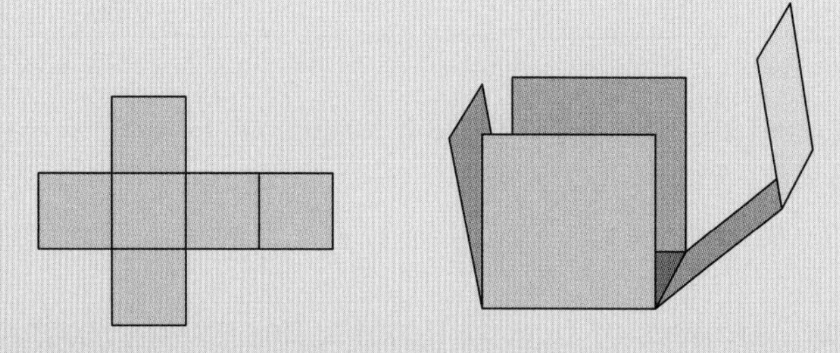

Pairs

There are 11 possible nets for a cube and your task is to find the other ten. This activity is best done with manipulatives.

**1** In pairs, build a cube with six interlocking squares. Carefully unfold it so that it looks like the net above and that it lies flat. That is your first net of a cube.

**2** Now, with your partner, find the remaining ten nets. Draw each net on graph paper.

**3** What are some of the common characteristics of the nets that you found?

**4** Once you have found all of the nets and have shown them to your teacher, you can go to the NCTM Illuminations website (nctm.org) and search for the 'Cube Nets' activity. Click on each net, compare it to your configurations and determine if it will create a cube. Once you have decided if the net will form a cube you can check to see if you are correct.

▶ Continued on next page

5 Go to nrich.maths.org and search for 'A Puzzling Cube'. Given the net of the cube, you have to determine the location of each face picture. Your teacher may print off the sheet with the different faces on it and a net to help you work through the problem.

## Investigation 4 – Surface area of 3D shapes

The *surface area* of a three-dimensional shape is the area of all of its faces added together. Creating a net is an efficient way to determine the surface area of a 3D shape, since you can flatten the shape to see all of its faces and then add the areas of the individual sections.

In pairs, determine the nets of each of the shapes in the table below, and hence write the formula for the surface area of each shape. To determine the surface area:

- draw the net of the 3D shape
- label the dimensions of each section
- find the area of each section
- total the areas of all of the sections; this will be the surface area of the shape.

Your teacher may provide you with manipulatives of 3D prisms that can be broken down into nets.

WEB LINK

Alternatively, search online for 'Annenberg Interactives Prisms'. Here, you can see how the net of each type of prism unfolds as a visual to help you determine the formula for surface area.

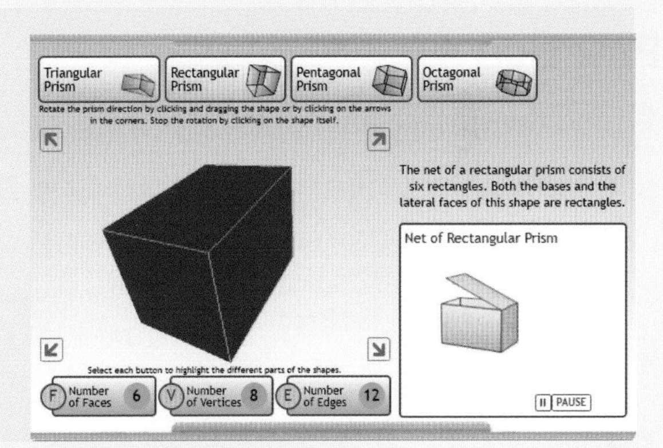


Create a table with the following headings, making sure there is enough room in each row to draw the net and label all dimensions clearly.

| 3D solid | Dimensions and shapes of sections needed to determine Surface Area (SA) | SA formula |
|---|---|---|
| Cube | | |
| Rectangular prism | | |
| Triangular prism | | |

▶ Continued on next page

## Did you know...?

An elephant uses its ears to cool its body temperature? Blood that flows through the ears is close to the outside of the body, and it can be cooled when the elephant flaps its ears. Having a large surface area means blood can be cooled more rapidly.

Similarly, radiators in your home release heat through small 'fins'. In order to achieve a larger surface area and keep the radiator relatively compact, many small fins are used. Smaller fins also use less material, allowing for a design that is both efficient and better for the environment.

## Reflect and discuss 6

How is finding the surface area of a three-dimensional shape similar to finding the area of a compound two-dimensional shape? Explain.

## Practice 4

**1** Find the surface area of these prisms.

**a**

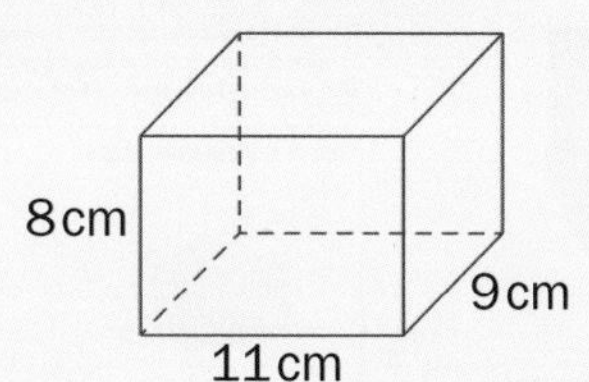


**b**

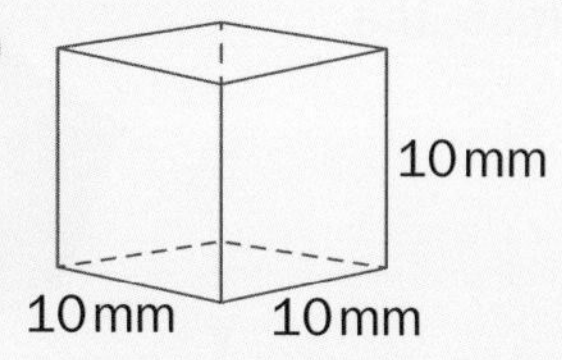


**c**

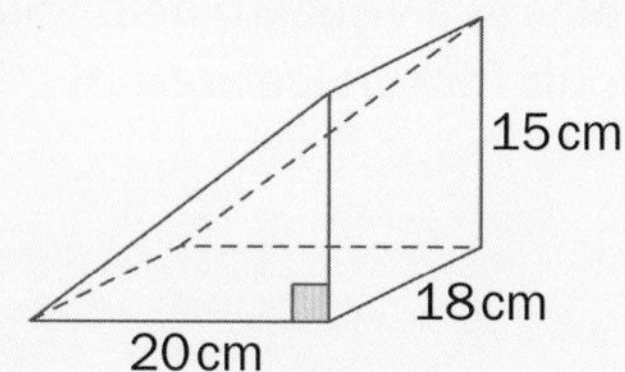


**d**

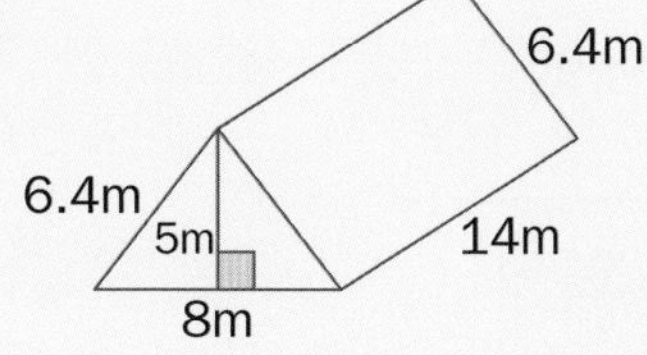


**e**

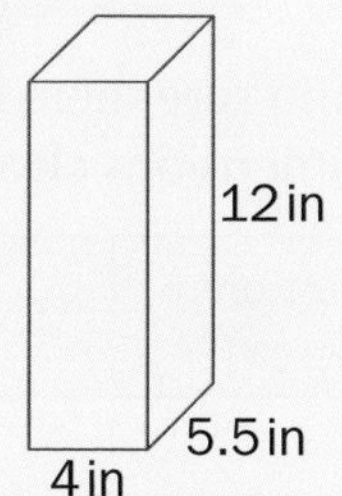


**f**

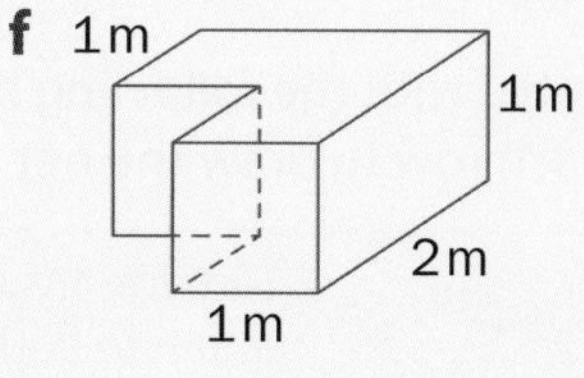


▶ Continued on next page

ATL2

**2** Small chocolate candies used to come in a box that measured 10 cm × 2 cm × 5 cm. The company changed the box to one that measures 11 cm × 3 cm × 6 cm.

**a** Calculate the change in volume and the change in surface area of the box.

**b** What reasons would a company use to justify a change in packaging?

**c** The original box contained 50 g of candies while the new box contains 45 g. Does this justify the change in packaging?

**d** Describe two environmental impacts of switching to a larger package.

**3** The handbag shown here is shaped like a triangular prism. The base of the handbag measures 28 cm across the front and 11 cm on the side. It has a vertical height (not including the straps) of 14 cm. Ignoring the strap and any overlap, how much material is needed to make the handbag? In other words, what is its surface area?

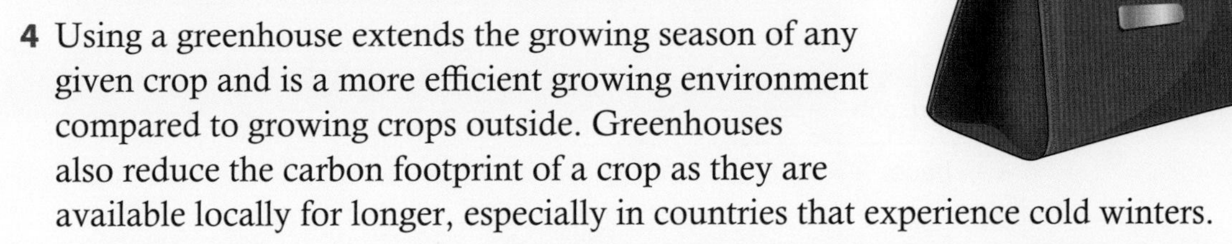

**4** Using a greenhouse extends the growing season of any given crop and is a more efficient growing environment compared to growing crops outside. Greenhouses also reduce the carbon footprint of a crop as they are available locally for longer, especially in countries that experience cold winters.

The Netherlands has some of the largest greenhouses in the world, with the most popular being a Venlo design structure, which is the shape of a traditional house built out of glass.

▶ Continued on next page

A commercial greenhouse, similar to the one in the photograph, has 10 peaked roof sections of equal dimensions. The depth of the greenhouse is 48 m, and the walls are 5 m high (from the ground to the base of the slanted roof sections). The slanted roof pieces are 2.2 m in length and they rise 1 m above the 5 m walls.

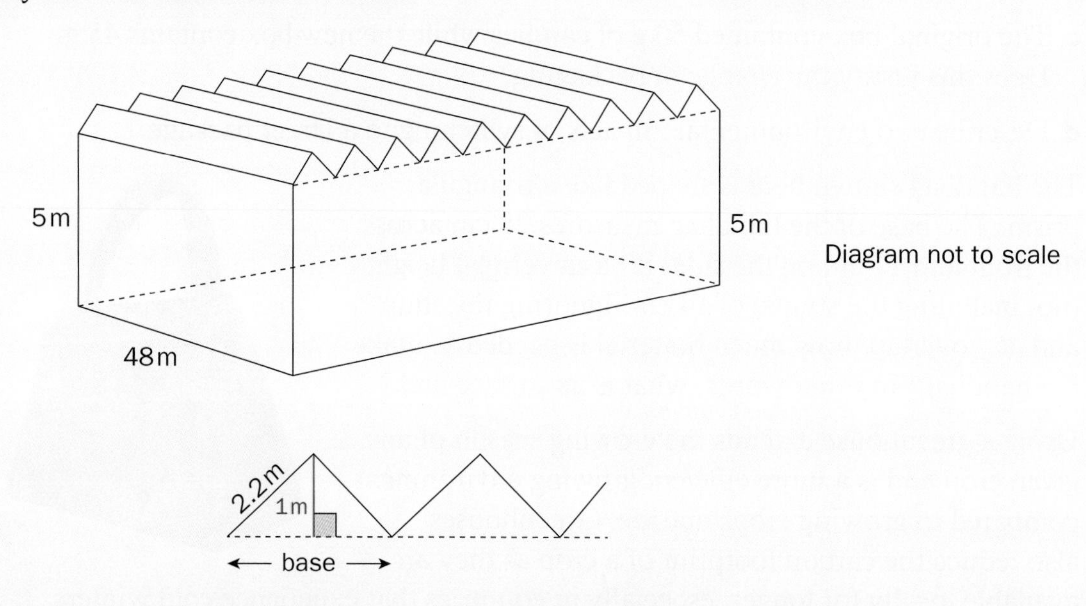


**a** If the base of the greenhouse is soil, and the construction is all glass windows, find the area of glass windows needed to construct the greenhouse. Hint: You will need to find the length along the front of the greenhouse. To do so, first find the base length of 1 peaked section, as shown in the diagram.

**b** The growing capacity of the greenhouse does not include the peaked roof sections. Find the volume of the growing capacity of this greenhouse.

**c** If solar panels were installed on half of the peaked roof pieces of this greenhouse, find the total area covered by the panels.

**d** To improve the efficiency of a greenhouse, some designs include insulation being buried 1m into the ground around the entire perimeter of the greenhouse. Find the area of insulation needed for this greenhouse.

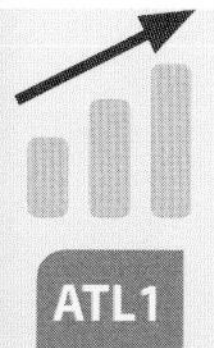

## Formative assessment – Clean teeth!

ATL1

Toothbrushes are typically packaged in plastic, which can take centuries to biodegrade. In an effort to reduce the amount of plastic in the world, an ecologically concerned company would like to start packaging their toothbrushes in recycled cardboard.

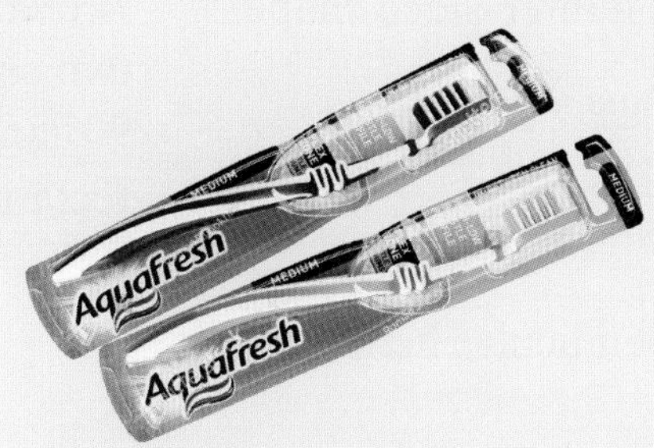


Your task is to determine whether a rectangular or triangular prism will use less cardboard, and reduce wasted space inside the packaging for one toothbrush. Using a standard toothbrush that does not have any attachments, batteries or lights, you will make a recommendation to the company as to how to package their product.

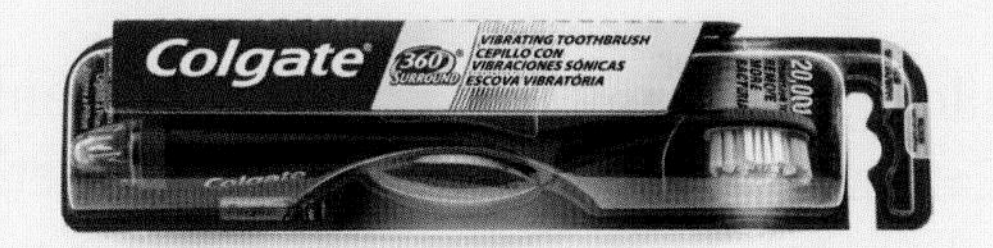


1 Design a geometric net for a rectangular prism and a triangular prism to contain the toothbrush. Make sure the toothbrush will actually fit inside, while minimizing the volume and surface area of the box.

2 Calculate the surface area and volume for each prism.

3 Explain the degree of accuracy of your results. Describe whether or not they make sense.

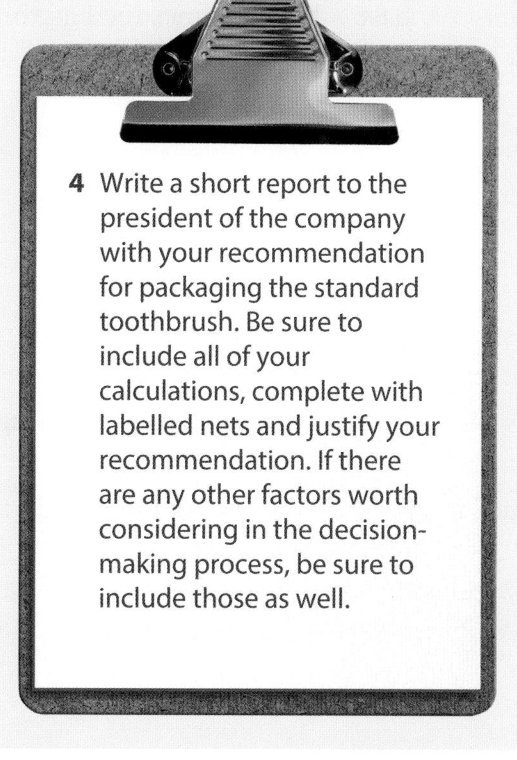

4 Write a short report to the president of the company with your recommendation for packaging the standard toothbrush. Be sure to include all of your calculations, complete with labelled nets and justify your recommendation. If there are any other factors worth considering in the decision-making process, be sure to include those as well.

# Unit summary

A triangle is a three-sided shape.

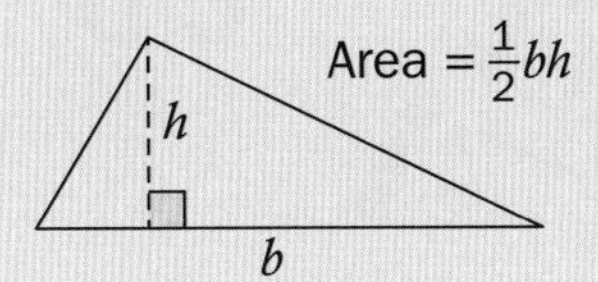


$h$ is the perpendicular height.

A rectangle is a quadrilateral with two sets of parallel sides and four 90° angles. Thus, a square is also a rectangle.

Area = $lw$

A parallelogram is any quadrilateral with two sets of parallel sides. Therefore, a rectangle is also a parallelogram.

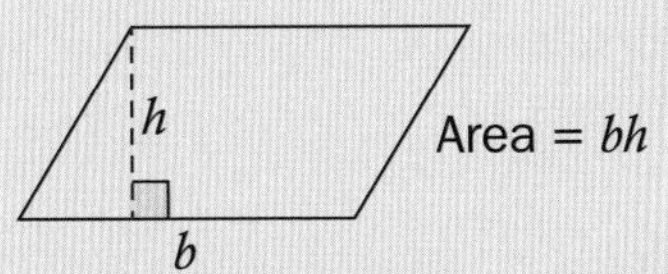


$h$ is the perpendicular height.

A rhombus is a quadrilateral with two sets of parallel sides and all the sides are the same length.

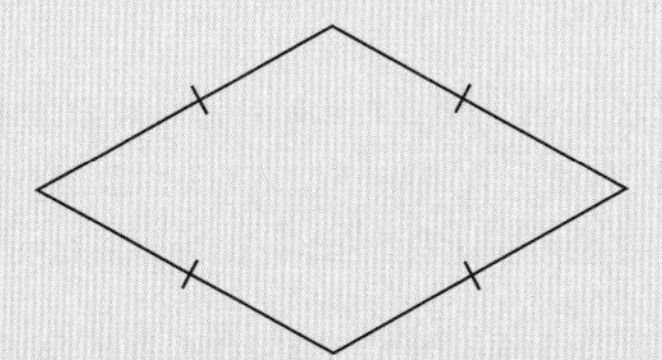

The volume of a rectangular prism is the area of the base (length × width) times the height. Volume is always expressed in units cubed, e.g cm$^3$.

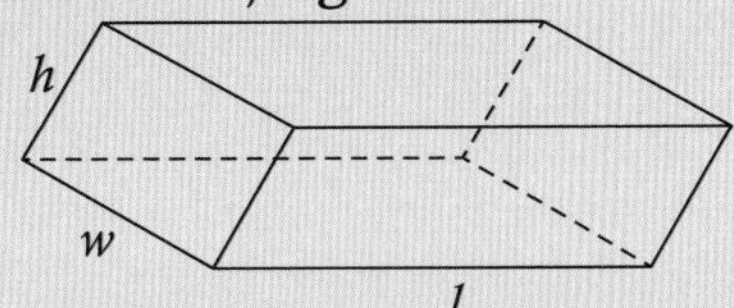


The volume of a triangular prism is the area of the base (one half the base × height) times the length.

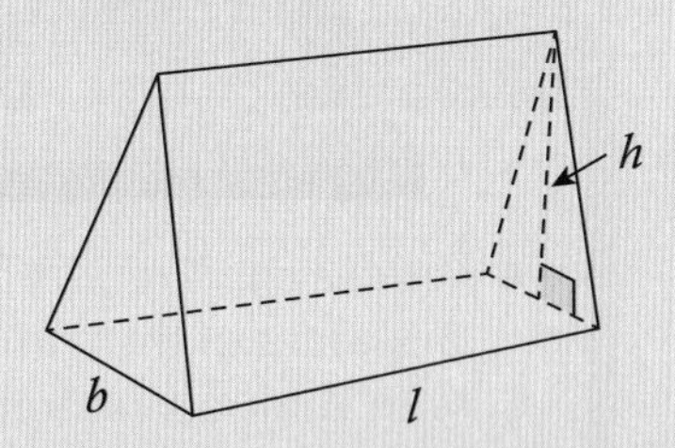


When a shape is made up of straight-line segments, finding the perimeter involves simply adding up the lengths of the sides.

Calculating the area of a compound shape usually means calculating the area of the rectangles and triangles that make up the shape.

Every 3D shape can be flattened to two dimensions. As long as all of the faces are connected somehow, we call these two-dimensional versions *nets*.The *surface area* of a 3D shape is the area of all of its faces combined together. Creating a net is an efficient way to determine the surface area of a 3D shape.

# Unit review

Launch additional digital resources for this chapter

Key to Unit review question levels:

Level 1–2 Level 3–4 Level 5–6 Level 7–8

1 Find the perimeter and area of these shapes.
Diagrams are not to scale, and all measurements are in inches.

a

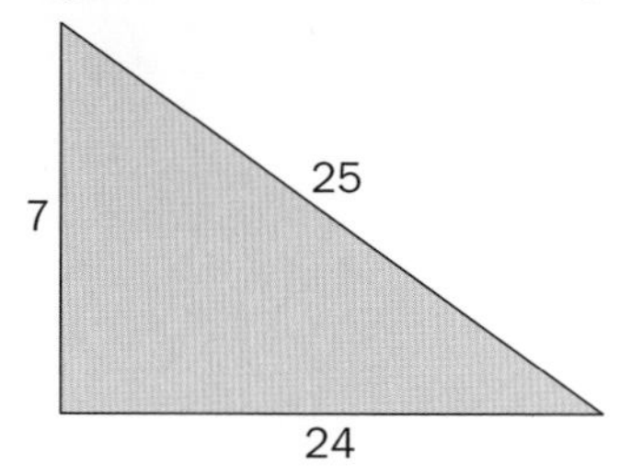


b

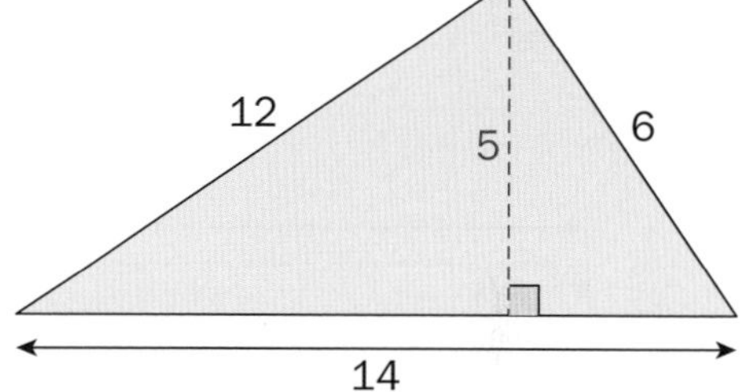


c

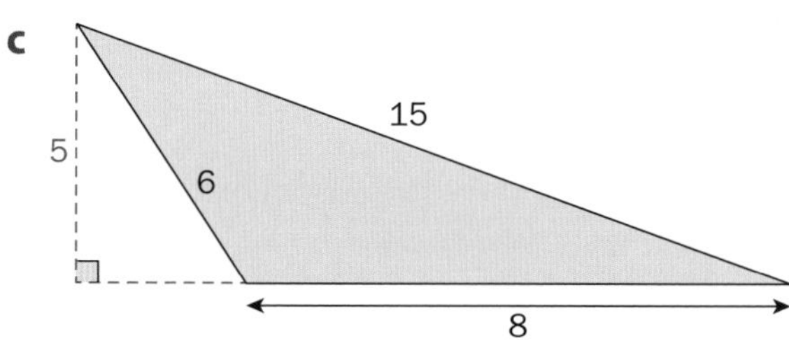


d

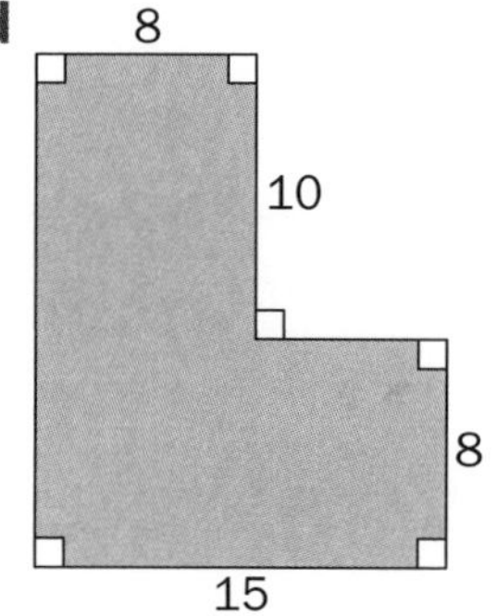


2 Find the volume of these 3D shapes. All measurements are in meters.

a

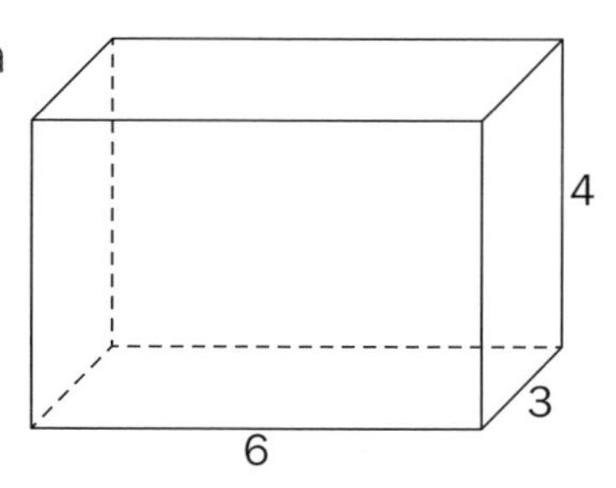


b

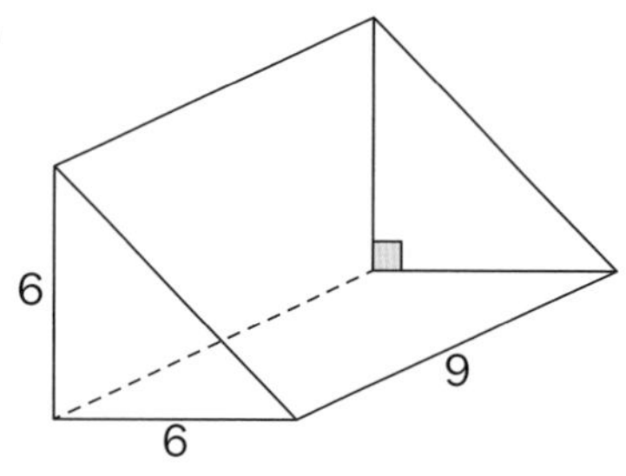


c

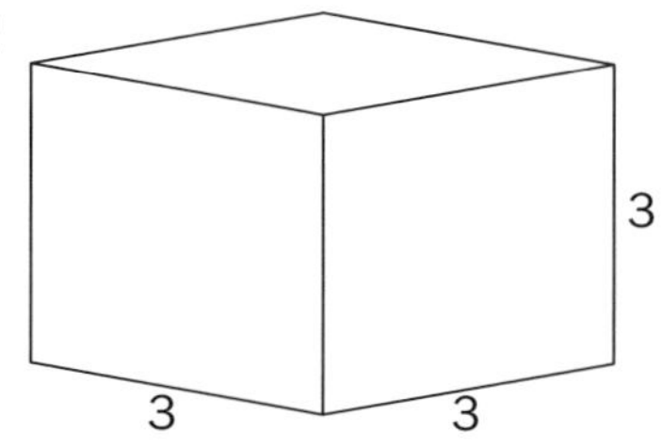


d

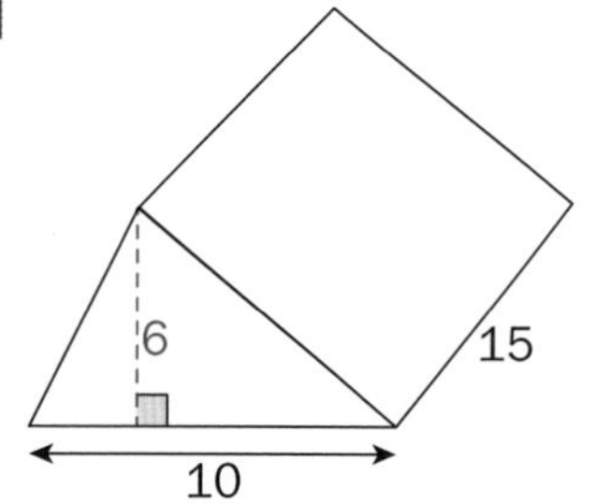

**3** Find the perimeter and area of these shapes. Diagrams are not to scale, and all measurements are in centimeters. Round your answers to the nearest hundredth where appropriate.

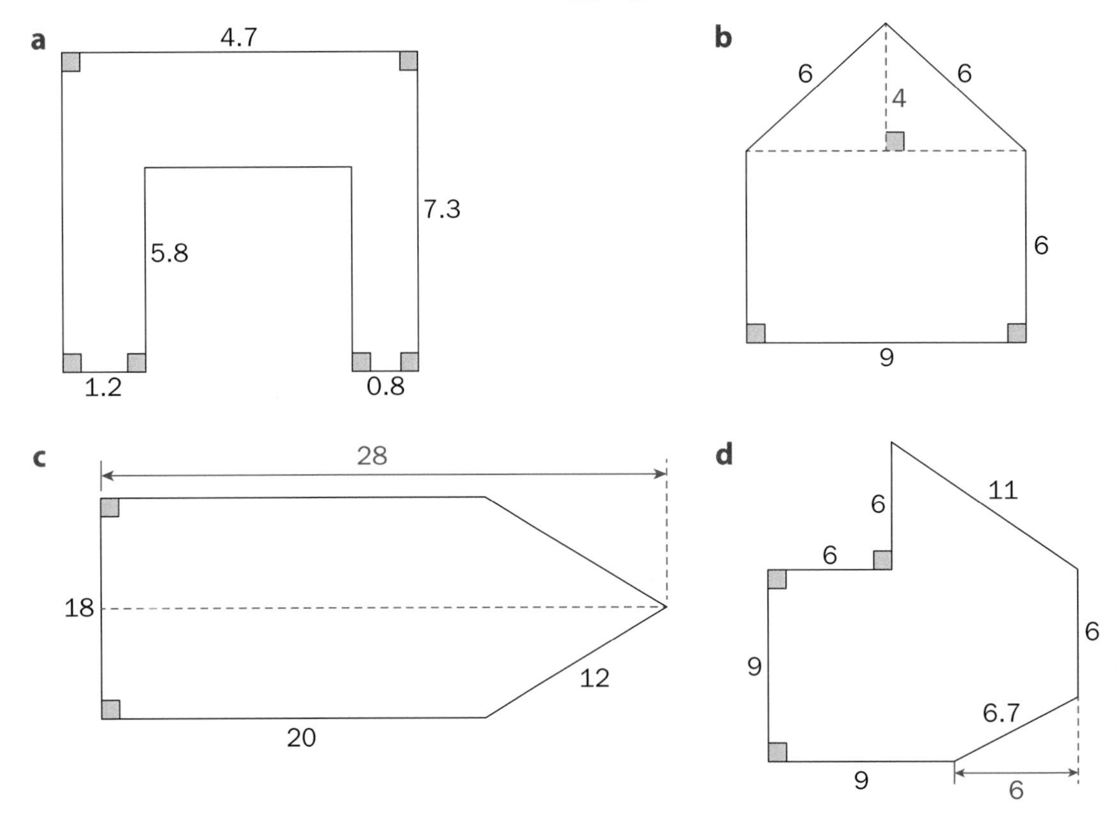


**4** Find the volume of these 3D shapes. All measurements are in feet.

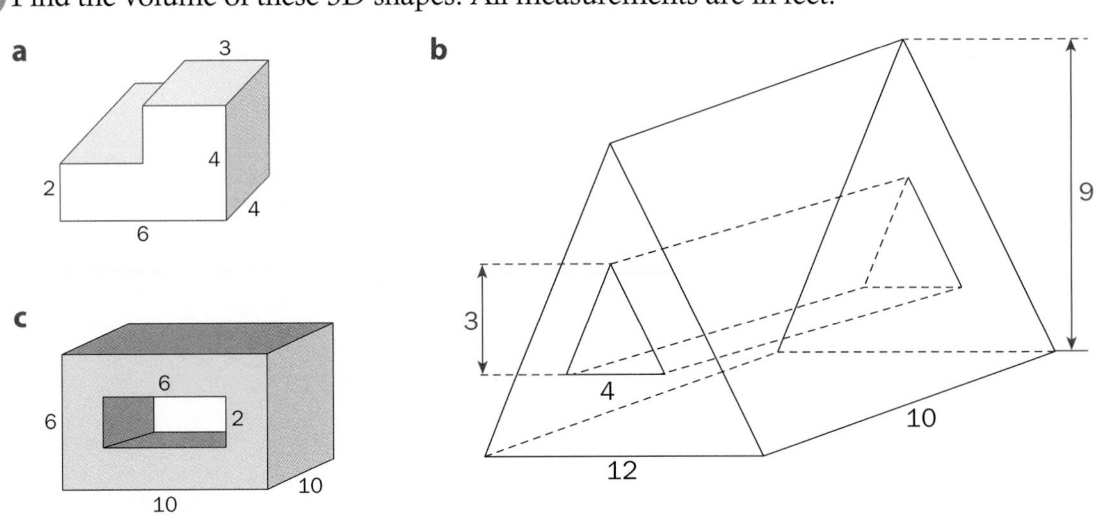

ATL2

5 Each solar panel on the roof below is square, with a side that measures 1.6 m.

**a** Find the perimeter and area of the solar panel arrangement.

**b** Ignoring the geometry of the roof for the moment, design a solar panel arrangement where the area of panels is the same as above, but the perimeter is less than it is now.

**c** Would your new arrangement fit on the roof in the picture?

**d** What limitations prevent people from converting to solar power for their energy needs?

6 On a large rectangular plot of land, three sustainable gardens are planted in rectangular sections. Each garden measures 20 m by 11 m. There is a 3 m gap between adjacent plots. There is also a 3 m border between each garden and the overall plot. If you want to fence in the entire rectangular plot, how much fencing will you need?

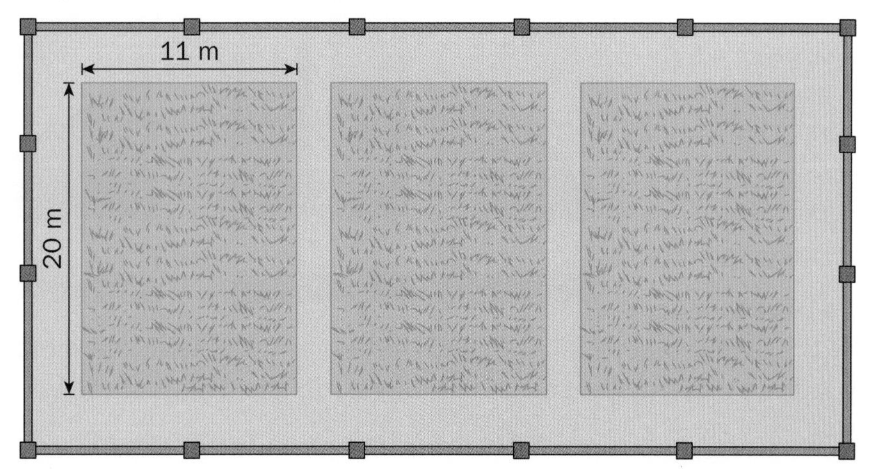

ATL2

**7** Tiny houses have become popular in the United States as a way of simplifying life, but also of being more ecologically friendly. With a living space of less than 46 $m^2$, tiny houses use less materials to build, require less energy and can be built on smaller amounts of land. The tiny houses below are two examples of designs from which customers can choose.

The house on the left has interior dimensions of $2.5 \text{ m} \times 18 \text{ m} \times 2.8 \text{ m}$. The triangular house has base dimensions of $4 \text{ m} \times 10 \text{ m}$, maximum height of 7 m and slant height of 7.3 m.

**a** Find the amount of space inside each house.

**b** Compare the environmental impacts (positive and negative) of each house.

**c** If you were building a tiny house, which design would you select? **Explain**.

**8** **a** Here is a sketch of some sun screen windows that block ultraviolet light. The two center windows, have parallel sides measuring 4.5 m and 3.5 m, respectively. The bottom (horizontal) side measures 1.5 m. What is the area of one of these center windows?

**b** The area of each outer window is equal to that of each center window. If the parallel sides of the outer windows measure 2.8 m and 0.4 m, what is the length of the bottom side?

ATL2 **9** The Saskatchewan glacier, in the Columbia icefields of Alberta, Canada, can be approximated by a rectangular prism with a length of 13 km, a width of 2.5 km and a depth of 1 km

**a** Find the area of the surface of the glacier that is exposed to air. (This is everything except the bottom surface.)

**b** Find the volume of ice in the glacier.

**c** If the length, width and depth are all cut in half, how would the volume of ice compare to the current amount? **State** the percentage decrease that this would represent.

**d** The Saskatchewan glacier is a primary source of water for the North Saskatchewan River, which is part of a major commercial route in Canada, as well as being the site of several hydroelectric dams. What impacts (both positive and negative) could there be from a melting Saskatchewan glacier?

**10** Each of these compost bins has a square base, measuring 1 m by 1 m. The front panel is 1.2 m high and the back panel is 1.5 m high.

**a** **Calculate** the volume of compost the set of three compost bins could hold when filled to capacity.

**b** Would it be realistic to fill them to capacity? What volume would be a more realistic amount to hold? **Explain** your reasoning.

★ **11** **a** **Show** that a rectangular prism with dimensions $4 \times 6 \times 12$ has a surface area that is numerically equal to its volume.

**b** There are nine other rectangular prisms with whole number sides that have a surface area numerically equal to their volume, one of which is a cube. Find the dimensions of the cube and one of the other rectangular prisms.

# Summative assessment

## ATL 1&2 Determine the best possible shape of a cereal box in order to minimize packaging

1 Get an empty box, for example a cereal box. Carefully pull it apart and flatten the box into its net.

2 Calculate the volume and surface area of the box.

3 The same volume of product could be packaged in a different size box, which would change its surface area (and the amount of material used to make the box.) Determine a set of dimensions (length, width, and height) that will keep the volume constant. Then calculate the new surface area. Fill in the first row of a table like the one below.

Note: Cereal has been chosen as it is popular in many countries, but your teacher may select another product to investigate that is more fitting in your country.

| Length | Height | Width | Volume | Surface Area |
|---|---|---|---|---|
| | | | | |
| | | | | |
| | | | | |

4 Find some different sets of dimensions that result in the same volume. Fill in the next few rows of your table, calculating the new surface area each time.

5 You can create a mathematical model to help you test possible dimensions and see the relationships between the different measurements (variables). You could continue trying different combinations by hand, but it would be best done using a spreadsheet. It will require the use of formulas entered into some of the cells.

6 Once you notice a pattern, determine the dimensions that will minimize the surface area of the box, hence reducing the amount of cardboard needed to make the box.

7 Sketch both the original box shape and the optimal box shapes, and label all dimensions.

8 Calculate the difference between the surface area of the original box and the surface area of the new optimal box.

9 There are many products that come in rectangular prism packaging. Choose one other product that is not cereal and determine if the packaging is optimal. If not, explain why this might be the case. Then, alter the design so that you are reducing the surface area while still maintaining the volume of the product. Make sure you clearly identify any constraints and explain the logic for your design, showing all calculations.

10 Describe the environmental implications of having as little packaging as possible.

11 Given these benefits, why is merchandise still sold in packaging that is far bigger than it needs to be? Do some research to find three reasons why an item might be sold in a package that is bigger than it needs to be. Be sure to cite your sources.

# Answers

## Unit 1

### Practice 1

**1** **a** Four-hundred-and-one billion, three-hundred-and-two million, fifty-nine thousand, six-hundred-and-seventy-seven
**b** Four-hundred-and-ninety-six thousand, eight-hundred-and-twelve thousand and three hundredths
**c** Two-hundred-and-forty-three ten-thousandths
**d** Eight-million-five-hundred-and-seventy-thousand-and-sixty-two hundred-millionths
**e** Three-hundred-and-two and five thousandths
**f** One and twenty-three thousandths
**g** Twenty-three thousand, four-hundred-and-seven and six tenths
**h** Two-thousand and one and two-hundred-and-ninety-five thousandths
**i** Nine-hundred million, ninety-thousand and ninety

**2** 827 096, 50 008, 275 069.00001, 14.06, 16 800 006.0007, 5000.4, 2.07, 0.012

**3** **a** **i** Six-hundred-and-forty thousand and seven-hundred
**ii** $(6 \times 100000) + (4 \times 10000) + (7 \times 100)$, $600000 + 40000 + 700$
**b** **i** Three-hundred-and-sixty-three thousand, one-hundred-and-five and twenty-one ten-thousandths
**ii** $(3 \times 100000) + (6 \times 10000) + (3 \times 1000) + (1 \times 100) + (5 \times 1) + (2 \times 0.01) + (1 \times 0.001)$, $300000 + 60000 + 3000 + 100 + 5 + 0.02 + 0.001$
**c** **i** Six-hundred-and-twenty-five ten-thousandths
**ii** $(6 \times 0.01) + (2 \times 0.001) + (5 \times 0.0001)$, $0.06 + 0.002 + 0.0005$,
**d** **i** Seven billion, four-hundred-and-ninety-two million, one-hundred-and-fifteen thousand, three-hundred and twenty-five
**ii** $(7 \times 1000000000) + (4 \times 100000000) + (9 \times 10000000) + (2 \times 1000000) + (1 \times 100000) + (1 \times 10000) + (5 \times 1000) + (3 \times 100) + (2 \times 10) + (5 \times 1)$,
$7000000000 + 400000000 + 90000000 + 2000000 + 100000 + 10000 + 5000 + 300 + 20 + 5$
**e** **i** Eighteen ten-thousandths
**ii** $(1 \times 0.001) + (8 \times 0.0001)$, $0.001 + 0.0008$

**4** **a** $(3 \times 100) + (8 \times 10) + (5 \times 1) + (9 \times 0.1) + (1 \times 0.001)$
$300 + 80 + 5 + 0.9 + 0.001$
**b** $(6 \times 1000) + (4 \times 100) + (7 \times 1) + (8 \times 0.1) + (3 \times 0.01)$
$6000 + 400 + 7 + 0.8 + 0.03$
**c** $(3 \times 100000) + (6 \times 10000) + (7 \times 1000) + (9 \times 100) + (8 \times 10) + (1 \times 1) + (2 \times 0.01)$
$300000 + 60000 + 7000 + 900 + 80 + 1 + 0.02$
**d** $(1 \times 1000000) + (4 \times 100000) + (6 \times 10000) + (7 \times 1000) + (3 \times 100) + (9 \times 10) + (9 \times 0.1) + (3 \times 0.01) + (2 \times 0.001)$
$1000000 + 400000 + 60000 + 7000 + 300 + 90 + 0.9 + 0.03 + 0.002$
**e** $(1 \times 1000) + (3 \times 1) + (2 \times 0.1)$
$1000 + 3 + 0.2$
**f** $(1 \times 10) + (2 \times 1) + (5 \times 0.1) + (7 \times 0.0001)$
$10 + 2 + 0.5 + 0.0007$
**g** $(4 \times 1) + (9 \times 0.001)$
$4 + 0.009$
**h** $(1 \times 0.01) + (4 \times 0.0001)$
$0.01 + 0.0004$

**5** **a** 252.47 **b** 4380.1 **c** 3073.005
**d** 20 507.0009 **e** 30 140.8 **f** 6 006 006

**6** **a** $(1 \times 10000) + (7 \times 1000) + (2 \times 100) + (3 \times 10) + (4 \times 1) + (1 \times 0.01)$
$10000 + 7000 + 200 + 30 + 4 + 0.01$
**b** $300 + 4 + 0.7 + 0.05$
304.75
**c** $(2 \times 1000) + (1 \times 10) + (2 \times 0.01) + (4 \times 0.001) + (5 \times 0.00001)$
2010.02405
**d** $(6 \times 100) + (8 \times 0.01)$
$600 + 0.08$
**e** $2000 + 50 + 0.03$
2050.03
**f** $60000 + 800 + 8 + 0.7 + 0.005$

## Practice 2

**1** **a** 1 3 **b** 3 2 **c** 2 27 **d** 1 19
**e** 1 40 **f** 1 25 **g** 2 0 **h** 3 39

**2** **a** 182 **b** 257 **c** 323 **d** 74
**e** 130 **f** 180 **g** 610 **h** 723

**3** Individual response

**4** **a** **b** **c** **d**
**e** **f** **g** **h**

**5** **a** 202 **b** 7159 **c** 643 **d** 1.61803 **e** 0.577218

**6** **a** 1 18 **b** 1 41 **c** 10 54 **d** 45

**7** **a** $67 - 49 = 18$ **b** $125 + 72 = 197$

**8** **a** **b** **c**
**d** **e** **f**

## Practice 3

**1** 1, 4, 9, 16, 25, 36, 49, 64, 81, 100, 121, 144, 169, 196, 225

**2** **a** 16 **b** 121 **c** 64 **d** 49
**e** 1 000 **f** 625 **g** 32 **h** 81

**3** **a** 3 **b** 9 **c** 10 **d** 18
**e** 80 **f** 15 **g** 50 **h** 4

**4** **a** 3 **b** 5 **c** 2 **d** 1

**5** **a** 5 **b** 7 **c** 4 **d** 12
**e** 2

## Practice 4

**1** **a** Yes **b** No **c** Yes **d** No
**e** No **f** Yes

**2** **a** Yes **b** Yes **c** No **d** No
**e** Yes **f** Yes

**3** **a** 1,2,4,6,8,9 **b** 1,2,4 **c** 1,3,5 **d** 1,2,4
**e** 1,2,4,8 **f** 1,2,4,5,6,8,9

**4** Individual response

**5** **a** Yes, $39 = 13 \times 3$
**b** Yes, $45 = 9 \times 5$ and $15 = 3 \times 5$ (no conversions scored)
**c** Individual response
**d** Individual response

## Practice 5

**1** **a** 12 **b** 16 **c** 27 **d** 1
**e** 7 **f** 8 **g** 12 **h** 60
**i** 25 **j** 16 **k** 1 **l** 2

**2** Individual response

## Practice 6

**1** **a** 12 **b** 24 **c** 192 **d** 96 **e** 595
**f** 594 **g** 80 **h** 38 352 **i** 5 940

**2** Individual response

## Practice 7

1 195

2 420

3 240

4 6 inches; 91 squares

5 **a** 18980 **b** 52

## Practice 8

1 **a** 86 **b** 24 **c** 26 **d** 159
**e** 560 **f** 60 **g** 2000 **h** 143
**i** 342 **j** 671 **k** 979 **l** 12300
**m** 120 **n** 31.4159 **o** 240 **p** 40010

2 210

3 5050

## Practice 9

1 **a** 20 **b** 0 **c** 1035 **d** 23
**e** 11 **f** 11 **g** 30 **h** 2
**i** –1 **j** 12

2 **a** ÷
**b** +
**c** ÷, ×

3 **a** $(18-6)\times 3+2=38$
**b** $(8+4)\div(2+2)=3$
**c** $(5+3)\times 6-10=38$
**d** $(16\div 8)+(4\div 4)=3$
**e** $12\div((8-5)\times 3)=1$
**f** $5\times(3+0)-6=9$

4 **a** $7\times 3-4\times 5=1$
**b** $(12-6)\times 3-10\div 5=4$
**c** $24\div(10-6)+2\times 5=16$
**d** $12+36\div 9\div 4=13$
**e** $6\times 8\div 12\times 2+7=15$
**f** $14+10-5-12\div 3=15$

5 **a** $(8-5)\times(1+7)=24$
**b** $8\times(3\times(8-7))=24$
**c** $(10+2)\times(6\div 3)=24$
**d** $(7\times 12)-(6\times 10)=24$
**e** $(2+6)\times(9-6)=24$

6 **a** 137.5
**b** Individual response

## Unit Review

1 **a** Thirty-two million, five-hundred-and-sixty thousand and forty-two
**b** Seventeen and eight hundredths
**c** One thousand, seven-hundred-and-forty-three and fourteen hundredths
**d** Two and seven-hundred-and-eighteen thousandths

2 **a** 10455 **b** 29.032 **c** 0.190645 **d** 0.018

3 **a** $(1\times 10000000)+(5\times 1000000)+(3\times 100000)+(4\times 10000)+(2\times 1000)+(8\times 100)+(7\times 1)$
$10000000+5000000+300000+40000+2000+800+7$
**b** $(4\times 100000)+(5\times 1000)+(6\times 100)+(9\times 10)+(6\times 1)+(7\times 0.1)$
$400000+5000+600+90+6+0.7$
**c** $(1\times 0.1)+(3\times 0.01)+(7\times 0.001)+(5\times 0.0001)$
$0.1+0.03+0.007+0.0005$

**d** (1 × 1 000 000 000) + (2 × 100 000 000) + (5 × 10 000 000) + (2 × 1 000 000) + (7 × 100 000) + (6 × 1 000) + (1 × 100) + (1 × 10) + (4 × 1)
1 000 000 000 + 200 000 000 + 50 000 000 + 2 000 000 + 700 000 + 6 000 + 100 + 10 + 4

**e** (2 × 0.00001) + (5 × 0.000001)
0.00002 + 0.000005

**4** **a** (1 × 10) + (2 × 1) + (7 × 0.1) + (5 × 0.01)
10 + 2 + 0.7 + 0.05

**b** (2 × 100) + (3 × 1) + (6 × 0.1)
200 + 3 + 0.6

**c** (1 × 10 000) + (2 × 1 000) + (4 × 1) + (2 × 0.1) + (5 × 0.001)
10 000 + 2 000 + 4 + 0.2 + 0.005

**d** (3 × 1) + (1 × 0.001) + (9 × 0.0001)
3 + 0.001 + 0.0009

**5** **a** 518.021 **b** 30620.07 **c** 9010.0108 **d** 406 020.03

**6** **a** (8 × 1 000) + (2 × 100) + (3 × 10) + (1 × 1) + (4 × 0.01)
8 000 + 200 + 30 + 1 + 0.04

**b** 4 000 + 30 + 0.5 + 0.004
4 030.504

**c** (8 × 1 000) + (8 × 100) + (7 × 0.1) + (8 × 0.001) + (2 × 0.00001)
8 800.70802

**d** (1 × 1 000) + (4 × 100) + (1 × 0.1) + (2 × 0.01)

**7** **a** 512 **b** 144 **c** 216 **d** 169
**e** 16 **f** 27 **g** 128 **h** 625

**8** **a** 86 **b** 24 **c** 26 **d** 159
**e** 560 **f** 60 **g** 2000 **h** 143

**9** **a** ⠼⠃⠓ **b** ⠼⠉⠋⠃ **c** ⠼⠁⠓⠉ **d** ⠼⠊⠚⠊⠃

**10** **a** 404 **b** 3.14 **c** 51.75 **d** 144.48

**11** **a** 1 39 **b** 2 30 **c** 11 44 **d** 17 16

**12** **a** 1 743 **b** 784

**13** **a** 5 814 **b** 25 773

**14** **a** 1 21 **b** 3 20 **c** 6 12 **d** 20 48

**15** **a** 74 **b** 185 **c** 163 **d** 300

**16** **a** 11 **b** 8 **c** 10 **d** 2
**e** 3 **f** 4 **g** 8 **h** 12

**17** **a** Yes **b** Yes **c** No **d** No
**e** Yes **f** No

**18** **a** No **b** Yes **c** Yes **d** Yes
**e** No **f** Yes

**19** **a** 1, 2, 4, 8 **b** 1, 2, 4, 8 **c** 1, 3, 5, 9 **d** 1, 2, 4, 5, 8
**e** 1, 2, 4, 8 **f** 1, 2, 3, 4, 5, 6, 8, 9

**20** **a** 7 **b** 8 **c** 48 **d** 6
**e** 12 **f** 18 **g** 24 **h** 30

**21** **a** 108 **b** 180 **c** 60 **d** 90
**e** 120 **f** 72 **g** 1155 **h** 84

**22** 2508 years

**23** **a** 16 **b** 15 **c** 1 **d** 121
**e** 150 **f** 10 **g** 8 **h** 1
**i** 100 **j** 2

**24** For example:

**a** $2 \times (11 + 5 - 4) = 24$

**b** $12 \times (10 \div (3 + 2)) = 24$

**c** $(3 + 9) \times (6 - 4) = 24$

**d** $(1 + 3) \times 4 + 8 = 24$

**25 a** No. Arturo simply worked from left to right rather than using the correct order of operations.

**b** Individual response e.g. $12 + 6 \times 4 - 12 = 24$

**c** For example: $(4 \times 3) \div (6 \div 12) = 24$ or $(12 \times 3) \div (6 \div 4) = 24$

**26** Individual response

**27** Individual response

**28 a** 6 **b** 45 **c** 22.5

# Unit 2

## Practice 1

**1 a** 25% **b** 47% **c** 30% **d** $33.3\dot{3}\%$

**2 a**

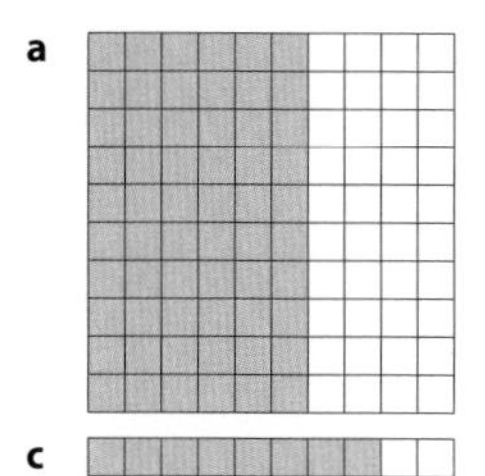

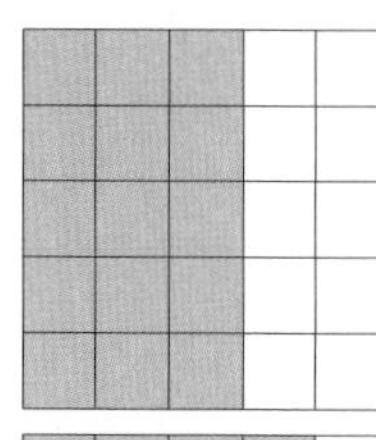

**b**

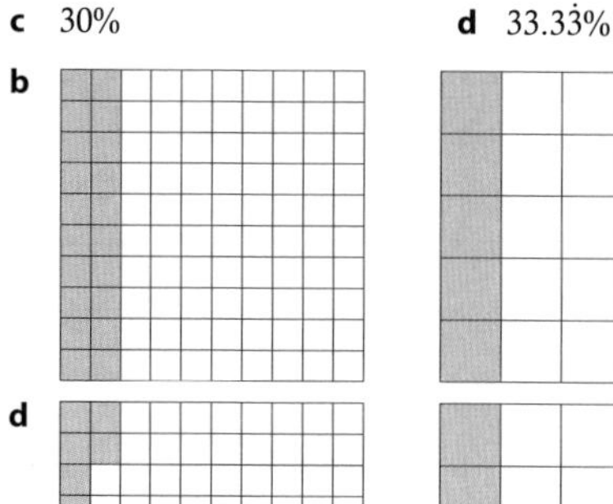

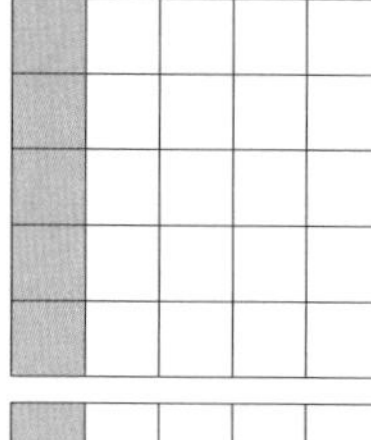

**c**

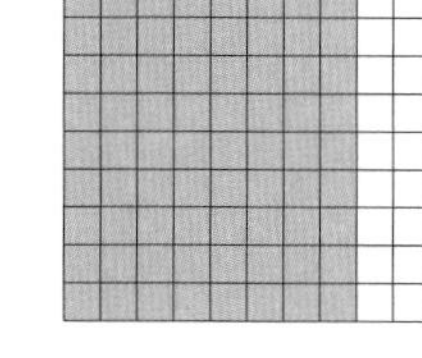

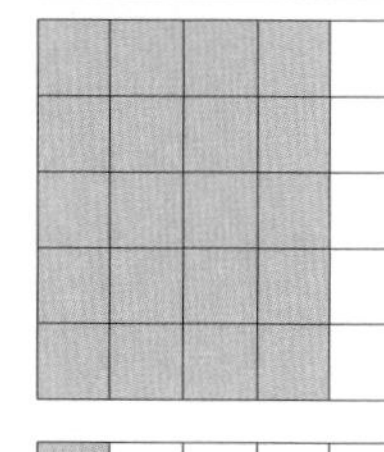

**d**

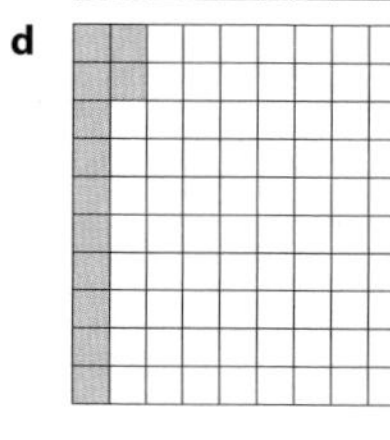

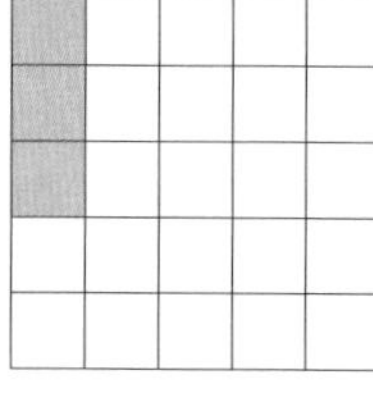

**e**

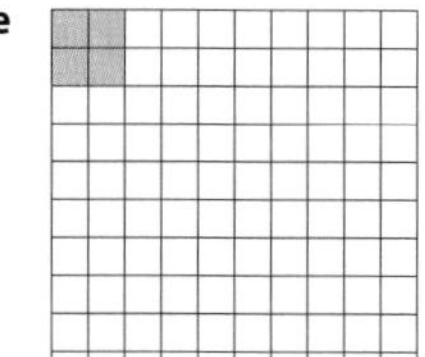

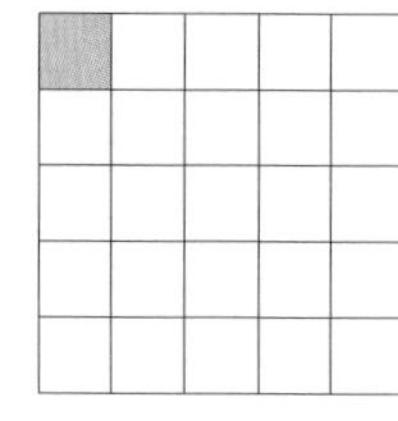

**3 a** 24 **b** 36 **c** 84
**d** 120 **e** 6 **f** 30

**4** Valid answers include:

**a** 12% is two lots of 6% or three lots of 4%
**b** 9% is one lot of 4% plus one lot of 5%
**c** 10% is two lots of 5% or one lot of 4% plus one lot of 6%
**d** 30% is five lots of 6% or six lots of 5% or five lots of 4% plus two lots of 5%
**e** 11% is one lot of 6% plus one lot of 5%
**f** 14% is one lot of 6% plus two lots of 4%
**g** 17% is two lots of 6% plus one lot of 5%

**5 a i** 8 million
**ii** 20 million
**iii** 10 million
**iv** 26 million
**b** Individual response

**6** 37.5%

## Practice 2

**1 a** 50% **b** 85% **c** 70% **d** 60%
**e** 62% **f** 75% **g** 44% **h** 15%
**i** 100% **j** 94%

**2 a** 50% **b** 40% **c** 15% **d** 10%
**e** 40% **f** 75% **g** 60% **h** 90%
**i** 75% **j** 60%

3 a 27.78% b 9.52% c 25.58% d 79.98%
e 94.74% f 45.28% g 35.87% h 81.97%
i 123.53% j 9.09%

4 80%

5 20%

## Practice 3

1 Note: for fractional form there is more than one correct answer

a $0.87 = \frac{87}{100} = 87\%$ b $0.5 = \frac{5}{10} = 50\%$ c $0.732 = \frac{732}{1000} = 73.2\%$

d $0.09 = \frac{9}{100} = 9\%$ e $0.006 = \frac{6}{1000} = 6\%$ f $1.2 = \frac{12}{10} = 120\%$

2 a 0.12 b 0.46 c 1.03
d 0.385 e 0.083 f 0.1225

3 a $\frac{2}{25} = 0.08 = 8\%$ b $\frac{4}{5} = 0.8 = 80\%$ c $\frac{35}{100} = 0.35 = 35\%$ d $\frac{4}{9} = 0.4\bar{4} = 44.4\bar{4}\%$

4 a $\frac{3}{10}$ b $0.7\bar{7}$ c 1.03
d $\frac{2}{9}$ e $\frac{24}{33}$ f They are all equal

5 a $\frac{42}{150} = 0.28 = 28\%$ b $0.4 = \frac{104}{260} = 0.40$

6 a 14%, $\frac{2}{11}$, 0.186, 36%, $\frac{2}{5}$, 0.6 b 0.05, 9%, $\frac{1}{9}$, 0.16, 21%, $\frac{7}{20}$

7 a 0.98, $\frac{31}{33}$, 91%, $\frac{45}{50}$, 0.89, 87% b $\frac{21}{70}$, 27%, $\frac{11}{55}$, 0.136, 0.09, 4%

8 a 10% of the three same sized bags of popcorn
b 40% of three of the same sized mangos
c The options are equivalent

9 150 million

10 1.79%

11 17 million to 102 million

12 a 235 000 b 71.28%

## Practice 4

1 a 75% b 75% c 25% d 20%
e 36% f 60% g 25% h 40%

2 a 40 b 40 c 120 d 152
e 260 f 350 g 200 h 60

3 a 25 b 40 c 10 d 80
e 100 f 140 g 30 h 3
i 80 j 70 k 4 l 3.84
m 33 n 36 o 18

4 $90

5 76.92%

6 $65 000 (per year)

7 $1344 (per week)

8 Individual response

9 26 667

## Practice 5

1 a Total costs
Shirt: $46
Flowers: $69
Ball: $34.50
Painting: $103.50
Book: $23
Umbrella: $17.25

**b** No, since $103.5 > 100$
**c** Any three which has total price summing to less than \$100 e.g. Shirt, Ball, Umbrella
**d** Shirt, Ball, Umbrella

**2** **a** **i** \$31
**ii** \$84
**iii** \$24
**b** **i** \$35.65
**ii** \$96.60
**iii** \$27.60

**3** \$31.05

**4** **a** 200 SYP **b** 220 SYP **c** 780 SYP
**d** 1820 SYP **e** 19.23% **f** Individual response

## Practice 6

**1** **a** 77.8% decrease **b** 167.6% increase **c** 327.4% increase
**d** 14.8% decrease **e** 900% increase **f** 77.3% decrease

**2** 8.8% decrease

**3** 58.8% decrease

## Unit Review

**1** **a** 47% **b** 38% **c** 65%
**d** 75% **e** 20%

**2** **a**

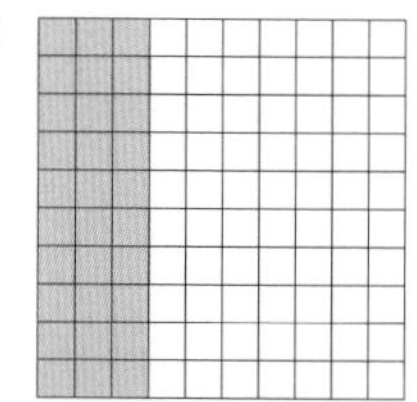

**b**

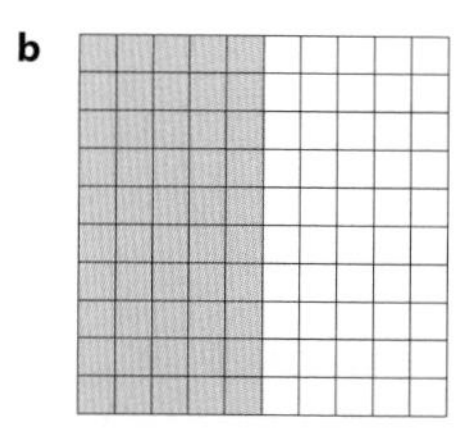

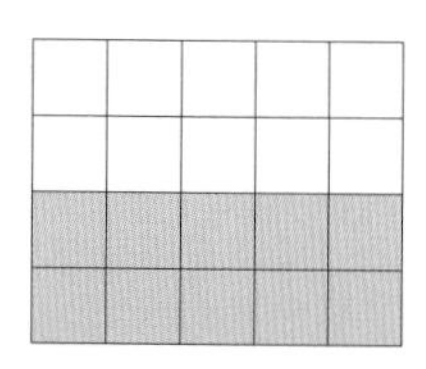

**c**

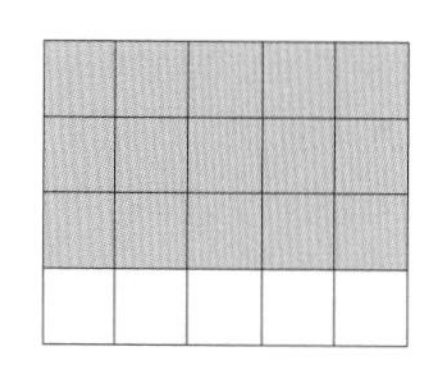

**d**

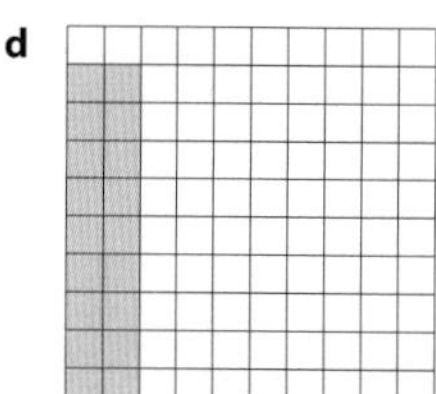

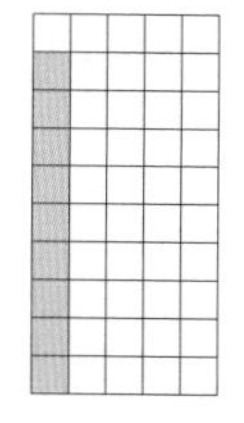

**e**

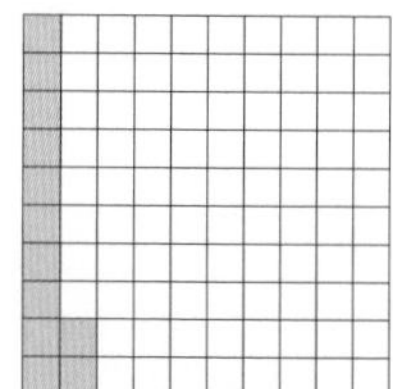

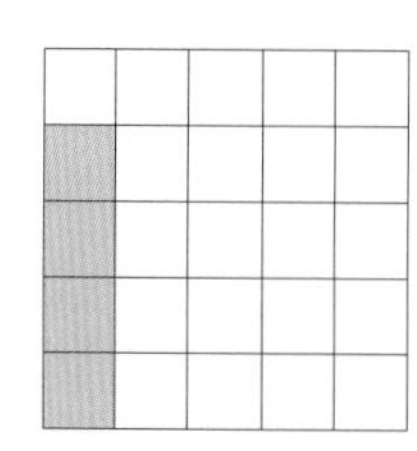

**3** Valid answers include:
**a** 9% : three lots of 3%
**b** 11% : one lot of 3% plus one lot of 8%
**c** 21%: one lot of 3% plus one lot of 8% plus on lot of 10%
**d** 18%: one lot of 8% plus one lot of 10%
**e** 16%: two lots of 8%
**f** 26%: two lots of 8% plus one lot of 10%
**g** 77%: nine lots of 3% plus five lots of 10%

**4** **a** 16 **b** 60 **c** 150
**d** 45 **e** 6 **f** 36

**5** **a** 90% **b** 85% **c** 200% **d** 25%
**e** 40% **f** 40% **g** 20% **h** 8.08%
**i** 100% **j** 363.64%

6 a 23% b 23% c 30% d 92%
e 37% f 61% g 16% h 174%
i 103% j 212%

7 a $\frac{7}{25}=0.28=28\%$ b $\frac{4}{5}=0.8=80\%$ c $\frac{13}{20}=0.65=65\%$ d $\frac{8}{25}=0.32=32\%$
e $\frac{11}{20}=0.55=55\%$ f $\frac{7}{15}=0.47=46.67\%$ g $\frac{23}{25}=0.92=92\%$ h $\frac{1}{5}=0.2=20\%$
i $\frac{3}{5}=0.6=60\%$ j $\frac{11}{10}=1.1=110\%$

8 a 0.81 b 0.92 c $\frac{2}{3}$
d 26% e 0.55

9 a $25\%=\frac{1}{4}$ and $\frac{12}{50}=0.24=24\%$ b $5\%=\frac{30}{600}=0.05$ and $0.5=50\%=\frac{1}{2}$

10 a 50% b 20% c 10%

11 10%

12 Canada

13 a 20% b 6 c $33.3\overline{3}$
d 54 e 160 f 75%

14 540 million

15 a 53.85% decrease b See comment c 3.5 million

16 10.4% assuming world population of 7.5 billion
546 million

17 a 30% decrease b 11.11% c 33.33%

## Unit 3

### Practice 1

1 a Rectangle of leaves 3 by 4, rectangle of leaves 3 by 5
b From one shape to the next, add a row of 3 leaves
c 3,6,9,12,15
We get to the next number of the sequence by adding 3
d Yes, this is a linear pattern because the difference between the terms is always the same
e $L+3$
f Yes, because 30 is a multiple of 3

2 a Cross of length 7 and height 7, cross of length 9 and height 9
b Cross of length 7 and height 7, cross of length 9 and height 9
Find the next number by adding 4
c Yes, this is a linear pattern because the difference between the terms is always the same
d 117
e $w+4$
f No. The $n$th term is $4n-3$. No whole number $n$ can make this expression equal to 40.

3 a Honeycomb shape with 13 hexagons; honeycomb shape with 16 hexagons
b 79
c Yes, the rule is $3n + 1$, so the number 409 will be the 136th term of the sequence.
d Let the $n$th term of the sequence be written as $a_n$. Then $a_n = a_{n-1} + 3$

4 a Individual response. Next two shapes should have 15 and 19 parts
b Individual response. Next two shapes should have 49 and 81 parts

5 Triangle

### Practice 2

1 Note: $n$ has been used throughout. Other letters are acceptable.
a $3n$ b $n+7$ c $5n$ d $n+12$
e $n-13$ f $n+60$ g $\frac{n}{2}$ h $n+17$
i $\frac{n}{43}$ j $\frac{8}{n}$ provided $n$ is non-zero k $5-n$ l $15n$
m $4n-26$ n $8n+3$ o $4(n+7)$ p $\frac{20}{n}-11$ provided $n$ is non-zero

2 a Subtract 1 from a number
b Add 11 to a number
c Multiply a number by 14
d Divide a number by 3
e Subtract 12 from a number
f Subtract a number from 20
g Divide 12 by a number
h Divide a number by 20 then add 7
i Multiply a number by 8 then subtract 9
j Times a number by 3 then subtract the result from 2

## Practice 3

1 a 50, $n+6$
b 88, $n-5$
c 81, $3n$
d 2, $\frac{n}{5}$
e 128, $4n$
f 53, $n+9$
g 12.5, $\frac{n}{2}$
h 40000, $10n$

2 a 16, 32, 64
b $512 = 2^9$ so division has occurred 9 times
c No. The difference between terms is not constant. (For those interested, this is an example of an *exponential* pattern)

3 a

| Animal | Heart rate (bpm) | Average approx. lifespan (years) | Lifetime heartbeats |
|---|---|---|---|
| Pig | 70 | 25 | **919 800 000** |
| Rabbit | 205 | 9 | **969 732 000** |
| Horse | 44 | 40 | **925 056 000** |
| Hamster | 450 | 3 | **709 560 000** |
| Monkey | 190 | 15 | **1 497 960 000** |
| Whale | 20 | 80 | **840 960 000** |
| Elephant | 30 | 70 | **1 103 760 000** |

b 995 261 142.9
c 126 to the nearest whole number
d No, lifetime heartbeats is approximately 2 207 520 000

4 a 3.5 cm, 7 cm, 10.5 cm, 14 cm, 17.5 cm
b $n+3.5$cm
c Yes, this is a linear pattern because the difference between the terms is always the same
d 98 000 000 cm

## Practice 4

1 a 10
b 60
c 13
d 36
e 7
f 17
g 65
h 58
i 17
j 23
k 75
l 0
m 1
n 13

2 a 52, 65, 78
b Yes, this is a linear pattern because the difference between the terms is always the same
c Multiply the cycle number by 13
d $13c$
e 260

3 a $r = 2n$
b Yes, this is a linear pattern because the difference between the segments is always the same
c 22 cm
d 15

4 a Divide by 5
b $\frac{t}{5}$
c 50 seconds

## Practice 5

1 a Rule: $4n$
Linear
b Rule: $2n$
Missing values: 8, 200
Linear

c Rule: $n+5$
Missing values: 9, 105
Linear

d Rule: $n+10$
Missing values: 14, 110
Linear

e Rule: $\frac{n}{2}$
Missing values: 200, 25
Linear

f Rule: $n-3$
Missing values: 34, 8
Linear

g Rule: $6n$
Missing values: 12, 24
Linear

h Rule: $n+8$
Missing values: 14, 23
Linear

i Rule: $2n+1$
Missing values: 9, 201
Linear

j Rule: $5n+2$
Missing values: 22, 252
Linear

k Rule: $3n-1$
Missing values: 11, 297
Linear

**2 a**

| Temp (°F) | Chirps per 14 seconds |
|---|---|
| 70 | 30 |
| 80 | 40 |
| 85 | 45 |

**b** $C = T - 40$, where $C$ = number of chirps per 14 seconds, and $T$ = temperature in Fahrenheit
**c** Yes, it is a linear pattern

**3 a**

| Temp (°C) | Chirps per 25 seconds |
|---|---|
| 19 | 45 |
| 21 | 51 |
| 24 | 60 |

**b** $C = 3T - 12$, where $C$ = number of chirps per 25 seconds, and $T$ = temperature in Celsius
**c** Yes, it is a linear pattern

## Practice 6

**1 a** $x = 14$ **b** $y = 4$ **c** $i = 18$ **d** $f = 34$
**e** $d = 200$ **f** $m = 30$ **g** $k = 35$ **h** $a = 25$

**2 a** $q = 15$ **b** $g = 8$ **c** $r = 32$ **d** $d = 50$
**e** $c = 8$ **f** $y = 25$ **g** $z = 35$ **h** $u = 6$
**i** $x = 8$

**3** Letters refer to the equation label, not the unknown quantity
2: a, g, i, l
9: b, d, e, h
16: c, f
38: j
40: k

**4 a**

| Day | Number of hexagons |
|---|---|
| 1 | 10 |
| 2 | 22 |
| 3 | 34 |
| 4 | 46 |
| 5 | 58 |
| 6 | 70 |
| 7 | 82 |
| 8 | 94 |
| 9 | 106 |
| 10 | 118 |

**b** $h = 10 + 12(d - 1) = 12d - 2$
**c** 358
**d**

**5 a** $f = 249 + t$
**b** 369
**c** 751

**6 a** The circumference of the tree (in cm) is the age of the tree (in years) multiplied by $\pi$
**b** Measure the circumference of the tree and divide the result by $\pi$
**c** $c = \pi a$
**d** 20 years
**e** 157.08

## Unit Review

**1 a** 160 320 (the next term is double the current term)
**b** 26 31 (the next term is the current term plus 6)
**c** 32 31 (alternating sequence with 31 and terms increasing by 2)
**d** 3 1.5 (the next term is half the current term)
**e** 60 70 (the next term is the current term plus 10)

**2** ?

**3 a** $\frac{28}{n}$ provided $n$ is non-zero **b** $n - 4$ **c** $n + 4$ **d** $17n$
**e** $12 + n$ **f** $13n$ **g** $n - 6$ **h** $2n + 3$

**4 a** Four times a number
**b** 63 divided by a number (that is non-zero)
**c** 15 less than a number
**d** 61 more than a number
**e** 7 added to 3 times a number

**5 a** 8 **b** 9 **c** 4

**6 a** $x = 15$ **b** $x = 8$ **c** $m = 21$ **d** $p = 24$
**e** $b = 7$ **f** $w = 20$ **g** $h = 9$ **h** $z = 40$
**i** $d = 150$ **j** $x = 8$ **k** $m = 19$ **l** $n = 49$
**m** $g = 38$ **n** $y = 11$ **o** $a = 114$ **p** $x = 24$

**7 a** From one shape to the next, add a ladybird to each endpoint
**b** The 10th term would have 1 ladybird in the centre, with two arms of 10 ladybirds each, for a total of 21 ladybirds.
**c** No, 78 will not be in the sequence because it is an even number, and this is the sequence of odd numbers starting with 3.
**d** If the current term is $n$ then the next term is $n + 2$

**8 a** The first shape in the sequence contains two clouds. Each subsequent shape contains three more clouds than the previous one, so the sequence is 2, 5, 8, 11, 14, ....
**b** The first shape contains 2 clouds. Each subsequent shape adds 3 more clouds, so the 20th term will contain $2 + 19(3) = 59$ clouds.
**c** No, the pattern rule is $3n - 1$, where $n$ is the term number, so no integer value of $n$ will give 46.
**d** If the current term is $n$ then the next term is $n + 3$

9 a The $n^{th}$ term is $n$ by $n$ lightning bolts
b Yes, the pattern rule is $t = n^2$, where $n$ is the term number. These are simply the square numbers: 1, 4, 9, 16, etc. so as 100 is the square of 10 it will be in the sequence.

10 Individual response

11 a $x = 3$ b 28 c $g = 8$ d $h = -22$

12 a $m = 75d$ b 40 days c Between $\frac{2}{9}$ and $\frac{1}{6}$

13 Snail = 20
Dragonfly = 10
Ladybug = 5

14 a

| Seconds | Eggs in the clutch |
|---|---|
| 3 | 7 |
| 4 | 8 |
| 5 | 9 |
| 6 | 10 |
| 7 | 11 |
| 8 | 12 |
| 9 | 13 |
| 10 | 14 |

b $e = 7 + (t - 3) = t + 4$
c 196 seconds
d 149 seconds

15 a $e = 2t$ b $\frac{5}{3}$ of a minute c 40 seconds

## Unit 4

## Practice 1

1 a Line through points $X$ and $Y$
b Line segment between $K$ and $L$
c Line from $P$ extending through $Q$
d Just a point $A$
e A line labelled with $n$
f Parallel lines, one through $B$ and $C$, the other through $F$ and $G$

2 a $\overleftrightarrow{BA}$ b $\overrightarrow{CD}$ c $\overline{EF}$ d $\overrightarrow{GH}$
e $\overleftrightarrow{IJ}$ f $\overline{KL}$ g $\overrightarrow{NM}$ h $\overrightarrow{PO}$ i $\overline{QR}$

3 Individual response

4 a Individual response
b Individual response

## Practice 2

1 a Acute b Right-angle c Obtuse d Acute
e Straight f Obtuse g Obtuse h Obtuse

2 a 60 degrees; acute b 126 degrees; obtuse c 35 degrees; acute d 89 degrees; acute

3 a 30 degrees; acute b 90 degrees; right c 175 degrees; obtuse d 54 degrees; acute
e 83 degrees; acute f 135 degrees; obtuse

4 Individual response; any angles intersecting with given angles are valid

5 Individual response; any two rays intersecting with angle 75 degrees are valid

6 Two line segments $\overline{CD}$ and $\overline{EF}$ intersecting perpendicularly. There are four right-angles at least one of which should be indicated by an empty box

7 Triangular face with angles made with the horizontal labelled '43.2 degrees' or equivalent

8 Drawing as described, with height of longer side = 26.2 cm (accurate to 1 d.p.)

9 **a** Individual response
**b** One angle of 50 degrees and two angles of 130 degrees
**c** Sum of all of the angles is 360 degrees
**d** Adding the angles along either of the two lines gives 180 degrees

## Practice 3

1 **a** **i** Vertically opposite: $a$ and $c$ or $b$ and $d$
**ii** Supplementary: $a$ and $b$ or $c$ and $d$
**b** **i** Vertically opposite: $e$ and $g$ or $f$ and $h$
**ii** Supplementary: $f$ and $g$ or $e$ and $h$
**c** **i** Vertically opposite: $j$ and $l$ or $k$ and $m$
**ii** Supplementary: $j$ and $k$ or $m$ and $l$
**d** **i** Vertically opposite: $p$ and $s$ or $q$ and $r$
**ii** Supplementary: $p$ and $q$ or $r$ and $s$

2 Two pairs of vertical angles and two pairs of supplementary angles at each of B,F,G

3 **a** Complement: 55 degrees; Supplement: 145 degrees
**b** Complement: 79 degrees; Supplement: 169 degrees
**c** Complement: 6 degrees; Supplement: 96 degrees
**d** Complement: 81 degrees; Supplement: 171 degrees
**e** Complement: 38 degrees; Supplement: 128 degrees

4 **a** $AXZ = 150°$, $BXZ = 30°$, $BXY = 150°$
**b** $PTR = 93°$, $RTQ = 87°$, $QTS = 93°$
**c** $PJK = 35°$, $KJQ = 90°$, $MJQ = 55°$, $MJL = 35°$
**d** $IGJ = 129°$, $IGF = 51°$, $FGH = 129°$, $GHM = 51°$, $MHL = 129°$, $LHK = 51°$
**e** $VRU = 50°$, $WRS = 50°$, $SRT = 40°$
**f** $BKR = 115°$, $EKC = 40°$, $MKC = 140°$
**g** $DPW = 100°$, $DPF = 80°$, $FPR = 100°$, $NRX = 40°$, $XRW = 140°$, $PRN = 140°$, $PWT = 120°$, $PWR = 60°$, $AWR = 120°$
**h** $WYR = 90°$, $FYC = 100°$

5 **a** $x = 37$ **b** $x = 115$ **c** $x = 28$
**d** $x = 15$ **e** $x = 63$

6 **a** Yes there are many, by the definition of vertical angles
**b** Individual response
**c** Individual response
**d** Individual response
**e** Individual response

## Practice 4

1 Alternate interior angles: Z-angles
Vertically opposite angles: X-angles
Complementary angles: are angles that add up to 90 degrees
Supplementary angles: are angles that add up to 180 degrees
Interior angles of a triangle:
Corresponding angles: F-angles

2 Individual response

3 **a** $d, e, g, i, j, n, p$ **b** $a, b, f, k, l, m, o$

4 **a** $a = b = 74$, $c = 106$
**b** $d = e = f = 133$
**c** $g = 55$, $h = 125$, $j = 55$, $k = 45$, $m = 55$
**d** $p = 50$, $q = r = 40$
**e** $s = 30$, $t = 20$
**f** $w = x = 115$, $y = 148$, $z = 32$

5 **a** $x = 60$ **b** $x = 35$ **c** $x = 110$

## Practice 5

1 Individual response

2 $a = 43, b = 75, c = 110, d = 120, e = 60, f = 63, g = 96, h = 81, i = 36$

**3** $a = 55, b = 43, c = 32, d = 63, e = 67, f = 108, g = 63, h = 117, i = 136$

$j = 100, k = 35, l = 80, m = 65, n = 57, o = 64, p = 116, q = 59, r = 160, s = 120, t = 40, u = 20$

**4** The angles $c$, $d$ and $e$ add to 180 degrees because they are the angles on a straight line, so $c + d + e = 180$. Angles $a$ and $d$ are alternate angles, so $d = a$. Angles $b$ and $e$ are alternate angles, so $e = b$. Substituting these values for $d$ and $e$ into the first equation gives $c + b + a = 180$, thus the angles in a triangle sum to 180 degrees.

**5** Individual response

## Unit Review

**1** **a** Line through $E$ and $F$
**b** Line segment between $X$ and $Y$
**c** Line from $R$ through $S$
**d** A point labelled $T$
**e** Perpendicular lines; a line through $A$ and $B$ perpendicular to a line through $J$ and $M$
**f** Parallel lines; a line through $T$ and $V$ parallel to a line through $Y$ and $Z$

**2** **a** $\overline{PQ}$ **b** $l \parallel m$ **c** $\overrightarrow{GH}$
**d** $\overline{PQ}$ **e** $\angle HGI$ or $\angle IGH$ **f** $\overline{AB} \perp \overline{MN}$

**3** **a** 47 degrees, acute
**b** 90 degrees, right-angle
**c** 56 degrees, acute
**d** 120 degrees, obtuse

**4** Individual response

**5** **b** 22 degrees **c** 56 degrees **f** 76 degrees

**6** **a** $x = 158$ **b** $x = 62$ **c** $x = 15$
**d** $x = 60$ **e** $x = 5$

**7** Sum of interior angles in a triangle is 180 degrees. There exists a right-angle, so the remaining two angles must sum to $180 - 90 = 90$

**8** **a** $x = 58$ **b** $y = 80$ **c** $p = 54, q = 36, r = 108$ **d** $x = 60, y = 50$

**9** **a** $x = 18$ **b** $x = 7$

**10** **a** $w = 56, x = 59, y = 65$
**b** $j = 65, k = 85, m = 65, n = 25$
**c** $p = 60, q = 30, r = 20, s = 160, t = 20$
**d** $p = 153, q = 63, r = 27, s = 81, t = 72, u = 153$
**e** $f = 142, g = 38, h = 142, j = 38, k = 52, m = 38, n = 142, p = 61, q = 61,$
**f** $A = 59, B = 111, C = 153, D = 37, E = 37, F = 58, G = 26, H = 96, J = 84, K = 37, L = 53$

**11** Individual response

# Unit 5

## Practice 1

**1** **a** There are four members of the family. Each member of the family eats 12 grapes
$\therefore 4 \times 12 = 48$
48 grapes are eaten
**b** After the sixth round, since $\frac{12}{2} = 6$
**c** After the third round, since $\frac{12}{4} = 3$
**d** Three quarters of the way through the seventh round
since $\frac{9}{16} \times 12 = 6\frac{3}{4}$
**e** $\frac{7}{12}$
**f** $\frac{23}{24}$
**g** Individual response

**2** **a** $\frac{5}{6} < \frac{8}{9}$ **b** $\frac{12}{25} < \frac{5}{9}$ **c** $\frac{3}{7} > \frac{21}{50}$ **d** $\frac{5}{24} < \frac{2}{5}$
**e** $\frac{3}{20} < \frac{7}{24}$

3 a Social Networks, Calling, Texting
b Individual response
c Individual response

4 $\frac{2}{11}, \frac{1}{5}, \frac{8}{33}, \frac{1}{4}, \frac{11}{40}, \frac{3}{9}$

5 a $\frac{2}{3} < \frac{5}{7}$ b $\frac{3}{5} > \frac{7}{12}$ c $\frac{9}{10} > \frac{5}{6}$ d $\frac{2}{7} < \frac{3}{8}$
e $\frac{8}{13} < \frac{5}{8}$ f $\frac{4}{9} < \frac{6}{13}$ g $\frac{5}{11} < \frac{10}{19}$ h $\frac{1}{4} > \frac{2}{9}$
i $\frac{3}{4} < \frac{6}{7}$ j $\frac{10}{13} > \frac{7}{11}$

6 Decimal form; Individual response

7 a Approximately 1.25 billion (assuming the world population is 7.5 billion)
b $\frac{15}{365} = \frac{3}{73}$
c $\frac{12}{80} = \frac{3}{20}$
d Road, Rail, Airplane, Boat
e Rail: 0.35 billion
Boat: 0.045 billion
Airplane: 0.06 billion
Road: 2.5 billion

8 a "Whole" here refers to the person's total ethnicity ancestry
b A whole can be divided into different sizes
c Morocco, Benin, Algeria, Cameroon, Spain, England, Italy, Portugal
d Same as in (c)

## Practice 2

1 a $\frac{4}{7} \times \frac{3}{5} = \frac{12}{35}$ b $\frac{7}{9} \times \frac{4}{7} = \frac{4}{9}$ c $\frac{4}{5} \times \frac{2}{5} = \frac{8}{25}$ d $\frac{5}{10} \times \frac{3}{7} = \frac{3}{14}$

2 a $\frac{7}{36}$ b $\frac{3}{4}$ c $\frac{1}{11}$
d $\frac{7}{4}$ e $\frac{5}{2}$ f $\frac{4}{3}$

3 a $\frac{1}{6}$ b $\frac{5}{18}$ c $\frac{5}{14}$
d $\frac{2}{35}$ e $\frac{3}{44}$ f $\frac{5}{6}$

4 a $\frac{1}{6}$ b $\frac{3}{8}$ c $\frac{2}{7}$
d $\frac{1}{6}$ e $\frac{1}{5}$ f $\frac{3}{7}$

5 a $\frac{7}{25}$ b $\frac{21}{250}$ c $\frac{3}{5}$ d Individual response

## Practice 3

1 a $\frac{9}{2}$ b $\frac{5}{3}$ c $\frac{21}{10}$
d 8 e $\frac{1}{4}$ f $\frac{11}{4}$

2 a $\frac{2}{3}$ b $\frac{32}{3}$ c $\frac{4}{7}$
d $\frac{1}{24}$ e 4 f $\frac{153}{50} = 3\frac{3}{50}$

3 Individual response

$5 \times \frac{1}{2} = \frac{5}{2}$

$5 \div \frac{1}{2} = 10$

4 a $\frac{3}{20}$ b $\frac{9}{20}$ c $\frac{1}{5}$

5 Fourth Grade: 20
Fifth Grade: 16
Sixth Grade: 14

6 a 14 b 8 c $\frac{7}{32}$

7 a $\frac{1}{2}$ b Individual response

## Practice 4

1 a $\frac{3}{7}$ b $\frac{5}{7}$ c $\frac{1}{3}$ d $1\frac{3}{7}$
e $\frac{3}{5}$ f $\frac{3}{11}$ g $1\frac{1}{6}$ h $\frac{17}{40}$
i $\frac{17}{20}$ j $1\frac{1}{15}$ k $-\frac{5}{18}$ l $1\frac{1}{24}$
m $\frac{37}{56}$ n $3\frac{11}{13}$ o $4\frac{5}{12}$ p $1\frac{1}{10}$

2 Three eighth-pages

3 $\frac{4}{15}$

4 a $\frac{5}{32}$ b $\frac{9}{32}$ c $\frac{9}{32}$ d Individual response

5 $\frac{18}{35}$

6 $\frac{3}{4}$

7 a This makes sense as the sum of one's fractional ethnic identities is equal to their total ethnic identity, which we denote as a whole (i.e. 1)
b $\frac{17}{100}$
c $\frac{83}{100}$

8 a It is possible that a single item can fall into multiple categories. Therefore, the categories are not disjoint and the sum of the teenagers across all categories is greater than the whole (i.e. the population of teenagers) since individuals can be double (or triple) counted
b Individual response

9 $\frac{4}{15}$

## Practice 5

1 a $\frac{1}{21}$ b $\frac{9}{4}$ c $\frac{9}{20}$ d $\frac{23}{12}$
e $\frac{1}{2}$ f $\frac{9}{4}$ g $\frac{79}{84}$ h $\frac{23}{30}$
i $\frac{1111}{100}$ j 1 k $\frac{29}{30}$ l $\frac{1}{2}$

2 a The measure includes four eighth notes and one half note. $\frac{1}{2}$ equals $\frac{4}{8}$; $\frac{4}{8}$ plus $\frac{4}{8}$ equals one whole note
b measure is a quarter note
c measure is a sixteenth note

3 $\frac{1}{3}$

4 $\frac{2}{3}\div\frac{4}{5}\div\frac{5}{1}$ and $\frac{2}{3}\times\frac{5}{4}\times\frac{1}{5}$

## Unit Review

1 a $\frac{3}{7}>\frac{2}{5}$ b $\frac{7}{10}>\frac{5}{8}$ c $\frac{4}{11}>\frac{2}{7}$
d $\frac{13}{20}>\frac{5}{9}$ e $\frac{5}{12}<\frac{9}{16}$ f $\frac{17}{21}<\frac{5}{6}$

2 a $\frac{1}{3}, \frac{5}{12}, \frac{1}{2}, \frac{17}{24}, \frac{5}{6}, \frac{7}{8}$ b $\frac{11}{25}, \frac{1}{2}, \frac{27}{50}, \frac{3}{5}, \frac{7}{10}, \frac{3}{4}$

3 a Disagreed, Neutral, Strongly Agreed, Agreed b $\frac{18}{25}$ c Individual response

4 a $\frac{6}{5}$ b $\frac{3}{17}$ c $\frac{3}{35}$ d $\frac{10}{11}$

5 a $\frac{8}{15}$ b $\frac{1}{36}$ c $\frac{20}{21}$ d $\frac{4}{15}$
e $\frac{31}{36}$ f $\frac{21}{40}$ g $\frac{161}{10}$ h $\frac{32}{11}$

6 a $2\times\frac{1}{4}$ cup b $4\times\frac{1}{4}$ cup or $3\times\frac{1}{3}$ cup c $1\times\frac{1}{4}$ cup $+ 1\times\frac{1}{3}$ cup d $3\times\frac{1}{4}$ cup $+ 2\times\frac{1}{3}$ cup

7 a $\frac{4}{11}<\frac{3}{8}$ so blue b $\frac{2}{7}<\frac{4}{13}$ so orange c India

8 Eric
a First line: incorrect conversion from improper to mixed fraction
Second to third line: incorrect diagonal cancellation
b, c $\frac{5}{14}\div 3\frac{4}{7}=\frac{5}{14}\div\frac{25}{7}=\frac{5}{14}\times\frac{7}{25}=\frac{1}{10}$

Amanda
a First line: you cannot perform diagonal cancellation when adding fractions
Second to third line: incorrect addition of fractions
b, c $\frac{5}{9}+\frac{6}{15}=\frac{25}{45}+\frac{18}{45}=\frac{43}{45}$

9 a $\frac{5}{18}$ b $\frac{1}{3}$ c $\frac{121}{64}$ d $\frac{2}{15}$

10 a $\frac{8}{5}$
b 2 sticks of butter
$\frac{4}{5}$ small, sweet onion, grated
1 cup yellow mustard
$\frac{6}{5}$ cup dark brown sugar
$\frac{3}{5}$ cup apple cider vinegar
2 tablespoons dry mustard
$\frac{4}{5}$ teaspoon cayenne pepper
$\frac{16}{5}$ bay leaves
Salt & pepper to taste
c Individual response

11 a $\frac{1}{5}$ b $\frac{5}{32}$ c $\frac{1}{50}$ Individual response d $\frac{2}{125}$

# Unit 6

## Practice 1

1 Categorical: whether you are right-handed or left-handed, colour of your favourite hat, the day of the week on which you were born, grade on a final exam
Ordinal: shirt size (S,M,L), your rating of a movie (1,2,3,4 or 5 stars), year of birth
Continuous: age, height
Discrete: results of an election, family income

2 a Discrete b Continuous c Discrete d Categorical
e Categorical f Discrete g Categorical h Continuous

3 Individual response

4 Individual response

## Practice 2

1 a Individual response
b There are 360 degrees in a circle. Therefore, for each percent, multiply by 3.6 and draw a sector using the corresponding angle (with a protractor).
c There is a very slight decrease in percentage of students in each MYP year from year 1 to year 5.
d No, it simply summarizes the raw data.
e Individual response

2 a There are more females than males in schools in Portland Bay
  b No. The graph only represents the overall statistic across all schools in Portland Bay. However, there could be some schools in which there are more males than females (or equal numbers of males and females)
  c No. The results are localised to Portland Bay and say nothing about schools across the country.
  d Individual response
  e Individual response

3 a No, it is not an organized picture representing the data, it is a summary of the raw data in tabular form
  b Individual response
  c Individual response
  d Individual response
  e Individual response

4 a Individual response
  b Individual response
  c Individual response
  d No. The graph only represents the overall collection of student responses and contains no information about the distribution of responses across the city.

5 a No, it is not an organized picture representing the data, it is a summary of the raw data in tabular form
  b (Individual response) In general, students overall agree that students are treated fairly by other students regardless of nationality or ethnicity. However, in general as students get older, they are less likely to strongly agree, and more likely to agree, disagree or strongly disagree.
  c (Individual response) As students get older, they are more likely to be socially aware of discrimination between nationalities and ethnicities, thus are less likely to strongly agree and more likely to acknowledge some extent of unfair treatment.
  d Individual response

6 a Individual response
  b Individual response

7 a Individual response
  b Individual response with respect to type of graph used
  c Individual response

## Practice 3

1 a Bar graph
  b In general, males are more literate than females. In countries where overall literacy rates are lower, there is a greater difference between the literacy rates of males and females
  c No. There is no reference to timescale or date in the graph
  d Individual response

2 a Individual response (the most correct answer arises from measuring, using a protractor, the angles corresponding to each sector and scaling accordingly to a percentage.
  b Individual response
  c Individual response
  d Individual response
  e Individual response

3 a Individual response
  b Individual response
  c Individual response
  d Individual response
  e Individual response
  f Individual response
  g Individual response
  h Individual response
  i Individual response

4 a Individual response
  b To be constructed after corrections, or could be described
  c Individual response
  d Individual response
  e Individual response

5 a No (Individual response)
  b (Individual response) No labelling of axes. The line joining the points suggests that the graph shows a continuous variable. Title says "percentage" but the vertical axis gives the count
  c To be constructed after corrections, or can be described
  d Individual response
  e Individual response

## Practice 4

1 a Water
  b 11
  c 26
  d 80
    Tea: 10, Coffee: 16, Soft Drink: 5, Water: 23, Juice: 11, Smoothie: 15
  e To be constructed after corrections
  f Individual response
  g Individual response

2 a Several times a day
  b Several times a day: 216 degrees
    Once a day: 36 degrees
    Almost constantly: 90 degrees
    Once a week: 14.4 degrees
    Less than once a week: 3.6 degrees
  c Individual response
  d Individual response
  e Individual response

3 a Individual response
  b Individual response
  c Individual response
  d Individual response
  e Individual response

4 a Individual response
  b The majority of bullying receives no intervention
  c Individual response

## Practice 5

1 a i Any estimate between 128-129 cm
    ii 140 cm
  b Individual response
  c Individual response
  d Individual response
  e Individual response
  f Individual response

2 a Football / Soccer
  b Field hockey, tennis, volleyball
  c 3
  d 10
  e 94
  f 4
  g Individual response
  h Individual response

3 a 125
  b To be completed after review or described
  c To be completed after review or described
  d Individual response
  e Individual response
  f Individual response
  g Individual response

4 a To be constructed after review or can be described
  b Individual response
  c Individual response
  d Individual response

## Practice 6

1 a Answers include, but are not limited to:
  - Brazil: Overall average increase in homicide rate over period (including specific data points). Fall-off in homicide rate towards the end of the period. Peak of around 33 homicides per 100 000 people around 2003.
  - Ecuador: Overall average increase in homicide rate over priod (including specific data points). Peak of around 23 homicides per 100 000 people around 2008.
  - Colombia: Significant rise in homicide rate from 1980 to around 1992, followed by significant fall in homicide rate from 1992 to 2010 (including specific data points, for example, the peak homicide rate of 80 people per 100 000 people around 1992.)

  b No, the $y$-axis range and scaling vary between the graphs

2 a Individual response
  b Overall increase in youth obesity overtime. Levelling off more recently.
  c Individual response
  d Individual response

3 a In general, there is an increase in violent crime among the cities from the beginning of the period to the early 1990s, and then a decrease thereafter until the end of the period
  b Individual response
  c This is not a good assumption to make from the data, because the data contains information on only six American cities and only for the period 1985–2012. Thus "... always been one of the safest cities in America." cannot be justified by this graph alone.
  d No; Individual response

4 a Answers include, but are not limited to:
  - Left-hand graph represents a numerical total whereas the right graph represents a percentage
  - The results for "Asia and the Pacific" and "Sub-Saharan Africa" vary between the graphs (i.e. for the left-hand graph the totals for "Asia and the Pacific" are higher than for "Sub-Saharan Africa" whereas on the right-hand graph, the percentages for "Sub-Saharan Africa" than for "Asia and the Pacific"

  b Comparing year-to-year, child labour was more frequent in "Asia and the Pacific" than it was in "Sub-Saharan Africa"
  c Comparing year-to-year, child labour was more frequent in "Sub-Saharan Africa" than it was in "Asia and the Pacific"
  d Individual response
  e Individual response

5 a Individual response
  b Individual response
  c Individual response
  d Individual response

6 a High school graduation percentages are most likely based on a lot of factors, including economic climate, school district budgets, immigration, national testing, etc. There is no evidence here to suggest that the results are correlated to new government.
  b Individual response (main point to be noted is that the number of books has no direct relationship with the percentage).
  c To be constructed after review, or can be described
  d Individual response
  e Individual response
  f Individual response
  g Individual response (main point to be noted is the Obama – and perhaps Bush - period)

## Unit Review

1 a Discrete
  b The distribution of the dwelling situations for the homeless
  c Individual response
  d Individual response

2 a Discrete
  b Individual response
  c Individual response

3 a Individual response
  b Individual response
  c Individual response

**4 a** Individual response
**b** It is not stated what units are used on the vertical axis. The text mentions "the raptor" but it is not clear whether this is. Also, we don't know the population size.
**c** Individual response

**5 a** Discrete
**b** Individual response
**c** Individual response
**d** Individual response
**e** Individual response

**6 a** 'Grade level' – ordinal
'Hours of homework' – Discrete
**b** 1 hour
**c** Increases by 0.75 hours
**d** Grades 1 and 2
**e** Grade 8
**f** Between grades 6 and 7 or between grades 9 and 10
**g** Individual response
**h** Individual response
**i** Individual response
**j** Individual response

**7 a** Individual response; can construct different graphs and/or describe them after review
**b** The students who use their cell phone more frequently in class tend to achieve worse grades (students should use data to support this justification)
**c** Individual response
**d** Individual response

**8 a** Pie charts with sectors as follows:
Basketball: Under \$25k = 58°, \$25k – \$49.9k = 76°, \$50k – \$74.9k = 68°, \$75k – \$100k = 54°, Over \$100k = 104°
Soccer: Under \$25k = 47°, \$25k – \$49.9k = 68°, \$50k – \$74.9k = 58°, \$75k – \$100k = 61°, Over \$100k = 126°
Swimming: Under \$25k = 29°, \$25k – \$49.9k = 58°, \$50k – \$74.9k = 43°, \$75k – \$100k = 75°, Over \$100k = 155°
Lacrosse: Under \$25k = 14°, \$25k – \$49.9k = 36°, \$50k – \$74.9k = 58°, \$75k – \$100k = 50°, Over \$100k = 202°
**b** Individual response
**c** Individual response
**d** Individual response (Lacrosse)
**e** Individual response

**9 a** High School Graduate
**b** Individual response
**c** At least a doctoral degree
**d** Individual response

**10 a** For the teens: circle graph with the following sectors.
0 = 7°, 1–10 = 79°, 11–20 = 40°, 21–50 = 65°, 51–100 = 65°, 101 or more = 104°.
For the adults: circle graph with the following sectors.
0 = 32°, 1–10 = 184°, 11–20 = 47°, 21–50 = 43°, 51–100 = 25°, 101 or more = 29°.
**b** Individual response
**c** Individual response
**d** Individual response
**e** Individual response
**f** Individual response

**11 a** Individual response
**b** Individual response
**c** Individual response
**d** Individual response
**e** Individual response

**12 a** Individual response (An urban area is a densely populated region that is equipped with infrastructure. A rural area is an area that is not considered to be rural.)
**b** Congo
**c** Nigeria
**d** Egypt
**e** Individual response
Individual response
**f** Individual response

## Unit 7

### Practice 1

1 a 6 and 8; b 7; c 10

2 a 60 $in^2$ b 45 $m^2$ c 126 $cm^2$ d 48 $cm^2$

3

| Triangle's Base (cm) | Triangle's Height (cm) | Triangle's Area (cm) |
|---|---|---|
| 10 | 12.6 | **63** |
| **18.21** | 14.5 | 132 |
| 8.4 | **50** | 210 |
| **10.34** | 16.25 | 84 |
| 10.5 | **9.14** | 48 |

4 a 10 $m^2$
b 10.825 $m^2$
c If the sun is not directly overhead, the effective shade is not necessarily the same area as the area of the triangular shade

### Practice 2

1 a Perimeter: 20 cm
Area: 19 $cm^2$
b Perimeter: 24 cm
Area: 30 $cm^2$
c Perimeter: 66 cm
Area: 99 $cm^2$
d Perimeter: 30 cm
Area: 46 $cm^2$
e Perimeter: 32 cm
Area: $32\sqrt{3}$ $cm^2$
f Perimeter: 50 cm
Area: 160 $cm^2$

2 a 30 $mm^2$ b 16 $mm^2$ c 798 $mm^2$ d 126 $mm^2$

3 a 120 $km^2$ b 20% c Individual response

### Practice 3

1 a 792 $cm^3$ b 2700 $cm^3$ c 280 $m^3$ d 3 $m^3$

2 a 7 cm b 10 cm c 12 cm

3

| Length (cm) | Width (cm) | Height (cm) | Volume ($cm^3$) |
|---|---|---|---|
| 10 | 12 | **5** | 600 |
| **2.5** | 4 | 9 | 90 |
| 3 | **4** | 25 | 300 |
| 7 | 4 | 11 | **308** |
| 8 | 20 | **3.5** | 560 |

4 a 540 $cm^3$ b 210 $cm^3$ c 330 $cm^3$ d Individual response

5 a 303.24 $cm^3$; isometric drawing of rectangular prism, with measurements in proportion
b 100.45 $cm^3$; isometric drawing of triangular prism, with measurements in proportion
c Individual response (the tetra packaging is more space efficient)
d Individual response

### Practice 4

1 a 518 $cm^2$ b 600 $mm^2$ c 1380 $cm^2$
d 331.2 $m^2$ e 272 $in^2$ f 14 $m^2$

**2** **a** Change in volume: $98\,cm^3$
Change in surface area: $74\,cm^2$
**b** Individual response
**c** No. The new box has a larger volume so does not justify a decrease in the mass of candy contained.
**d** Individual response

**3** $1304.3\,cm^2$

**4** **a** Two sides, each measuring $5 \times 48$. Surface area $= 2 \times 5 \times 48 = 480\,m^2$.
20 slanted panels, each measuring $2.2 \times 48$. Surface area $= 2.2 \times 48 \times 20 = 2112\,m^2$.
Using Pythagoras, the base length of each peaked roof is $2 \times \sqrt{2.2^2 - 1^2} = 3.92\,m$.
There are ten peaked roof sections, so the length along the front (and back) of the greenhouse is $10 \times 3.92 = 39.2$ m.
Thus, the surface area of the front and back is $2 \times 5 \times 39.2 = 392\,m^2$.
The entire surface of glass windows is the combined surface area: $480 + 2112 + 392 = 2984\,m^2$.
**b** Growing capacity $= 48 \times 5 \times 39.2 = 9408\,m^3$.
**c** Surface area of the 20 slanted panels $= 2112\,m^2$ (from part a), so half of this is $1056\,m^2$.
**d** $174.4\,m^2$

## Unit Review

**1** **a** Perimeter: 56 in
Area: $84\,in^2$
**b** Perimeter: 32 in
Area: $35\,in^2$
**c** Perimeter: 29 in
Area: $20\,in^2$
**d** Perimeter: 66 in
Area: $200\,in^2$

**2** **a** $72\,m^3$ **b** $162\,m^3$ **c** $27\,m^3$ **d** $450\,m^3$

**3** **a** Perimeter: 35.6 cm
Area: $18.65\,cm^2$
**b** Perimeter: 33 cm
Area: $72\,cm^2$
**c** Perimeter: 82.08 cm
Area: $432\,cm^2$
**d** Perimeter: 53.7 cm
Area: $153\,cm^2$

**4** **a** 72 $feet^3$ **b** 480 $feet^3$ **c** 480 $feet^3$

**5** **a** Perimeter: 57.6 m
Area: $76.8\,m^2$
**b** Individual response
**c** Individual response
**d** Individual response

**6** $1224\,m^2$

**7** **a** Volume of left house: $126\,m^3$
Volume of right house: $140\,m^3$
**b** Individual response
**c** Individual response

**8** **a** $6\,m^2$ **b** 3.75 m

**9** **a** $63.5\,km^2$
**b** $32.5\,km^3$
**c** The new volume would be an eighth of the current volume. This is an 87.5% decrease in the volume
**d** Individual response

**10** **a** $4.05\,m^3$ **b** $3.6\,m^3$, Individual response

**11** **a** Surface area $= 2 \times (4 \times 6 + 4 \times 12 + 6 \times 12) = 288\,units^2$
Volume $= 4 \times 6 \times 12 = 288\,units^3$
So numerically the surface area and volume are equal
**b** A cube of side length 6 will have a volume of $6 \times 6 \times 6 = 216\,units^3$, and a surface area of $2 \times (6 \times 6 + 6 \times 6 + 6 \times 6) = 216\,units^3$.
One such rectangular prism whose surface area is numerically equal to its volume is one that measures $4 \times 6 \times 12$.

# Index

**V**